Managing a Sustainable and Low-Carbon Society

Managing a Sustainable and Low-Carbon Society

Editors

Taoyuan Wei
Qin Zhu

MDPI • Basel • Beijing • Wuhan • Barcelona • Belgrade • Manchester • Tokyo • Cluj • Tianjin

Editors

Taoyuan Wei
Climate Economics
CICERO Center for
International Climate
Research
Oslo
Norway

Qin Zhu
School of Social Development
and Public Policy
Fudan University
Shanghai
China

Editorial Office
MDPI
St. Alban-Anlage 66
4052 Basel, Switzerland

This is a reprint of articles from the Special Issue published online in the open access journal *International Journal of Environmental Research and Public Health* (ISSN 1660-4601) (available at: www.mdpi.com/journal/ijerph/special_issues/Sustainable_Low_Carbon).

For citation purposes, cite each article independently as indicated on the article page online and as indicated below:

LastName, A.A.; LastName, B.B.; LastName, C.C. Article Title. *Journal Name* **Year**, *Volume Number*, Page Range.

ISBN 978-3-0365-7943-6 (Hbk)
ISBN 978-3-0365-7942-9 (PDF)

Contents

About the Editors vii

Preface to "Managing a Sustainable and Low-Carbon Society" ix

Lars Lindholt and Taoyuan Wei
The Effects on Energy Markets of Achieving a 1.5 °C Scenario
Reprinted from: *Int. J. Environ. Res. Public Health* **2023**, *20*, 4341, doi:10.3390/ijerph20054341 . . . 1

Shuang Liang, Xinyue Lin, Xiaoxue Liu and Haoran Pan
The Pathway to China's Carbon Neutrality Based on an Endogenous Technology CGE Model
Reprinted from: *Int. J. Environ. Res. Public Health* **2022**, *19*, 6251, doi:10.3390/ijerph19106251 . . . 21

Wanlin Yu and Jinlong Luo
Impact on Carbon Intensity of Carbon Emission Trading—Evidence from a Pilot Program in 281 Cities in China
Reprinted from: *Int. J. Environ. Res. Public Health* **2022**, *19*, 12483, doi:10.3390/ijerph191912483 . 43

Qian Zhang and Qizhen Wang
Digitalization, Electricity Consumption and Carbon Emissions—Evidence from Manufacturing Industries in China
Reprinted from: *Int. J. Environ. Res. Public Health* **2023**, *20*, 3938, doi:10.3390/ijerph20053938 . . . 63

Weiwu Wang, Huan Chen, Lizhong Wang, Xinyu Li, Danyi Mao and Shan Wang
Exploration of Spatio-Temporal Characteristics of Carbon Emissions from Energy Consumption and Their Driving Factors: A Case Analysis of the Yangtze River Delta, China
Reprinted from: *Int. J. Environ. Res. Public Health* **2022**, *19*, 9483, doi:10.3390/ijerph19159483 . . . 85

Po Yun, Chen Zhang, Yaqi Wu and Yu Yang
Forecasting Carbon Dioxide Price Using a Time-Varying High-Order Moment Hybrid Model of NAGARCHSK and Gated Recurrent Unit Network
Reprinted from: *Int. J. Environ. Res. Public Health* **2022**, *19*, 899, doi:10.3390/ijerph19020899 . . . 111

Usman Mehmood, Ephraim Bonah Agyekum, Solomon Eghosa Uhunamure, Karabo Shale and Ayesha Mariam
Evaluating the Influences of Natural Resources and Ageing People on CO_2 Emissions in G-11 Nations: Application of CS-ARDL Approach
Reprinted from: *Int. J. Environ. Res. Public Health* **2022**, *19*, 1449, doi:10.3390/ijerph19031449 . . . 131

Haiying Liu, Wenqi Guo, Yu Wang and Dianwu Wang
Impact of Resource on Green Growth and Threshold Effect of International Trade Levels: Evidence from China
Reprinted from: *Int. J. Environ. Res. Public Health* **2022**, *19*, 2505, doi:10.3390/ijerph19052505 . . . 145

Xiaoxue Liu, Fuzhen Cao and Shuangshuang Fan
Does Human Capital Matter for China's Green Growth?—Examination Based on Econometric Model and Machine Learning Methods
Reprinted from: *Int. J. Environ. Res. Public Health* **2022**, *19*, 11347, doi:10.3390/ijerph191811347 . 165

Jianming Wang, Xincheng Yang, Yini Xi and Zhengxia He
Is Green Spread? The Spillover Effect of Community Green Interaction on Related Green Purchase Behavior
Reprinted from: *Int. J. Environ. Res. Public Health* **2022**, *19*, 6571, doi:10.3390/ijerph19116571 . . . 193

Jianfeng Guo, Kai Zhang and Kecheng Liu
Exploring the Mechanism of the Impact of Green Finance and Digital Economy on China's Green Total Factor Productivity
Reprinted from: *Int. J. Environ. Res. Public Health* **2022**, *19*, 16303, doi:10.3390/ijerph192316303 . **217**

Zhanjie Wang, Yongfeng Ma, Shasha Wang and Yongjian Wang
Research on the Heterogeneity Threshold Effect of Foreign Direct Investment and Corporate Social Responsibility on Haze Pollution
Reprinted from: *Int. J. Environ. Res. Public Health* **2023**, *20*, 4802, doi:10.3390/ijerph20064802 . . . **235**

Aiqun Wang, Ming Zhang and Shuya Zhou
Air Pollution, Environmental Violation Risk, and the Cost of Debt: Evidence from China
Reprinted from: *Int. J. Environ. Res. Public Health* **2022**, *19*, 3584, doi:10.3390/ijerph19063584 . . . **249**

About the Editors

Taoyuan Wei

Dr. Taoyuan Wei, Senior Researcher at CICERO Center for International Climate Research, has been responsible for macroeconomic modeling and analysis in interdisciplinary research projects related to energy, climate change, and the environment. He is a leading energy economist on the rebound effects of energy efficiency improvement, one of the key measures commonly suggested for climate change mitigation and adaptation. His Ph.D. dissertation was on accounting for income arising from nonrenewable resources. He has also worked on input-output surveys, accounting, and analysis at the National Bureau of Statistics of China. So far, Dr. Wei has published nearly 60 peer-reviewed articles in international journals.

Qin Zhu

Dr. Qin Zhu is a professor of economic demography at the School of Social Development and Public Policy, Fudan University, China. His areas of expertise cover demography, social policy, and low-carbon economics. In recent years, His research interests focus on population aging, the energy and environmental effects of demographic change, and population big data. He has published more than 60 papers and 2 monographs, the latest of which is titled "Population Aging and Carbon Emissions: Impacts of Labor Supply and Consumption Pattern". His research has been awarded the first prize for outstanding achievements in population science in China.

Preface to "Managing a Sustainable and Low-Carbon Society"

Modern society faces various global and regional challenges, including climate change and sustainable development. In order to deal with these challenges, society must be managed properly to follow a sustainable and low-carbon pathway. To advance the studies on this topic, 13 articles have been included in a Special Issue of the "International Journal of Environmental Research and Public Health". This is a reprint of the 13 articles published in the Special Issue. The themes covered by the 13 articles include the driving forces of carbon emissions, the carbon market, implications of regional and global low-carbon pathways, green finance, green growth, and air pollution. Various methods have been used in these studies, such as computable general equilibrium (CGE) models, econometric models, and machine learning methods. Most of the articles provide new insights based on evidence from China while the others focus on several regions and the global economy.

Both co-editors, Taoyuan Wei and Qin Zhu, were honored to be invited to serve as guest editors for this special issue. In the process of organizing the Special Issue, many people were involved. We are grateful for all the authors contributing their wonderful articles. Thanks also to the editorial team of the journal led by Ms. Pearl Yang for their excellent assistance. Finally, we would acknowledge that our editorial work was partially supported by the grants from Research Council of Norway (No. 303486) and the Humanities and Social Science Foundation of the Ministry of Education of China (No. 18YJA840025).

Taoyuan Wei and Qin Zhu
Editors

International Journal of
Environmental Research and Public Health

MDPI

Article

The Effects on Energy Markets of Achieving a 1.5 °C Scenario

Lars Lindholt [1] and Taoyuan Wei [2,*]

[1] Research Department, Statistics Norway, 0177 Oslo, Norway
[2] CICERO Center for International Climate Research, 0349 Oslo, Norway
* Correspondence: taoyuan.wei@cicero.uio.no

Abstract: Net zero emission scenarios are aligned with the criteria for the Paris Agreement to keep global warming below 1.5 °C. By soft-linking an energy model with a macroeconomic model, we create a similar pathway to the net zero emission scenario from the International Energy Agency (IEA) to 2050 both of demand for fossil fuels and total CO_2 emissions. Soft-linking entails that we insert endogenous variables from one model into the other model. We implement measures such as CO_2 taxes, improved energy efficiency, more renewables in electricity production and other sectors, easier substitution between electricity and fossil fuels for final users, and drastically limiting future production of oil, gas and coal. Our conclusion is that net zero is possible by introducing very strict measures, e.g., a high rate of energy efficiency improvement, far above what has been achieved in the past. While our partial equilibrium energy model, similar to the IEA model, overlooks the potential rebound effects, i.e., more energy used by consumers due to lower prices caused by energy efficiency improvement, our macroeconomic model does capture the rebound effects and has to implement stricter supply-side measures to reduce fossil fuel use to achieve the 1.5 °C scenario.

Keywords: net zero; climate change; mitigation; energy model; integrated assessment; CGE model; fossil fuel; energy transition

Citation: Lindholt, L.; Wei, T. The Effects on Energy Markets of Achieving a 1.5 °C Scenario. *Int. J. Environ. Res. Public Health* **2023**, *20*, 4341. https://doi.org/10.3390/ijerph20054341

Academic Editor: Paul B. Tchounwou

Received: 4 January 2023
Revised: 22 February 2023
Accepted: 23 February 2023
Published: 28 February 2023

1. Introduction

To keep global temperature increase below 1.5 or 2 °C compared to pre-industrial levels, the global CO_2 emissions will need to become net zero and potentially below zero. As indicated by the Intergovernmental Panel on Climate Change (IPCC) in its special report on 1.5 °C: "From a physical science perspective, limiting human-induced global warming to a specific level requires limiting cumulative CO_2 emissions, reaching at least net zero CO_2 emissions, along with strong reductions in other greenhouse gas emissions" [1]. Jones et al. [2] suggest that, essentially, the global warming stops in the case of net zero CO_2 emissions, implying that current choices can avoid the worst impacts of global warming in the future. Net zero energy systems, where residual CO_2 emissions are offset by removals, are crucial to achieve economy-wide net zero emissions (see e.g., [3]).

The Paris Agreement from 2016 proposes an ambitious target of limiting global warming to well below 2 °C, preferably to 1.5 °C, compared to pre-industrial levels. This was strengthened to below 1.5 °C at the Glasgow meeting in 2021. There is a strand of studies using integrated models that analyze how 1.5 °C can be reached, e.g., the IPCC-scenarios presented in Masson-Delmotte et al. [1]. In 2021, a report released by the IEA [4] described a roadmap to achieve net zero emissions for the global energy sector by 2050 (NZE hereafter), which is necessary for a 1.5 °C world.

The purpose of this study is to examine how global energy markets might be affected during the transition period of achieving net zero emissions in 2050 based on two soft-linked complementary models. Soft-linking entails that we insert endogenous variables from one model into the other model. The IEA [4] presents the most-cited and well-known NZE pathway for energy sectors (we refer to this scenario as NZE IEA). Brecha et al. [5] have assessed various scenarios, including those from the IPCC reports, and find that the

NZE IEA scenario is aligned best with the requirement specified in the Paris Agreement when it comes to having the likelihood to limit global warming to 1.5 °C. To the best of our knowledge, we have not seen any studies that use energy models to test whether the NZE IEA pathway is achievable and which relevant measures to apply to reach such a pathway. By soft-linking an energy model with a macroeconomic model, this study creates a similar pathway to 2050 and analyzes how such a pathway impacts the energy markets.

We introduce various mitigation measures simultaneously to achieve NZE in 2050 in our models. Among them, we consider CO_2 taxes, improved energy efficiency, more renewables in electricity production and other sectors, easier substitution between electricity and fossil fuels for final users, and drastically limiting future production of fossil fuels. The effect of our various measures is to reduce fossil fuel consumption and corresponding emissions to around 90 per cent from the reference scenario by 2050. Even if it is possible to reach large reductions by implementing such strict measures, it would in practice be very challenging, if not impossible. By presenting the NZE scenarios simulated by our models, the paper aims to discuss the magnitude of the various measures that are necessary to obtain NZE and whether this is realistic or not.

The remainder of the article is organized as follows. In the next section, we present a brief literature review to justify our study. In Section 3, we introduce our models and describe how the models are used to simulate reference and NZE scenarios. In Section 4, we present the results of the effects on energy markets of the simulated NZE scenarios. In Section 5, we further discuss the issues related to the simulation results. We conclude in the last section.

2. Literature Review

A 1.5 °C scenario refers to a simulated pathway where there is at least a 50 per cent probability for the global warming to be no more than 1.5 °C in at least 2100 and possibly above 1.5 °C in some years before 2100 [1]. Such scenarios have been developed by various institutions, including BP [6], Equinor [7], the IEA [4], and Shell [8], besides the integrated assessment model (IAM) scenarios [1,9]. Among these scenarios, the NZE IEA scenario is identified to align best with the requirement specified in the Paris Agreement in terms of the likelihood to limit global warming to 1.5°C [5]. However, we know of no other model-based studies that test and verify the achievability of the NZE IEA pathway.

The NZE IEA pathway [4] illustrates a marked transformation of the energy system from fossil fuels to renewable energy, rapid improvement of energy efficiency, and wide electrification in the economy. However, the IEA energy model is a partial equilibrium model and assumes exogenous key determinants of energy demand such as gross domestic product (GDP). In addition, the NZE IEA pathway introduces considerable energy efficiency improvement to reduce energy demand, but excludes the potential rebound effects on energy consumption due to lower energy prices [10–12]. It is necessary to address what additional measures are needed to offset the rebound effect in a 1.5 °C world.

Various measures can contribute to reducing emissions. Considering the limited resources available for society, degrowth is proposed to reduce GHG emissions and contribute to emissions neutrality [13]. However, degrowth implies welfare losses of certain groups and is difficult to implement, which makes the other option, "decoupling", attractive, i.e., breaking the link between economic growth and environmental pressures like GHG emissions [14]. Decoupling provides a win–win perspective to control global warming without sacrificing economic growth. As the NZE IEA scenario assumes GDP growth at an annual rate of around 3 per cent [4], all the measures adopted in the scenario can be regarded as decoupling measures. Given the considerable barriers to implement these decoupling measures, the NZE IEA scenario was described as achievable but challenging [4].

The energy market of particularly fossil fuels in such a 1.5 °C scenario may markedly differ across scenarios simulated by different models, depending on the assumptions in the models on the adopted mitigation measures [1]. Hence, the energy market in the NZE

IEA scenario can differ from other models that achieve the same global level of emission reductions, and this will be tested by this study.

3. Methods and Materials

The two models we have soft-linked are a computable general equilibrium (CGE) model of the world economy, GRACE [15], and a partial equilibrium model of the global energy markets, FRISBEE [16]. Descriptions of both models are presented in [17]. Both models have been used to analyze issues related to climate policies (e.g., [16,18]). We introduce policy measures by changing central parameter values in each model, so that an NZE scenario is simulated to roughly follow the pathway of both demand for fossil fuels and total CO_2 emissions from fossil fuels in the NZE IEA.

In studies that rely on CO_2 prices as the only climate policy measure, as is often the case in global climate policy modelling, they can be interpreted as the marginal cost of abatement [19]. In practice, other measures than CO_2 prices are necessary. Hence, we follow NZE IEA and the IPCC special report [1] to include various energy policies and accompanying measures designed to reduce emissions such as CO_2 prices, renewable electricity production, and efficiency improvements.

By soft-linking the models, we insert endogenous variables from one model into the other model. In this study, the endogenous regional GDP and global crude oil price generated from GRACE are used as exogenous variables for FRISBEE to simulate the 1.5 °C scenario, so that FRISBEE captures the effects on GDP and global oil price of various mitigation measures introduced in GRACE. We do not harmonize any other issues for both models in order to keep the individual advantages of each model, as explained below.

Energy market models are, in many cases, better suited to identify various energy goods and the associated costs and investments, but do not address the impacts on the overall economy due to, e.g., interactions between markets of energy and those of other goods. On the other hand, CGE models are generally better suited to study overall economic impacts indicated by, e.g., GDP growth rates, market price effects, and structural change. These differences imply that the two modelling approaches can contain different instruments to achieve emission reductions. In addition, the effects of the same measures in the models may create diverging results. While we soft-link two top-down models in this study, it seems more common to soft-link a bottom-up energy system model and a CGE model [20,21].

3.1. Model Description

FRISBEE is a recursive, dynamic partial equilibrium model simulating the global energy markets with 2012 as the start year. Prices are in terms of 2012 USD and exchange rates are assumed constant over time. The model is sequentially solved year by year. The energy goods in the model cover coal, gas, oil, and bioenergy, and, further, electricity (and heat) generated from feedstock of either the fossil or non-fossil fuels, assisted by a sector of transformation and distribution. Global demand equals supply for each energy good. Demands for secondary energy goods in households and industry are modeled as log-linear functions of prices and income. In addition, autonomous energy efficiency improvement (AEEI) is implemented in the model. FRISBEE elaborately models the global oil market and regional gas markets while, in less detail, modelling the world markets for electricity, coal, and renewables. The oil price in the world market is exogenous as the residual demand is satisfied by OPEC as the difference between world demand and Non-OPEC supply at the prevailing price. FRISBEE also elaborately describes oil and gas investments and production, explicitly accounting for discoveries, reserves, and field development.

GRACE is a multi-sectoral, multi-regional recursive dynamic CGE model for the global economy. The model is calibrated to mimic the global economy in 2014. All economic values are stated in 2014 USD and exchange rates are assumed constant over time. Like FRISBEE, the GRACE model is sequentially solved year by year. A regional economy consists of 15 production activities including agriculture, forestry, fishery, three manufacturing sectors, services, three transport sectors, and five energy sectors (coal, oil, gas, refined oil and electricity). Electricity is generated from feedstock of either fossil fuels or non-fossil energy, assisted by a transformation and distribution sector. Global demand equals global supply for each good. In GRACE, production of primary energy is described by nested-CES functions. At the top level, energy output is a CES combination of a value-added intermediate aggregate and natural resources. At the second level, the value-added intermediate aggregate is a Leontief combination of intermediates and the value-added aggregate (a CES combination of capital and labor).

As the functional forms of the demand are different in the two models (CES in GRACE; Cobb Douglas in FRISBEE), it is difficult to directly compare the modelling output. However, we do not harmonize the functional forms as this can strengthen the analyses. We emphasize that, other things being equal, substitution elasticities closer to one in GRACE reduce the difference between the two models (as the substitution elasticities between the various energy goods are all equal to one in FRISBEE).

3.2. Reference Scenarios

The IEA's Stated Policies Scenario (STEPS) stands out as a reference scenario for energy markets, energy security, and emissions and explores the implications of announced climate targets as well as existing energy policies [22]. In STEPS, primary energy demand grows by a quarter to 2040. Hence, there is no peak in global energy-related CO_2 emissions until 2040, as the effects of an expanding economy and population on energy demand outweigh the need for a more efficient and lower-emissions energy system. By soft-linking GRACE and FRISBEE, the reference scenario simulated by each model aligns with the regional energy development of STEPS in [22], but we do not strive for a perfect match. STEPS ends in 2040, but we simulate our models to 2050 as we will compare policy scenarios with NZE IEA up to 2050.

Both models assume regional GDP growth rates, population development, and CO_2 prices (and, to some extent, other policy-related variables) as in STEPS in [22]. Due to lack of data, we performed various estimations and calculations regarding regional CO_2 prices and GDP growth rates, where pricing of CO_2 emissions is by emissions trading systems or taxes although GRACE and FRISBEE only cover CO_2 (see [17,23]). If the simulated demand (or supply) of the various energy goods in different regions is far off the levels in STEPS, we adjust central parameter values.

In the reference scenario simulated by GRACE, population development is set exogenously as in STEPS. The GDP path of STEPS is obtained through adjustment of factor-augmented technological changes with respect to capital and labor. Fossil fuel consumption of STEPS is approximated by adjustment of the endowment of reserves (or natural resources) available for production in these sectors and fossil fuel tax/subsidy rates for final users besides efficiency improvement of energy for final use. CO_2 emissions from fossil fuels are then calculated by fossil fuel use multiplied with given emission factors (i.e., CO_2 emissions per unit of fossil fuel use by sector and region). Hence, the population and GDP growth in the reference scenario of GRACE are the same as in STEPS. Figures A1 and A2 in Appendix A show that consumption of fossil fuels and emissions of CO_2 in the reference scenario of GRACE align quite well with STEPS.

In FRISBEE, assumptions on population development, GDP growth rates, and CO_2 prices in each region are taken from Cappelen et al. [4], which is also based on STEPS in IEA [22]. While renewable energy (incl. nuclear) in the power sector is endogenous in GRACE, it is exogenous in FRISBEE. We include the regional volumes of non-fossils in the power sector from STEPS in FRISBEE. Then, if the simulated demand (or supply) of the various energy goods in the different regions are far off targets in STEPS, we mainly adjust the income elasticities and the parameters for Autonomous Energy Efficiency Improvement (AEEI). For oil, we, in addition, include the exogenous oil price taken from IEA [22]. Figures A3 and A4 in Appendix A show that consumption of fossil fuels and emissions of CO_2 in the reference scenario of FRISBEE also align quite well with STEPS.

3.3. NZE Scenario in IEA

The NZE IEA scenario seems stricter than the scenarios in the IPCC special report, in terms of reaching net zero of total CO_2 emissions in 2050. The CO_2 emissions include emissions from burning fossil fuels and non-renewable wastes and emissions from fuel transformation and industrial processes, with the subtraction of CO_2 removals from carbon capture, utilization, and storage (CCS). Focusing on the burning of fossil fuels, we study the effects on energy markets in the two models to roughly follow the pathway of total CO_2 emissions in NZE IEA.

This study does not simulate removals of emissions for simplicity. Instead, we assume an exogenous pathway of CCS to remove CO_2 over time, following IEA (2021), allowing for more emissions from fossil fuels in the two models equivalent to the exogenous amount of CO_2 removals, at the same level as in NZE IEA. Although a given path of greenhouse gas (GHG hereafter) emissions can imply different temperature changes depending on which climate model is used, the 1.5 °C scenario in this study is defined as roughly following the pathway of total CO_2 emissions from fossil fuels in NZE IEA. A discussion of which measures to use in GRACE and FRISBEE to make the scenarios align with NZE IEA can be found in Cappelen et al. [24].

3.4. The NZE Scenario in GRACE

We introduce six types of measures for all regions to reduce CO_2 emissions (Table 1). All measures other than CO_2 taxes are introduced from 2020.

Table 1. The measures introduced in GRACE to achieve a net zero scenario.

Measure	Description
CO_2 taxes	Regional CO_2 taxes following NZE IEA are introduced [24]
Lower cost of power generation	Non-thermal electricity generation costs are reduced by 3.0, 1.5, and 0.5 per cent over the periods 2020–2030, 2030–2040, and 2040–2050, respectively.
Upper limit of thermal power	The generation of thermal power in a year after 2020 is not allowed to be more than in the previous year.
Increased substitution between fossils and power	Substitution elasticities between the use of fossil fuels and electricity are increased over time.
Improved energy efficiency	Energy-augmented technology for all final energy users increases by 1.5 per cent yearly in 2020–2030 and 1.0 per cent in 2030–2050 over the level in the reference scenario.
Reducing fossil reservoirs	The natural resources used in fossil fuel production in 2050 dramatically reduce to become only 10 per cent for oil and 5.5 per cent for coal and gas of the levels in the reference scenario.

Based on information available from [4], CO_2 taxes by year and region were derived from NZE IEA [24]. We introduce the CO_2 taxes as differences between the CO_2 prices in NZE and STEPS, since our reference scenario simulated by GRACE does not introduce explicit CO_2 prices.

In NZE IEA, much more renewable energy is used to generate electricity than in STEPS. The unit cost of renewable electricity is expected to decline over time. The assumed costs in NZE IEA are generally lower than in STEPS, in both absolute values and changes over time. A decline rate of 1 per cent of the costs means 1 per cent decline of all inputs to generate a given amount of electricity. In GRACE, this is interpreted as a 1 per cent improvement in Hicksian neutral technology, meaning if we keep the inputs of labor and capital constant, the generated electricity increases by 1 per cent.

Four technologies are modelled in GRACE to generate electricity fueled by coal, gas, oil and non-fossils. We assume that the cost for non-fossil-fueled electricity generation declines yearly by 3.0 per cent during 2020–2030, 1.5 per cent during 2030–2040, and

0.5 per cent after 2040, further than in the reference scenario, corresponding to a yearly rate of 1.7 per cent over the whole period of 2020–2050.

In NZE IEA, no electricity is assumed to be generated from unabated fossil fuels in 2050 and only 2 per cent of electricity generation is from fossil fuels with CCS. As GRACE has not included CCS so far, we introduce a modest restriction on electricity generated from fossil fuels either with or without CCS, i.e., we set upper limits for thermal electricity generation. Electricity generated from fossil fuels is assumed to be less than the previous-year level from 2030.

In NZE IEA, the renewable energy (either electricity or other types) is assumed to account for most of the final use in 2050. To achieve this, one must invest massively in infrastructure to facilitate the final energy users in replacing fossil fuels with renewable energy, which implies that considerable barriers to use renewable energy will gradually be removed. This makes it easier for the final users to substitute fossil fuels with renewable energy.

It has been argued that a high elasticity of substitution between clean and dirty inputs in an economy is crucial for sustainable and green growth [25]. In the specifications of nested CES production functions, [26] find that the substitution elasticity between clean and dirty energy inputs can significantly exceed one (around 2 in the electricity generation sector and 3 in the non-energy sectors), contrary to the findings in earlier interfuel substitution literature.

Hence, in GRACE, we allow for gradually larger substitution possibilities between fossil fuels and renewables by changing two parameters. Substitution elasticities between electricity and fossil fuels gradually become higher for final users (including producers and consumers) from 2020 to 2050, and so does the substitution elasticity between electricity generated from fossil fuels and renewable energy sources. For final users, the values of the substitution elasticities are assumed to increase from 0.5 (or 0.4) for producers (or households) in 2020 by 6.5 per cent yearly until 2040 and then the yearly rate is gradually lowered to 3.25 per cent in 2050, when the elasticities become 2.8 for producers and 2.2 for households. For the substitution elasticities between thermal and other types of power generation (1.386 for base load and 0.472 for peak load in 2020), the rates are 5.5 per cent until 2040 and are gradually lowered later until reaching 2.75 per cent in 2050. In 2050, the substitution elasticities become 6.0 for base load and 2.0 for peak load.

In GRACE, energy efficiency is implemented directly as a change over time in the parameter of energy-augmented technology in production functions and household consumption functions. Hence, it differs from the widely used energy intensity (energy use per unit GDP). If such a change is a 1 per cent increase, then it is interpreted as a 1 per cent decrease in energy input (or energy use) to produce a given amount of output (or welfare level), while all other inputs remain constant.

In NZE IEA, the energy intensity is reduced yearly by 4.2 per cent from 2020 to 2030 and 2.7 per cent from 2030 to 2050, compared to a decrease of 2.3 per cent in STEPS. Following the pattern of the energy intensity over time in the NZE IEA scenario, we assume that the parameter of energy-augmented technology for all final energy users of all regions in NZE GRACE increases by 1.5 per cent yearly until 2030 and 1.0 per cent between 2030 and 2050 compared to the level in STEPS. As a result, to produce a given amount of output, the assumed technology changes mean that the energy needed in 2050 is only 35 per cent of the energy needed in 2020 in NZE GRACE, other things being equal.

In NZE IEA, it is suggested to stop establishing new coal mines and mine extensions, and in addition to stop approving new oil and gas fields for development from 2021. This implies that the fossil reservoirs available for production are gradually reduced over time. Hence, this is interpreted as the gradual reduction of natural resources for fossil fuel production. In GRACE, we assume that the natural resources for fossil fuel production are gradually reduced from 2020 to become only 10 per cent for oil and 5.5 per cent for coal and gas of the 2050 level in the reference scenario. This measure alone pushes up the cost and price of fossil fuel production, resulting in less supply and demand.

3.5. The NZE Scenario in FRISBEE

We apply the following strategy for all three fossil fuels to establish the NZE scenario in FRISBEE. First, we implement the regional endogenous GDP growth rates from NZE GRACE discussed in the last section. We introduce the CO_2 taxes by year and region derived from the NZE IEA scenario [24]. Further, we introduce exogenous amounts of non-fossils (renewables and nuclear) in the power sector. We have the regional volume of non-fossils per year and region for the Sustainable Development Scenario (SDS) in IEA [27]. We also have total global volume of non-fossils in NZE per year. We adjust the regional figures in SDS upwards with global non-fossil NZE/global non-fossils SDS. This makes electricity demand in FRISBEE align relatively close to the large increase in electricity demand in NZE IEA. In addition, we increase the efficiency improvement in demand for fossil fuels by 4 per cent to roughly align with IEA [4]. They apply a reduction in energy intensity (energy use/GDP) of more than 4 per cent per year between 2020 and 2030 and almost 3 per cent in the 2030-2050 period. Further, we apply the endogenous oil price from NZE GRACE. We also run model simulations with a halt in new investments in gas reserves and also reduce coal supply from 2021, which are key milestones in NZE IEA. Lastly, we consider if a further increase in energy efficiency is necessary to reduce demand even more. Hence, the measures we take into consideration are CO_2 taxes (the same as that introduced in GRACE), increased renewables in the power sector (following NZE IEA), increased AEEI by 4 per cent for oil consumption and 8 per cent for coal and gas, the endogenous global oil price from GRACE, a stop in gas investments, and a lower supply for coal.

4. Results

4.1. The Energy Markets in the NZE Scenario in GRACE

Figure 1 shows that oil demand in GRACE follows closely to the demand in NZE IEA but is somewhat higher over the simulated period, ranging from around 200 to 400 Mtoe. For the IEA, we only have data for specific years. In 2050, oil demand is reduced by 88 per cent from the reference scenario, lower than the reduction of 94 per cent in NZE IEA from STEPS (see Table 2). Gas demand in GRACE is always lower than in NZE IEA, above all in 2030 when it is 22 per cent below the NZE IEA target. Gas demand in GRACE ends up 93 per cent lower than the level in the reference scenario in 2050, somewhat lower than the 95 per cent in NZE IEA. Coal demand in GRACE is almost the same as NZE IEA in 2020 but is higher in the subsequent years. In 2050, coal demand is 93 per cent lower than in the reference scenario. This is lower than the reduction of 98 per cent in NZE IEA from STEPS (see Table 2).

Table 2. Reduction in predicted final demand and CO_2 emissions from reference scenarios in 2050; per cent.

	NZE FRISBEE	NZE GRACE	NZE IEA *
Oil	89	88	94
Gas	89	93	95
Coal	93	93	98
CO_2	91	91	95

* Calculated from extrapolation for 2040-50. We add the amount of fossil fuel use with CCS in NZE IEA to allow for additional emissions in the two models. With only unabated use of gas and coal, the emission reduction in NZE IEA would have been closer to 100 per cent, especially when it comes to coal.

The relatively higher demand for coal and oil in GRACE leads to higher CO_2 emissions from fossil fuels than in NZE IEA (Figure 2). In 2030, emissions are almost the same as in NZE IEA due to lower gas demand offsetting the effect of higher coal and oil demand. In 2050, emissions are 91 per cent below the reference scenario. This is lower than the reduction of 95 per cent in NZE IEA (see Table 2).

Figure 1. Comparison of primary fossil fuel consumption predicted in NZE IEA and NZE GRACE. Markers show the predicted values for the years available from IEA [4].

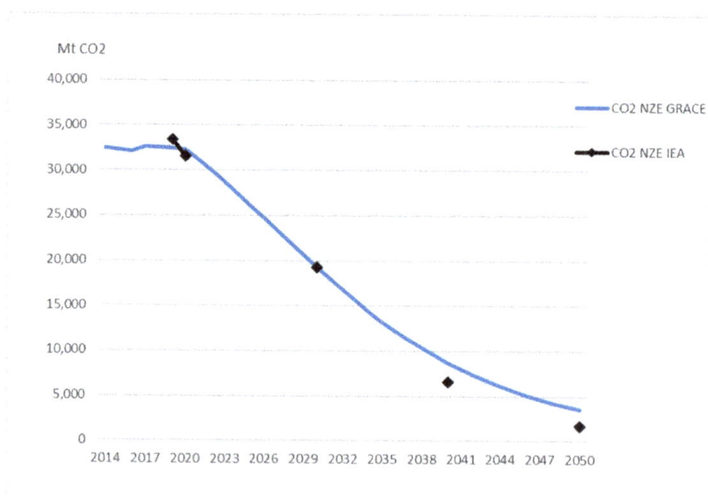

Figure 2. Comparison of CO_2 emissions predicted in NZE IEA and NZE GRACE. Markers show the predicted values for the years available from IEA [4].

Table 3 shows the effect of each measure on oil, gas, and coal demand. Introducing CO_2 taxes reduces oil demand only by 2 per cent in 2050 as oil can hardly be substituted by other energy goods at this level of substitution possibilities, not even for electricity in transport. The effect of the two measures directed towards power generation contributes only marginally to the reduction in oil demand, which is due to limited substitution possibilities, assumed as oil is only to a small extent used in power generation on a global scale. With increased substitution possibilities between oil and electricity, oil demand becomes 14 per cent lower than in the reference scenario. With improved efficiency of energy used by final users, oil demand is further reduced to become 21 per cent lower than in the reference scenario. When we assume reduced natural resources for oil supply, oil demand declines further to 88 per cent below the reference level in 2050.

Table 3. Reduction in predicted primary energy demand in 2050 due to each additional measure in NZE GRACE.

Measures Adopted in GRACE	2050		
	Oil	Gas	Coal
Consumption in reference scenario	4797 mtoe	5114 mtoe	3699 mtoe
CO_2 taxes	2%	25%	10%
+ Cost changes in power generation	2%	25%	11%
+ Upper limits on thermal power	2%	30%	15%
+ Increased substitution between fossil fuels and renewables	14%	56%	62%
+ Energy efficiency	21%	60%	66%
+ Reduced natural resources for fossil fuel production	88%	93%	93%

Surprisingly, introducing CO_2 taxes has a much larger effect on gas demand than oil and reduces gas demand by 25 per cent in 2050, mainly because of relatively higher carbon taxes per monetary unit (USD) for gas users in most regions. The carbon tax in NZE GRACE is the same for each ton of CO_2 emissions in a region. As the carbon content of gas is lower, the carbon tax per ton gas is lower than that per ton coal. However, in the reference scenario simulated by GRACE, the gas price is already lower than the coal price in 2050 since the stated policies of governments encourage gas consumption rather than coal consumption. As a result, the carbon tax per USD gas becomes higher than that per USD coal. The two measures directed towards power generation further reduce gas demand by 5 percentage points. Easier substitution between electricity and fossil fuels for final users makes gas demand 56 per cent lower than in the reference scenario in 2050. Improved energy efficiency reduces gas demand further by only 4 percentage points from the reference level. Reduced natural resources for gas supply further reduces gas demand by 33 percentage points from the reference scenario.

Introducing CO_2 taxes reduces coal demand by 10 per cent from the reference scenario, partly because coal is not easy to replace in the main coal-use regions such as, e.g., China, although coal is more carbonaceous than gas and is hit more by the CO_2 tax. The joint effect of the two measures directed towards power generation reduces coal demand further by 5 percentage points. Increased substitution possibilities between electricity and fossil fuels leads to considerable reduction in coal demand by a further 43 percentage points from the level in the reference scenario. Improved energy efficiency contributes to reducing demand by 4 percentage points. When we, in addition, restrict natural resources availability for coal production, total demand in 2050 is reduced by 93 per cent from the reference level.

GDP in NZE GRACE is always higher than in NZE IEA from 2020 to 2050, and the difference slightly increases from 2040 (see Figure 3) due to, e.g., the ignorance in GRACE of the COVID-19 pandemic. This partially explains why the CO_2 emissions in NZE GRACE are greater than in NZE IEA. However, GDP in NZE GRACE is lower than in the reference scenario (STEPS) and the difference increases over time. Notice that the model does not consider the cost of implementing the various measures and the simulated GDP might to a large extent underestimate the potential GDP losses to achieve the 1.5 °C scenario. In the measures we introduced, only CO_2 taxes and reduced fossil fuel resources lead to GDP losses, while we do not consider the costs of the extensive use of renewable energy and the high pace of energy efficiency improvement.

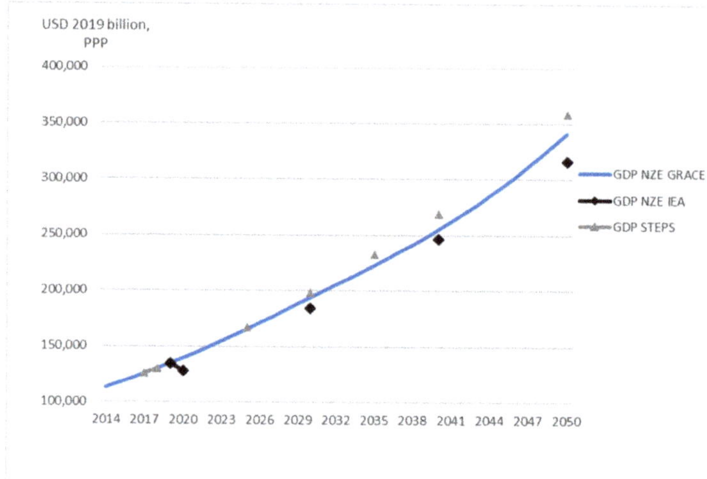

Figure 3. Comparison of world GDP predicted in STEPS, NZE IEA, and NZE GRACE. The 2050 GDP in STEPS is calculated from extrapolation for 2040-2050. Markers show the predicted values for the years available from IEA [4].

Gas is the least carbonaceous fossil fuel compared to coal and oil. Figure 4 shows the sectoral shares of global gas demand in 2050 in the reference scenario and NZE GRACE. The shares of gas demand increase in four sectors, power generation, other transport than air and water, gas production, and crude oil production, compared to the shares in the reference scenario. The reason is that it is relatively harder to replace gas with renewables in these sectors.

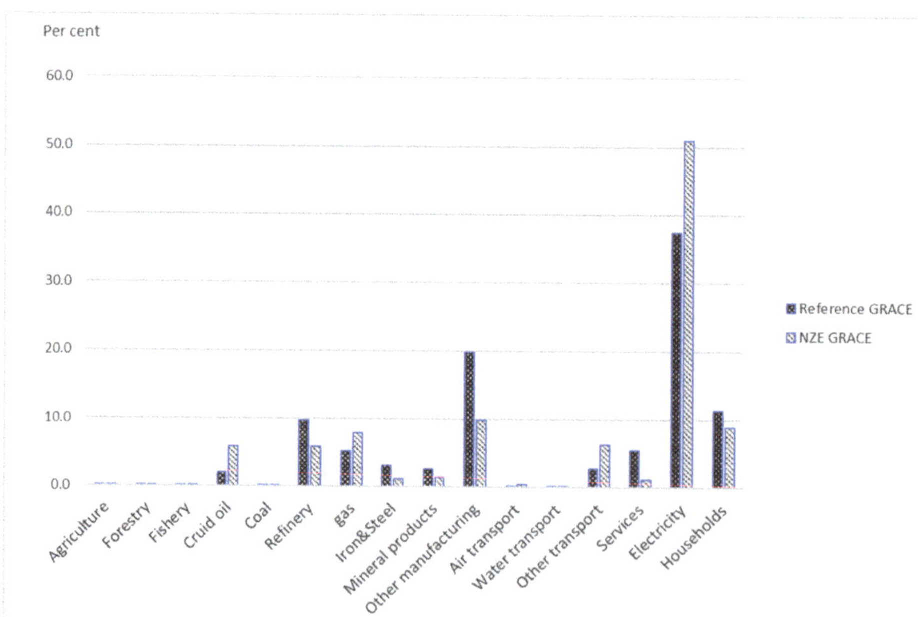

Figure 4. Sectoral shares of global gas demand in 2050 predicted by GRACE.

4.2. The Energy Markets in the NZE Scenario in FRISBEE

Figure 5 shows the aggregated effects on energy markets. Oil demand in FRISBEE is somewhat higher than demand in NZE IEA until 2040 (note that we do not consider short-term effects such as the decline in oil demand in 2020 that was due to the COVID-19 pandemic). In 2050, oil demand is reduced by 89 per cent from the reference scenario, almost at the same level as NZE GRACE and somewhat lower than the reduction of 94 per cent in NZE IEA from STEPS (see Table 2). We emphasize that the differences between our model results and NZE IEA in 2050 also are shown in Figures 1 and 5.

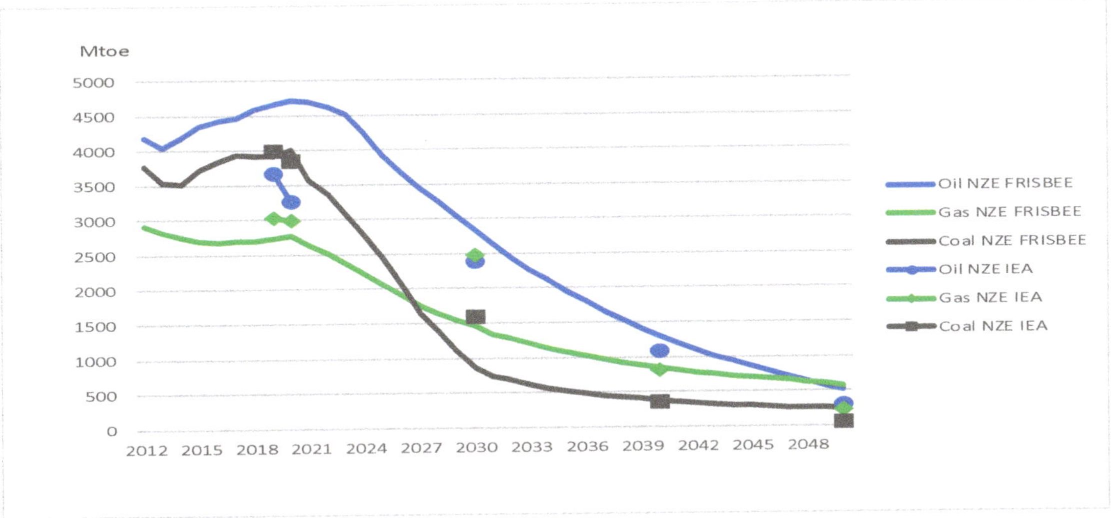

Figure 5. Comparison of primary fossil fuel consumption predicted in NZE IEA and NZE FRISBEE. Markers show the predicted values for the years available from IEA [4].

Gas demand in FRISBEE is lower than NZE IEA in 2030, but at the same level in 2040. Gas demand in FRISBEE ends up 89 per cent lower than the level in the reference scenario in 2050, somewhat lower than the reduction of 93 per cent and 95 per cent in NZE GRACE and NZE IEA, respectively (see Table 2). The gas supply in FRISBEE is affected by, e.g., prior investments (which again is a function of the extensive reserve database). Hence, it is difficult to further adjust demand-side parameters to reduce demand even more (as it must equal supply).

Coal demand in FRISBEE is also lower than in NZE IEA in 2030, but at the same level in 2040. Further, in 2050, coal demand is 93 per cent lower than in the reference scenario. This is at the same level as in NZE GRACE and somewhat lower than the reduction of 98 per cent in NZE IEA from STEPS (see Table 2).

The effects on fossil fuels in FRISBEE lead to a decline in CO_2 emissions, as shown in Figure 6. In 2030, emissions are lower than in NZE IEA due to lower coal and gas demand (even if oil demand is somewhat higher), but at the same level a decade later. In 2050, emissions are 91 per cent lower than in the reference scenario. This is at the same level as NZE GRACE and somewhat lower than the reduction of 95 per cent in NZE IEA (see Table 2).

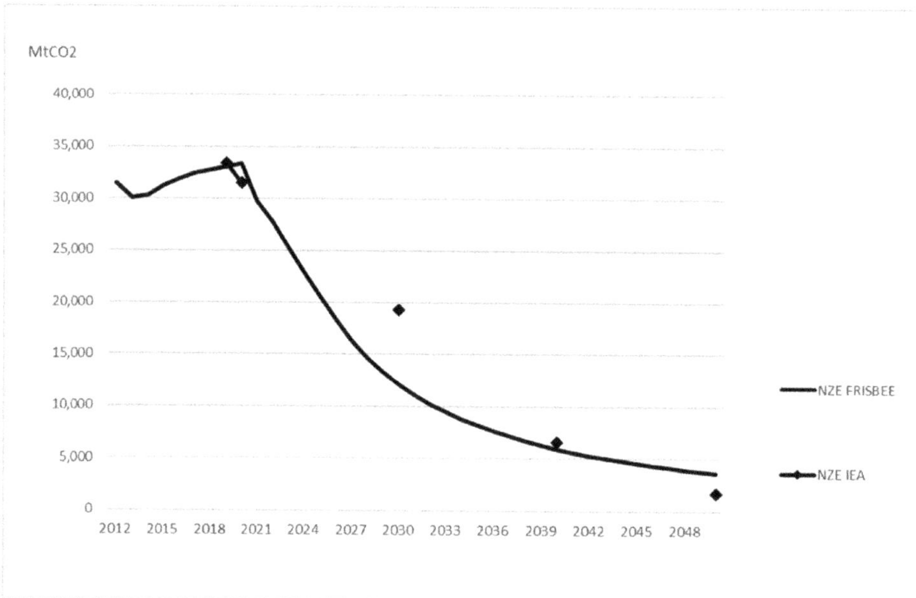

Figure 6. Comparison of CO_2 emissions predicted in NZE IEA and NZE FRISBEE. Markers show the predicted values for the years available from IEA [4].

Table 4 shows the effects of each measure on oil, gas, and coal demand in 2050. We emphasize that demand for all energy goods in either households or industry are log-linear functions of end-user prices, income, and AEEI, i.e., an increase in AEEI of x per cent leads to a decline in demand of x per cent for given prices and income.

Table 4. Reduction in predicted primary energy demand in 2050 due to each measure in NZE FRISBEE.

	2050		
	Oil	**Gas**	**Coal**
Reference scenario	4703 mtoe	5318 mtoe	3475 mtoe
CO_2 taxes and increased renewables in power sector	14%	33%	53%
+ AEEI = 4% (Includes increased bioenergy in households and industry.)	67%	80%	77%
+ Endogenous oil price	89%		
+ Stopping gas investments (Oil demand is not influenced by investments (as explained in the text).)		85%	
+ Lower supply for coal			84%
+ AEEI = 8%		89%	93%

Let us first take a closer look at oil demand. We see from Table 4 that introducing CO_2 prices and increased renewables in power production reduces oil demand only by 14 per cent in 2050. The reason is primarily that oil is hardly used in power production on a global scale. With an AEEI of 4 per cent, which is roughly in line with improvements in energy intensity in NZE IEA and NZE GRACE, demand is further reduced. In addition, we insert the endogenous oil price from NZE GRACE. This oil price is around four times higher than the oil price in the reference scenario in FRISBEE, and leads to a reduction in oil demand by 89 per cent from the reference scenario.

We emphasize that stopping new oil investments from 2021 does not change oil demand. OPEC satisfies the residual between demand and non-OPEC supply at the exogenous oil price, that being the reference oil price in STEPS or the endogenous oil price in NZE GRACE. A credible defense of the price target requires surplus capacity. In our model, we therefore assume that OPEC will always invest sufficiently to maintain a capacity surplus of 10 per cent. As demand and non-OPEC supply are determined independently of each other, stopping non-OPEC investment in new reserves has no effect on total supply (=demand) (In NZE FRISBEE, OPEC manages to keep the required surplus capacity from 2021, when they increase production as non-OPEC supply declines).

In the scenario with the endogenous oil price from NZE GRACE, OPEC is thrown out of the market if non-OPEC is allowed to continue to invest. Hence, we show the effect of stopping investment in the scenario with AEEI of 4 per cent and with the reference oil price. Figure 7 shows that the world oil supply declines by almost 70 per cent from 2021 to 2050. This is mainly due to a relatively steep decline in OPEC production and, to some extent, a moderate decrease in non-OPEC supply. OPEC satisfies the residual demand between global supply (=demand) and non-OPEC production over the period. Stopping investment in non-OPEC countries leads to a large reduction in their supply. However, world supply does not change as OPEC increases production to keep demand at the prevailing oil price.

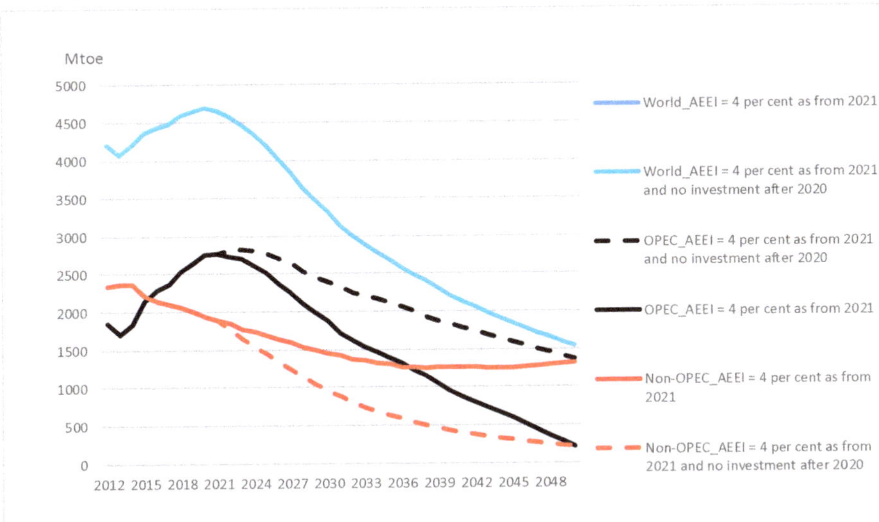

Figure 7. Oil supply; world, OPEC, and non-OPEC in various scenarios.

If we assume that OPEC, instead of increasing supply to keep production constant, stops investments in non-OPEC when AEEI is 4 per cent, this reduces oil supply and consumption by 90 per cent from the reference scenario in 2050. Thus, a more realistic energy efficiency improvement of 4 per cent and a reference oil price would then be sufficient to let oil demand be relatively close to the NZE IEA target. However, we implement the high endogenous oil price from NZE GRACE and demand is reduced by 89 per cent from the reference scenario.

Introducing CO_2 prices and increased renewables in power production reduces gas demand by 33 per cent, mainly because gas is used in power production. By introducing a yearly AEEI of 4 per cent, gas demand is reduced to 80 per cent of the reference level in 2050. Stopping all new investments as from 2021 clearly also has an impact on gas demand as the reduction now is 85 per cent. By applying an AEEI of 8 per cent, demand declines further to 89 per cent of the reference scenario in 2050. We might add that stopping investment before introducing the AEEI of 4 per cent would make this measure stand out as relatively more important than the impression one gets from Table 4. Figure 8 shows world gas supply in various scenarios. When we move from a situation with an AEEI of 4 per cent to a situation where we, in addition, stop new investments, supply is only reduced by around one-third in 2050. We see that, if we, in addition, increase energy efficiency improvements, there are still enough profitable proven reserves today to sustain production until 2050 at around one-fifth of today's level. Production is increasingly concentrated in resource-rich countries due to the large size and slow decline rates of their existing fields. In 2050, a large part of global gas is produced in the Middle East and Russia.

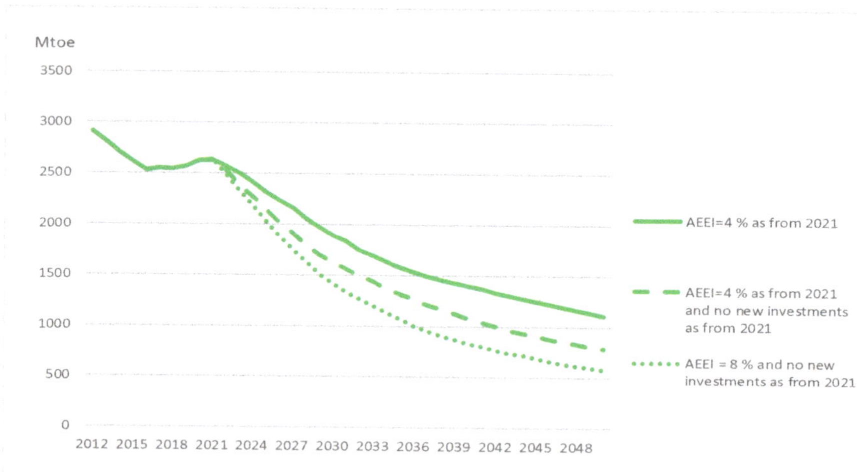

Figure 8. World gas supply in various scenarios.

Introducing CO_2 taxes and increased renewables in power production has a much larger effect on coal than on gas, as coal demand declines by 53 per cent from the reference scenario. Coal is also used in power production, but the main reason is that coal, in addition, is more carbonaceous than gas and thus is hit more by the CO_2 tax. By introducing a yearly AEEI of 4 per cent, coal demand declines by 77 per cent from the reference level. By suppressing coal supply further, demand is 84 per cent lower in 2050. By applying an AEEI of 8 per cent, coal demand declines further to 93 per cent from the reference scenario in 2050.

5. Discussion

According to [4], technology alone is not enough to reach NZE in 2050. The active support of people is crucial. The report does not describe this as behavioral changes, but a mixture of low carbon technologies and people's engagement, such as "buying an electric vehicle or insulating a loft". However, it is emphasized that behavioral changes—meaning adjustments in everyday life that reduce, e.g., excessive energy consumption—are also needed. IEA [4] argues that this is especially important in richer parts of the world where energy-intensive lifestyles are the norm. Behavioral changes include cycling or walking instead of driving, turning down heating, and going on holiday nearer to home. Total accumulated CO_2 emissions in the NZE between 2021 and 2050 are around 4 per cent less than they would be without such behavioral changes.

Neither GRACE nor FRISBEE include behavioral changes or changes in preferences. Behind demand and supply lies traditional maximization of utility and profit. However, if we include behavioral changes to the extent that the IEA does, reduction in emissions in 2050 in both models might increase closer to the level of NZE IEA, as Table 2 shows. Further, we included both energy use with CCS and unabated use in our simulations when making the comparison in Table 2. Hence, if CCS were included in our models, reduction in energy use and emissions from the reference scenario could be closer to 100 per cent. Further, opposed to the IEA, we did not include hydrogen in our scenarios.

Even with the introduction of CCS and energy alternatives such as hydrogen, strict measures are necessary to reduce emissions to achieve the demand required by a 1.5 °C scenario. A major worldwide push to increase energy efficiency is an essential part of the efforts to reduce energy use and emissions. We introduced an efficiency improvement of 8 per cent per year in FRISBEE for gas and coal, but we see from Table 2 and the discussion related to the table that large reductions in emissions can be obtained by an AEEI of 4 per cent. For oil, an AEEI of 4 per cent is enough. Both NZE GRACE and NZE IEA operate with reductions in energy intensity. Both apply a reduction in energy intensity (energy use/GDP) of around more than 4 per cent per year between 2020 and 2030 and almost 3 per cent in the 2030–2050 period. According to IEA [4], 4 per cent is about three times the average intensity decline rate achieved over the last two decades. To accomplish such an increase would, of course, be a very challenging task.

The IEA has been criticized for applying exogenous growth in GDP in their scenarios. Mohn [28] emphasizes that empirical models of energy economics and climate change should open for the indigenization of economic activity. For the IEA, this could allow for variation in energy prices and policies to imply corresponding variation in economic growth between the different scenarios. IEA [4] uses similar exogenous GDP growth in both NZE and the reference scenario. GRACE endogenizes the growth in NZE, which is inserted in NZE FRISBEE and leads to a somewhat lower growth compared to the STEPS scenario. However, we did not take into consideration all costs connected to the climate policies, e.g., efficiency improvement, and how these costs may affect future growth. Empirical estimates of such costs vary considerably and depend on sectoral and regional environment [11].

We apply the endogenous oil price from NZE GRACE in the NZE FRISBEE scenario. This oil price is four times higher than the reference oil price in STEPS, reaching over 400 USD (2012 prices) per barrel in 2050. Is this realistic? IMF [29] has shown that the NZE scenario can be consistent with both increasing and declining oil prices. When they only consider demand-side policies, oil prices could decline to 20 USD in 2030. When reductions in oil production are driven by supply-side measures, like the one we implement in NZE GRACE, this would result in a strong upward pressure, taking prices to roughly 190 USD a barrel in 2030. The latter oil price is somewhat higher than our price in 2030 in the NZE scenarios.

In the simulation of GRACE, the lower availability of natural resources in fossil production is crucial for reducing CO_2 by pushing up the cost of fossil fuel production. On the other hand, we must introduce relatively high levels of efficiency improvement in FRISBEE for gas and coal while allowing extraction to phase out gradually over time to approach the emission reduction level of NZE IEA. As a result, the gas and coal prices are relatively low in FRISBEE and can potentially result in more gas and coal consumption, a well-known phenomenon of rebound effects [11,12]. As FRISBEE is a partial equilibrium model, the model does not consider the potential economy-wide rebound effects on energy consumption of energy efficiency improvement. However, the rebound effects are taken into account in the macroeconomic model GRACE and thus lead to more energy consumption, which has to be controlled by supply-side measures, e.g., the lower availability of natural resources in fossil production in our case.

When we introduce CO_2 taxes, increased use of renewables, and an efficiency improvement of around 4 per cent, emissions are reduced more in FRISBEE than in GRACE in 2050. In NZE GRACE, lower availability of oil resources, but also to some extent gas and coal resources, are paramount to reach the desired emission targets. In NZE FRISBEE, it is to some extent important that gas investment in new reserves is stopped from 2021 and that coal supply is reduced. Oil investments decline rapidly due to the high oil price implemented from NZE GRACE. One might question if we can see signs of a halt in development of new oil, natural gas, and coal reserves or a reduction in investments. IEA [30] concludes that, even if the pace of growth in renewable investment has accelerated since 2020, today's fossil fuel spending is too high for a pathway aligned with limiting global warming to 1.5 °C.

Another milestone in NZE IEA is in power generation, where overall net-zero-emissions electricity is achieved by 2040 globally. This requires no additional coal power stations, with generation from unabated plants phased out by 2040 globally. Unabated natural gas power generation must fall by 90 per cent globally by 2040 from 2020. Global Energy Monitor [31] shows that a large amount of new coal-fired power capacity is still being built. In addition, Wilson et al. [32] conclude that these milestones on the route to NZE are at odds with the current trajectory of our energy system, with significant expansion planned in both fossil fuel reserves and fossil fuel power. They are also at odds with the financial flows supporting this expansion, despite financial institutions with 130 trillion USD in assets committing at COP26 in Glasgow in 2021 to align their portfolios with net zero by 2050. For fossil fuel production, the case studies in Wilson et al. [32] show that oil and gas companies' financing are continuing to expand fossil fuel reserves, allocating approximately 10 per cent of capital expenditures (CAPEX) to exploration-related activities.

6. Conclusions

A net zero pathway to 2050 in the energy sector was created in this study by soft-linking an energy model with a macroeconomic model. Based on a reference scenario assuming the same population growth, GDP growth rates, and CO_2 prices, both models were modified to roughly follow the same pathway of both demand for fossil fuels and total CO_2 emissions from the combustion of fossil fuels as in the NZE scenario from IEA [4]. We apply an endogenous GDP effect in our NZE scenarios, contrary to, e.g., NZE IEA where GDP growth is exogenous. To achieve our NZE scenarios, various mitigation measures were implemented simultaneously in our models, including CO_2 taxes, improved energy efficiency, more renewables in electricity production and other sectors, easier substitution between electricity and fossil fuels for final users, and drastically limiting future production of fossil fuels.

We find that, to achieve the net zero pathway, the energy model needs relatively high levels of efficiency improvement and the macroeconomic model needs relatively strict supply-side measures to reduce production of fossil fuels. This is likely related to the potential role of economy-wide rebound effects on energy consumption of energy efficiency improvement. Such rebound effects are simulated by the macroeconomic model but cannot be taken into consideration by the partial equilibrium energy model.

We conclude that net zero is possible by introducing very strict measures in a modelling world. However, the rate of, e.g., energy efficiency improvement needed is far above what has been achieved in the past. Further, even if renewables are on the rise, the world does not seem to be on a trend of sufficiently declining fossil fuel use, as investment in both new reserves and new fossil power plants continues. This reminds us what is needed for a 1.5 °C world, although it will be very challenging, if not impossible, to make different governments agree on such measures.

Author Contributions: All authors contributed to the study design. L.L. conducted simulation and analysis based on FRISBEE modeling. T.W. performed simulation and analysis based on GRACE modeling. L.L. wrote the first draft of the manuscript and all authors commented on previous versions of the manuscript. All authors have read and agreed to the published version of the manuscript.

Funding: This study was made possible with financial support from Research Council of Norway (grant no. 303486).

Institutional Review Board Statement: Not applicable.

Informed Consent Statement: Not applicable.

Data Availability Statement: We declare that the data and materials presented in the article are reliable.

Acknowledgments: Thanks to Knut Einar Rosendahl, Ådne Cappelen, Solveig Glomsrød, and three anonymous reviewers for valuable and constructive comments.

Conflicts of Interest: The authors have no relevant financial or non-financial interest to disclose.

Appendix A. More Details on Reference Scenarios

How does the reference scenario in GRACE compare with STEPS? Fossil fuel consumption and CO_2 emissions are slightly different from STEPS (Figure A1) due to sectoral and regional differences between STEPS and the reference scenario of GRACE. We see from Figure A1 that oil demand is generally marginally lower in GRACE than in STEPS and, in 2040, demand is 2 per cent lower. Demand for coal and gas is almost the same as in STEPS. Figure A2 shows that CO_2 emissions in GRACE are marginally higher than in STEPS—emissions are 3 per cent higher in 2040.

How does the FRISBEE reference scenario align with STEPS? We see from Figure A3 that oil demand generally is marginally higher in FRISBEE than in STEPS up to 2030, but is marginally lower in 2040. Demand for gas is generally lower than in STEPS and is almost 9 per cent below the level in STEPS in 2040 (A simple extrapolation of gas demand in STEPS after 2040 shows that demand in FRISBEE is 12 per cent higher in 2050). The gas price is endogenous in FRISBEE and supply is affected by, e.g., prior investments (which again is a function of the extensive reserve database). Hence, it is difficult to adjust demand-side parameters to change demand (as it equals supply). Coal demand is generally marginally higher than in STEPS up to 2030, but is slightly lower in 2040 (We also adjust demand for electricity and bioenergy to the final end-users households and industry). Figure A4 shows that CO_2 emissions in FRISBEE ends up marginally lower than in STEPS in 2040.

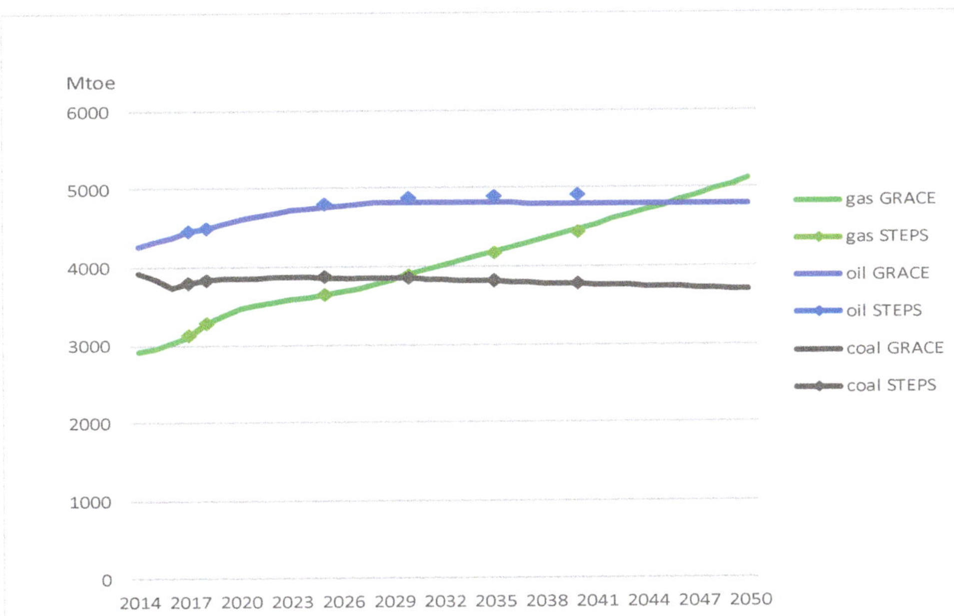

Figure A1. Comparison of fossil fuel consumption in STEPS and the reference scenario in GRACE.

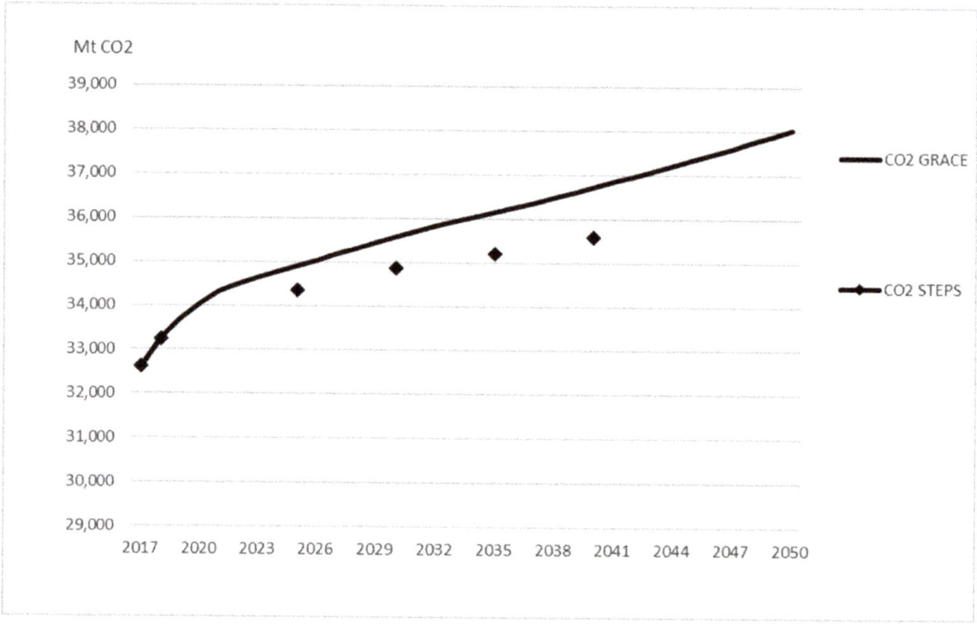

Figure A2. Comparison of CO_2 emissions from fossil fuels in STEPS and the reference scenario in GRACE.

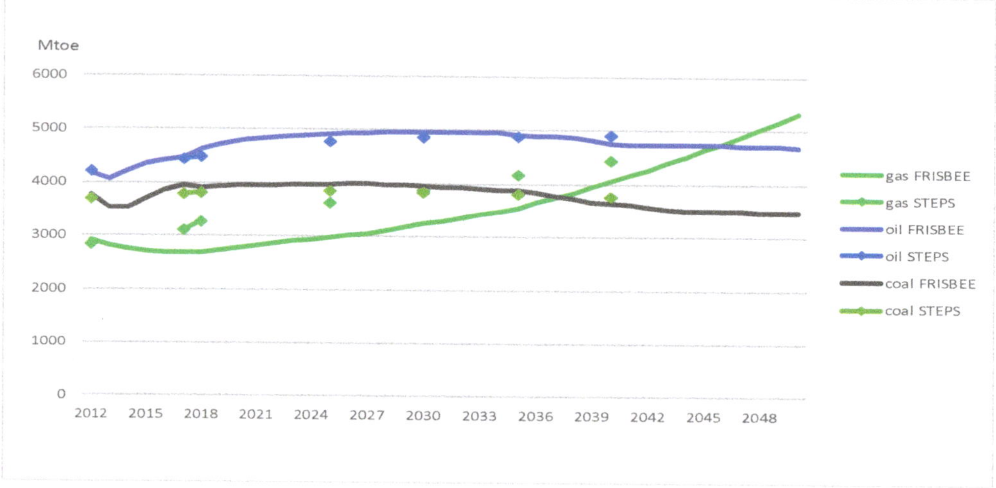

Figure A3. Comparison of primary fossil fuel consumption in STEPS and the reference scenario in FRISBEE.

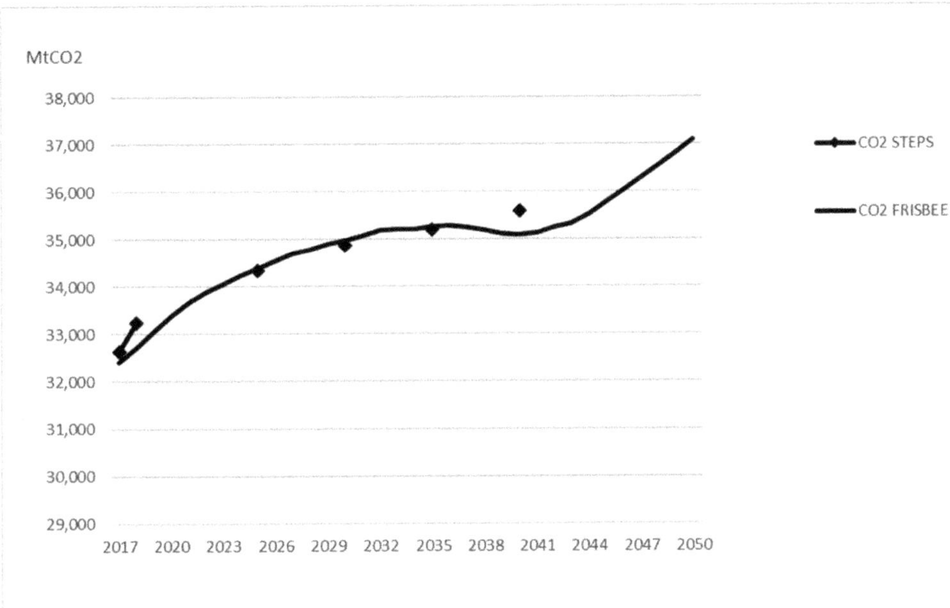

Figure A4. Comparison of CO_2 emissions from fossil fuels in STEPS and the reference scenario in FRISBEE.

References

1. Masson-Delmotte, V.; Pörtner, H.O.; Roberts, D.; Skea, J.; Shukla, P.R.; Pirani, A.; Moufouma-Okia, W.; Péan, C.; Pidcock, R.; Connors, S.; et al. Global Warming of 1.5 °C. An IPCC Special report on the impacts of global warming of 1.5 °C above pre-industrial levels and related global greenhouse gas emission pathways. In *The Context of Strengthening the Global Response to the Threat of Climate Change, Sustainable Development, And Efforts to Eradicate Poverty*; 2018; Available online: https://www.ipcc.ch/sr15/ (accessed on 27 February 2023).
2. Jones, C.D.; Frölicher, T.L.; Koven, C.; MacDougall, A.H.; Matthews, H.D.; Zickfeld, K.; Rogelj, J.; Tokarska, K.B.; Gillett, N.P.; Ilyina, T.; et al. The Zero Emissions Commitment Model Intercomparison Project (ZECMIP) contribution to C4MIP: Quantifying committed climate changes following zero carbon emissions. *Geosci. Model Dev.* **2019**, *12*, 4375–4385. [CrossRef]
3. Energy and Climate Change. Special issue on Net-Zero Energy Systems. *Energy Clim. Chang.* **2021**, *2*, 100066. [CrossRef]
4. IEA. *Net Zero by 2050: A Roadmap for the Global Energy Sector*; International Energy Agency, OECD: Paris, France, 2021.
5. Brecha, R.J.; Ganti, G.; Lamboll, R.D.; Nicholls, Z.; Hare, B.; Lewis, J.; Meinshausen, M.; Schaeffer, M.; Smith, C.J.; Gidden, M.J. Institutional decarbonization scenarios evaluated against the Paris Agreement 1.5 °C goal. *Nat. Commun.* **2022**, *13*, 4304. [CrossRef] [PubMed]
6. BP. Energy Outlook 2020 Edition. 2020. Available online: https://www.bp.com/content/dam/bp/business-sites/en/global/corporate/pdfs/energy-economics/energy-outlook/bp-energy-outlook-2020.pdf (accessed on 15 February 2023).
7. Equinor. Energy Perspectives 2020. 2020. Available online: https://www.equinor.com/sustainability/energy-perspectives-previous-reports (accessed on 15 February 2023).
8. Shell. The Energy Transformation Scenarios. 2021. Available online: www.shell.com/transformationscenarios (accessed on 15 February 2023).
9. Huppmann, D.; Rogelj, J.; Kriegler, E.; Krey, V.; Riahi, K. A new scenario resource for integrated 1.5 °C research. *Nat. Clim. Chang.* **2018**, *8*, 1027–1030. [CrossRef]
10. Saunders, H.D. Recent evidence for large rebound: Elucidating the drivers and their implications for climate change models. *Energy J.* **2015**, *36*, 23–48. [CrossRef]
11. Saunders, H.D.; Roy, J.; Azevedo, I.M.L.; Chakravarty, D.; Dasgupta, S.; de la Rue du Can, S.; Druckman, A.; Fouquet, R.; Grubb, M.; Lin, B.; et al. Energy Efficiency: What Has Research Delivered in the Last 40 Years? *Annu. Rev. Environ. Resour.* **2021**, *46*, 135–165. [CrossRef]
12. Wei, T.; Liu, Y. Estimation of global rebound effect caused by energy efficiency improvement. *Energy Econ.* **2017**, *66*, 27–34. [CrossRef]
13. Lenzen, M.; Keyβer, L.; Hickel, J. Degrowth scenarios for emissions neutrality. *Nat. Food.* **2022**, *3*, 308–309. [CrossRef]

14. OECD. Indicators to Measure Decoupling of Environmental Pressure from Economic Growth. 2002. Available online: http://www.oecd.org/dataoecd/0/52/1933638.pdf (accessed on 15 February 2023).
15. Aaheim, A.; Orlov, A.; Wei, T.; Glomsrød, S. *GRACE Model and Applications*; CICERO Report 01; CICERO: Oslo, Norway, 2018.
16. Lindholt, L.; Glomsrød, S. Phasing out coal and phasing in renewables—Good or bad news for arctic gas producers? *Energy Econ.* **2018**, *70*, 1–11. [CrossRef]
17. Cappelen, Å.; Glomsrød, S.; Lindholt, L.; Rosendahl, K.E.; Wei, T. *Stress-Testing the Norwegian Economy: The Effects of the 1.5 °C Scenario on Global Energy Markets and the Norwegian Economy*; Report 7; Statistics Norway: Oslo, Norway, 2021.
18. Carattini, S.; Kallbekken, S.; Orlov, A. How to win public support for a global carbon tax. *Nature* **2019**, *565*, 289–291. [CrossRef] [PubMed]
19. NGFS (Network for Greening the Financial System). Technical Documentation V2.2. 2021. Available online: https://www.ngfs.net/sites/default/files/ngfs_climate_scenarios_technical_documentation__phase2_june2021.pdf (accessed on 27 February 2023).
20. Krook-Riekkolaa, A.; Berg, C.; Ahlgren, E.O.; Söderholmd, P. Challenges in top-down and bottom-up soft-linking: Lessons from linking a Swedish energy system model with a CGE model. *Energy* **2017**, *141*, 803–817. [CrossRef]
21. Cunha, B.S.L.; Garaffa, R.; Gurgel, A.C. *TEA Model Documentation, FGV EESP—FGVAGRO n.00*; Working Paper Series 520; Sao Paulo School of Economics: São Paulo, Brazil, 2020; Available online: https://hdl.handle.net/10438/28756 (accessed on 27 February 2023).
22. IEA. *World Energy Outlook 2019*; International Energy Agency, OECD: Paris, France, 2019.
23. Wei, T. Smoothing a reference pathway based on discrete exogenous projections. *MethodsX* **2022**, *9*, 101605. [CrossRef] [PubMed]
24. Cappelen, Å.; Glomsrød, S.; Lindholt, L.; Rosendahl, K.E.; Wei, T. *Description of a 1.5 °C Scenario with Chosen Measures*; CICERO Report: Oslo, Norway, 2022.
25. Acemoglu, D.; Aghion, P.; Bursztyn, L.; Hemous, D. The environment and directed technical change. *Am. Econ. Rev.* **2012**, *102*, 131–166. [CrossRef] [PubMed]
26. Papageorgiou, C.; Saam, M.; Schulte, P. Substitution between Clean and Dirty Energy Inputs: A Macroeconomic Perspective. *Rev. Econ. Stat.* **2017**, *99*, 281–290. [CrossRef]
27. IEA. *World Energy Outlook 2022*; International Energy Agency, OECD: Paris, France, 2022.
28. Mohn, K. The gravity of status quo: A review of the IEA's World Energy Outlook. *Econ. Energy Environ. Policy* **2018**, *9*, 63–81. [CrossRef]
29. IMF. *World Economic Outlook—War Sets Back the Global Recovery*; International Monetary Fund: Washington, DC, USA, April 2022.
30. IEA. *World Energy Investment 2022*; International Energy Agency, OECD: Paris, France, 2022.
31. Global Energy Monitor. Building an Open Guide to the World's Energy System. 2022. Available online: https://globalenergymonitor.org/ (accessed on 27 February 2023).
32. Wilson, C.; Limburg, A.; Caldecott, B. *Implications of the International Energy Agency Net Zero Emissions by 2050*; Scenario for Net Zero Committed Financial Institutions, Briefing Paper, Oxford Sustainable Finance Group; Smith School of Enterprise and the Environment University of Oxford: Oxford, UK, 2022.

International Journal of
Environmental Research and Public Health

MDPI

Article

The Pathway to China's Carbon Neutrality Based on an Endogenous Technology CGE Model

Shuang Liang [1,2], Xinyue Lin [1,2], Xiaoxue Liu [3,*] and Haoran Pan [1,2]

1 School of Economics and Resource Management, Beijing Normal University, Beijing 100875, China; alice_liang@mail.bnu.edu.cn (S.L.); lin_xinyue@mail.bnu.edu.cn (X.L.); hrpan@bnu.edu.cn (H.P.)
2 Center for Innovation and Development Studies, Beijing Normal University, Zhuhai 519087, China
3 School of Economics, Beijing Technology and Business University, Beijing 100048, China
* Correspondence: liuxx@th.btbu.edu.cn

Abstract: Global warming resulting from greenhouse gas emissions has been a worldwide issue facing humanity. Simultaneously, governments have the challenging task of striking a judicious balance between increased economic growth and decreased carbon emissions. Based on the energy-environment-economy triple coupling (3E-CGE) model, we endogenously integrate climate-friendly technologies into the model's analysis framework through logic curves and refine and modify the CGE model's energy use and carbon emission modules. We conduct a scenario simulation and sensitivity analysis on carbon tax, carbon-trading, and climate-friendly technological progress, respectively. The results reveal that carbon tax and carbon trading contribute to reducing carbon emissions in the short-term but achieving the goals of peak carbon and carbon neutrality will cause the collapse of the economic system. In the long-term, climate-friendly technologies are key to achieving the dual carbon goal; the development of such technologies can also stimulate economic development. The best path for China to achieve its dual carbon goals and economic development in the next 40 years involves effectively combining the carbon tax, carbon trading, and a climate-friendly technological progress. Specifically, China can begin trading carbon in high-emissions industries then impose industry-wide carbon taxes.

Keywords: carbon neutral; endogenous technological progress; computable general equilibrium analysis; economic development; carbon tax; carbon trading

Citation: Liang, S.; Lin, X.; Liu, X.; Pan, H. The Pathway to China's Carbon Neutrality Based on an Endogenous Technology CGE Model. *Int. J. Environ. Res. Public Health* **2022**, *19*, 6251. https://doi.org/10.3390/ijerph19106251

Academic Editors: Taoyuan Wei and Qin Zhu

Received: 26 March 2022
Accepted: 18 May 2022
Published: 20 May 2022

1. Introduction

Recently, global natural disasters caused by climate change have become more frequent, with serious social and economic effects. Climate change has significantly impacted China's natural ecosystem and socio-economic system. Fossil energy shortages, dramatic price increases, and ecological damage are bottlenecks to future economic growth. In the 40 years of reform during China's opening-up period, the nation's rapid economic growth has been accompanied by heavy resource and environmental costs for development due to an over-reliance on resources and factor inputs. On 22 September 2020, China pledged to the world at the general debate of the 75th UN General Assembly that China would scale up its nationally determined contributions, adopt more vigorous policies and measures, and that China aimed to peak carbon dioxide (CO_2) emissions before 2030 and achieve carbon neutrality before 2060. To achieve this "30–60" target, the Chinese government should adopt various approaches to reduce carbon emissions, including various policy-related mechanisms.

Many realistic development needs suggest that achieving peak carbon and carbon neutrality involves extensive, profound economic and social systemic change, and that China's energy- and consumption-related, industrial, and regional structures will experience significant adjustments in the next 40 years. Therefore, China urgently requires a more realistic carbon-neutral model for scenario simulations to indicate the nation's path toward economic transformation. The relevant carbon-emissions models can be divided into 3 categories: (1) top-down models, such as the computable general equilibrium (CGE), dynamic integrated climate-economy, and regional integrated climate-economy models. These analyze the impact of reducing emissions from a macroeconomic perspective, providing useful information on the effects of climate environmental policy implementation [1–3]. (2) Bottom-up models, such as the Program of Energy and Climate Economics (PECE-LIU2020) and the low emissions analysis platform energy model, focus on production-level processes and technologies. These simulate energy consumption and technologies under different target scenarios from the local equilibrium perspective, providing enhanced insights to examine emissions-reduction sectors and initiatives [4]. (3) Combining the former 2 approaches can result in a comprehensive evaluation model. Such models emerged in the 1990s, predominantly as a result of climate change research, and represented multidisciplinary, policy-oriented application approaches developed for integrated evaluations, such as the China greenhouse gas emissions scenario-analysis model, the regional air pollution information and simulation model from the International Institute for Applied System Analysis, the Stockholm Environment Institute's clay-and-sand model, and the Imperial College of Science, Technology, and Medicine's all-scale atmospheric model [5,6]. Such models combine macro- and micro-economic sectors and can examine the economic losses that the economy and society must sustain while meeting specific emissions-reduction targets [7].

Given the imperative concern in implementing a pathway toward carbon neutrality in China at a minimum cost, this paper will first reveal the macroeconomic effects of implementing a carbon policy and technological progress in a different way, using an endogenous technology (3E-CGE) model. From a temporal perspective, this research aims to understand how developing climate-friendly technologies will profoundly impact China's economic structure. Second, this research will comprehensively study a combination of carbon-based policy and climate-friendly technology in the context of China's unique socio-economic, energy, and political contexts.

This study also provides the following contributions: first, we establish a 3E-CGE model, for the first time, with dynamic characteristics of the economy-energy system-carbon emission linkage. The model can portray the trajectory of carbon emissions, the required policies and investments in climate-friendly technology development, and the systemic impacts on the national economy over the next 40 years. Second, a logistic curve is introduced in the model to describe the cycle of technological improvement. The model with endogenous technological progress is closely associated with the investment in the energy sector, which can more clearly reflect the cause of technological development and its effect on the economy-energy-environment system. Third, we simulate the trade-off effects of decreased carbon emissions policy and increased technological progress, and provide the most appropriate development path for China, that is to use a combination of carbon tax and carbon trading policy instruments, while steadily developing climate-friendly technologies.

The remainder of this paper is organized as follows. Section 2 reviews current literature and outlines prior contributions. Section 3 integrates the carbon and technology progress modules with the energy sector to develop a dynamic CGE model. Section 4 discusses peak carbon emissions and the GDP growth rate, and three scenarios will be established and simulated. Section 5 considers the simulation results to analyze the strategies for carbon implementation. Section 6 concludes by proposing several corresponding policy suggestions for China to better realize the "30–60" target and low-carbon transformation.

2. Literature Review

The current literature has various paths of discourse regarding carbon neutrality and carbon emission reduction. For example, researchers have focused on the development of energy technologies to reduce carbon emissions [8–11], as well as carbon taxes and carbon trading [12,13]. A few scholars have also begun to study the long-term dynamics of carbon prices [14]. Further, some CGE models have attempted to simulate development toward carbon neutrality, but most of the existing CGE models consider technology as an exogenous factor and do not adequately describe the role of climate-friendly technologies in the context of carbon neutrality goals. Therefore, this paper considers these previous studies' results in an attempt to solve this problem.

First, existing literature primarily adopts a "bottom-up" model that focuses on describing the development path of energy technologies. Relatively few studies have addressed carbon neutrality's impact on China's macro-economy and micro-economy at the industry level. The Institute of Climate Change and Sustainable Development at Tsinghua University [15] proposed that China should further increase its overall efforts to reduce emissions by promoting breakthroughs in zero- or negative-emissions technologies, strengthening carbon sink-absorption and carbon-removal technologies, and achieving net-zero emissions of all greenhouse gases as soon as possible. Adair Turner et al. [16] comprehensively assessed China's zero-carbon society to argue that the realization of China's carbon neutrality vision requires the complete decarbonization of the power-generation sector; maximum electrification of all economic sectors; and the large-scale application of hydrogen and biomass energies as well as carbon capture, utilization, and storage technologies.

Their study details the development path of an energy system to achieve carbon neutrality in China, but only focuses on such economic growth indicators as GDP, consumption, investments, and import and export trades. Academic discussions on carbon emissions and economic growth have long focused on testing whether the environmental Kuznets curve hypothesis is tenable [17–21]. However, this curve cannot accurately predict the future relationship between the economy and environment, and any study of carbon neutrality in China should not be limited to an economic growth perspective. How can we better optimize China's economic structure? Existing literature has yet to comprehensively analyze industrial restructuring, development modes among non-energy industries, and changes in consumption patterns under carbon neutrality goals; further, any attempts in prior literature are inconsistent. For example, Liu and Cai [22] studied the impacts of technological progress, changes to the industrial structure, and price changes on the intermediate consumption level in the national economy through direct consumption-coefficient and intermediate demand consumption matrices as applied to three major industrial sectors.

Second, the path toward carbon neutrality in China is still unclear, and academic community is still exploring the most feasible solution. According to the Coase theorem, the government can correct negative externalities by internalizing environmental costs into the production and consumption costs of relevant emitters through policy measures or market actions. At present, generally accepted carbon pricing methods primarily include carbon taxes and trading; however, most related studies on these topics are short-term [23–32]. For example, Galeotti and Larsen [17] found that carbon taxes have not performed well in Norway, as a lower energy intensity and changes in the nation's energy structure led to a 14% decrease in carbon dioxide emissions, while a carbon tax reduced carbon dioxide emissions by only 2%. Mardones and García [13] found that implementing a carbon tax (US$5 to US$131 per ton) led to no significant decrease in agricultural emissions. Yang et al. [33] used the CGE model to study different carbon tax prices' impacts on China's provincial economy. The results reveal that levying a Chinese yuan (CNY) 40 carbon tax effectively reduced CO_2 emissions and negative effects.

The latest related research using the CGE model still focuses on examining economic impacts and decreases in carbon emissions through implementing appropriate carbon tax rates [34]. Nong et al. [35] analyzed the impacts of one carbon emissions-trading system on the environment and economy to demonstrate that such a system was effective in Vietnam, as it successfully decreased emissions at a low cost. Choi et al. [36] used the CGE model to analyze the usefulness of South Korea's carbon-trading policies and observed that the best carbon price to promote emissions trading is US$9.14 per ton. Some existing studies on the long-term dynamics of carbon prices have not been conducted in a carbon neutrality context. Ntombela et al. [14] used a dynamic CGE model to assess carbon tax policies' potential impacts on South Africa's agriculture and food sectors, among others. The results indicate that implementing a carbon tax by 2035 will reduce carbon dioxide emissions by 33% compared to a baseline scenario but would result in welfare losses of US$5.9 billion. Zhou et al. [37] used the economic-energy-environment CGE model to analyze the impacts of different carbon tax rates on China's economy and agriculture from 2020 to 2050. They noted that a carbon tax's short-term effects in reducing carbon dioxide emissions are superior to the long-term effects of reducing carbon intensity and improving energy efficiency.

However, some scholars have illustrated that the carbon tax has no significant impacts, but it also has dual emissions-reduction effects. Shi et al. [38] evaluated carbon trading's impact on China through the CGE model to reveal that a carbon-trading mechanism can effectively reduce both carbon and energy intensity and promote processes to conserve energy and reduce emissions in China. This mechanism also has certain simultaneous, negative impacts on economic output. Vera and Sauma [39] pointed out that while a carbon tax as a tax system can promote energy conservation and decreased emissions, it will inevitably redistribute wealth to a certain extent, affecting macroeconomic costs and social welfare. Lin and Jia [34] analyzed the carbon tax system's effects on energy, the environment, and the economy to discover that the carbon tax rate follows the "law of increasing marginal emissions reduction." Considering the driving factors to decrease emissions, the general consensus at this stage is that technological progress, improved efficiency, and adjustments to the energy structure are the main driving factors to decrease carbon dioxide emissions [40].

Third, most studies of existing CGE models regard climate-friendly technologies as a given exogenous factor [41,42], which does not fully illustrate the role of climate-friendly technology given the goal of carbon neutrality, thus failing to establish an effective link between the economy, energy, and the environmental system. Climate change research has mostly incorporated the theory of exogenous technological progress, with improved energy efficiency used to represent technological progress. The flaw in this theory is that technological progress is regarded as a black-box operational process unaffected by price-induced and innovative activities, which cannot truly reflect the process of technological progress. Another application of exogenous technological progress theory in the energy technology field is a purported "backstop" technology. This typically refers to a technology that has been developed but has not yet entered the market or will enter the market at a certain period in the future, thus changing the energy technology progress. Similarly to automatic energy efficiency index (AEEI), Löschel [43] argues that this assumption is unsatisfactory, because it cannot accurately predict a technology's future details and costs in the future.

The recent endogenous economic growth theory posits that long-term economic growth comes from the positive external effects of knowledge accumulation, and the economic system can influence this accumulation of knowledge by adjusting R&D investments, thereby promoting technological progress within the system [44–46]. Generally, models based on exogenous technological progress theory assume that technological progress is a definite time trend, while models based on endogenous technological progress theory regard knowledge accumulation as a form of capital accumulation that represents technological progress [47]. The application of endogenous technological progress theory in climate change takes many forms. Existing literature using endogenous technologies to study low-carbon and economic growth includes dynamic input-output models and CGE models [48]. Pan [49,50] regards the input-output coefficient as the result of combining a specific technology in an early stage and an existing specific technology and believes that technological progress is the process of alternately updating old and new technologies. The research designed a dynamic input-output model and introduced technological progress and diffusion as endogenous variables; R&D investments were used to drive new technological progress along a logical curve until maturity and decay. Further, fixed assets installation investments were used to diffuse new technologies and eliminate old technologies. The result of this alternate adjustment of new and old technologies within the industry promoted an industrial transformation, which then led to changes in the entire industrial structure.

This method was subsequently applied to study the replacement of fossil and non-fossil energy technologies in China's power industry, as low-carbon energy technologies will significantly change China's power industry structure and economic structure in the future. Wang et al. [51] constructed an endogenous technological progress CGE model to study China's climate change issues. The model focuses on technological progress in energy and environment modules. According to the theory of endogenous technological progress, the model separates the R&D matrix from the intermediate input sector, adds a knowledge capital input to the factor sector, and increases R&D investments in the final demand column. Kristkova et al. [52] introduced public R&D investment as a sector in the CGE model to study its effect on agricultural productivity and food security.

Current works on endogenous technological progress involving climate change and economic growth primarily considers R&D investments—or specifically, separates the R&D department—or conducts research with R&D investments as one element. This method not only creates spillover effects and inconsistency between social and private benefits, but also controls the development path of technological progress; thus, it is more suitable for short-term research [53]. Judging from a 40-year simulation of China's developing carbon neutrality, technological progress has significantly impacted both carbon neutrality and economic growth. Therefore, a more suitable long-term endogenous technological progress model is needed to explore China's decreased carbon emissions and path of development.

Generally, research using the CGE model to simulate carbon tax and trading policies and evaluate their role in decreasing carbon emissions has been relatively mature, and several discussions have included the Kuznets curve of carbon emissions and economic growth. However, these studies lack systematic, long-term characteristics, and rarely involve the mechanisms of economic structural adjustments and their specific effects on carbon dioxide emissions. Most of these also ignore endogenous technological progress' impacts on the economy-environment system. Additionally, relevant research on China's carbon neutrality is still in an early stage, and primarily focuses on practical approaches, technologies, and standards. Detailed empirical research is required, and especially studies of carbon neutrality targets' impact on economic transformations using a macroeconomic model. Therefore, this study is based on realistic economic theories, methods, and data from China, which has proposed an integrated energy system comprised of regional economic and social data. This system can be used to develop an integrated energy-environment-economy CGE (3E-CGE) model for China that evolves an endogenous, climate-friendly technological innovation process to study climate change policies. This study will focus on the trajectory of carbon emissions and the policies and investments required for climate-friendly technological developments. It will also examine the systemic impacts on the national economy, including changes to the industrial structure and energy system, and adjustments to consumption patterns from 2020 to 2060. Ultimately, this work will explore the optimal development path for China under "peak carbon" and "carbon neutrality" constraints.

3. Model and Methods

3.1. The Model's Basic Structure

We construct a computable general equilibrium model of China's energy and carbon dynamics that includes the following modules: production, energy, revenue and expenditures, trade, carbon, dynamic, climate-friendly technology, and closure (Figure 1). The model describes a closed-loop system, with each module interlocking and interacting through price and output variables. In the model, "PS" and "CC" denote the industry and product dimensions, respectively. Here, we only briefly describe the model structure.

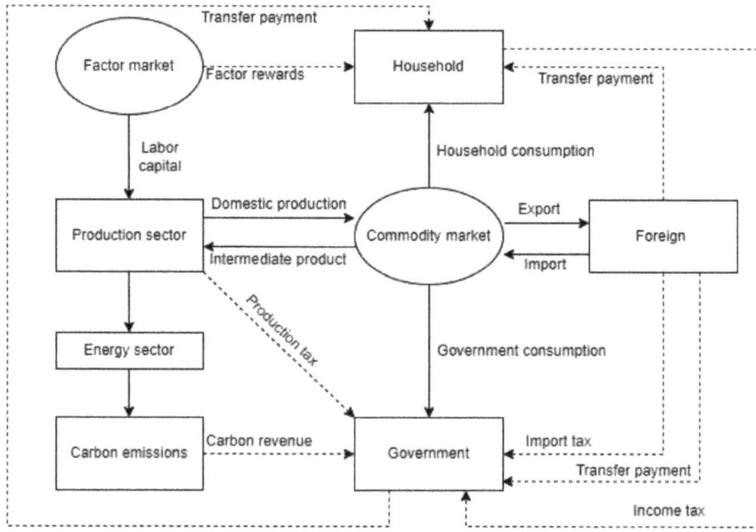

Figure 1. The CGE model framework diagram.

3.1.1. Production Module

The production module describes the relationship between the product input and output in the Chinese production sector. This model assumes that the market is completely competitive, sectoral output is determined by market equilibrium conditions, and production decisions are made in accordance with the principle of cost minimization. To reflect and address the more complex substitutive relationship between multiple inputs, the production module uses a multi-layer, nested design form (Figure 2). The first layer of nesting is solved by the intermediate input and composite elements through the CES function. The production function of the combination of intermediate input and added value is as follows:

$$X(PS) = AP(PS)\cdot\left[\beta(PS)\cdot U(PS)^{\rho(PS)} + \beta\prime(PS)\cdot V(PS)^{\rho(PS)}\right]^{\frac{1}{\rho(PS)}} \quad (1)$$

where AP (PS) is the scale coefficient of the sector; U (PS) and V (PS) represent the intermediate goods input aggregation and collected bundle of factors from sector PS, respectively; β (PS) and β′ (PS) denote the shared parameter representing intermediate goods and factors from sector PS, respectively; ρ (PS) is the substitution parameter; and X (PS) is the production function from combining intermediate input and added value. The second layer consists of two parts. The first compounds the intermediate input U(PS) through the LT function, and the second compounds the capital-labor-land element bundle V(PS) through the CES function. The formula is as follows:

$$U(PS) = \sum_{CC} ut(CC, PS)\cdot XX(CC, PS) \quad (2)$$

$$V(PS) = AV(PS)\cdot\left[\alpha(PS)\cdot L(PS)^{\rho 2(PS)} + \alpha\prime(PS)\cdot K(PS)^{\rho 2(PS)}\right]^{\frac{1}{\rho 2(PS)}} \quad (3)$$

where XX (CC, PS) represents the intermediate production input; ut (CC, PS) denotes the shared parameter of commodity CC used by sector PS in the LT function; AV (PS) is the total factor productivity; α(PS) and α′ (PS) are the input and shared parameters of labor factor L (PS) and capital factor K (PS) as used by sector PS in the CES function, respectively; and ρ2 (PS) is the substitution coefficient.

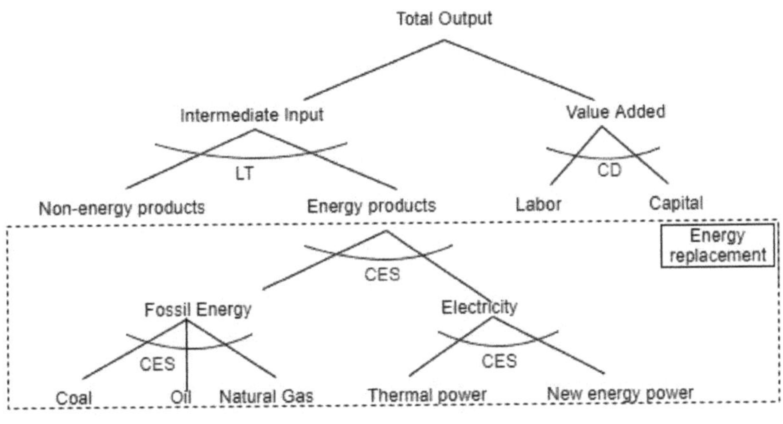

Figure 2. Production structure diagram.

The third layer describes the energy substitution portion. First, the energy product bundle is formed by combining both energy and power products through the CES function.

$$\mathrm{UEE(PS)} = \mathrm{APEE(PS)} \cdot \left[\gamma(\mathrm{PS}) \cdot \mathrm{UEN(PS)}^{\rho 3(\mathrm{PS})} + \gamma\prime(\mathrm{PS}) \cdot \mathrm{UEP(PS)}^{\rho 3(\mathrm{PS})}\right]^{\frac{1}{\rho 3(\mathrm{PS})}} \quad (4)$$

In the previous formula, UEE (PS) represents the sector's energy cluster production input; UEN (PS) and UEP (PS) represent the sector's fossil energy and power cluster production inputs, respectively; APEE (PS) denotes the scale coefficient; γ (PS) and γ'(PS) are the shared parameters of sector PS using both fossil energy and power products in the CES function; and ρ3(PS) is the substitution coefficient.

The fourth layer is the production structure of fossil energy products and power products. The energy products on the left side of Figure 2 are composed of coal, oil, and natural gas through the CES function, while the power products on the right are composed of thermal and new energy power.

$$\mathrm{UEN(PS)} = \mathrm{APEN(PS)} \cdot [\sum_{\mathrm{EN}} \varphi(\mathrm{EN,PS}) \cdot \mathrm{QXEN(EN,PS)}^{\rho 4(\mathrm{PS})}]^{\frac{1}{\rho 4(\mathrm{PS})}} \quad (5)$$

$$\mathrm{UEP(PS)} = \mathrm{APEP(PS)} \cdot [\sum_{\mathrm{EP}} \theta(\mathrm{EP,PS}) \cdot \mathrm{QXEP(EP,PS)}^{\rho 5(\mathrm{PS})}]^{\frac{1}{\rho 5(\mathrm{PS})}} \quad (6)$$

where QXEN(EN,PS) and QXEP(EP,PS) represent the sector's investment in fossil energy and power products, respectively; APEN(PS) and APEP(PS) are the scale coefficients; φ(EN,PS) denotes the shared parameter of coal, oil, and natural gas; θ(EP,PS) represents the shared parameter of thermal and new energy power product inputs; and ρ_4(PS) and ρ_5(PS) are substitution coefficients.

3.1.2. Revenue and Expenditure Module

The revenue and expenditure module includes two main bodies—residents and the government—which use the Cobb-Douglas utility function to maximize utility under the constraints of the income function. Residents' income comes from labor income, capital remuneration, and government transfer payments, and is used for consumption or savings after paying income taxes. Government revenue comes from production, consumption, value-added, import, and income taxes; the government's expenditures include the purchase of goods, transfer payments, and government savings.

3.1.3. Trade Module

The goods in the commodity supply come from domestic production and imports, which are used for final demand and intermediate consumption. To achieve the lowest consumption cost, rational consumers will optimize a combination of domestic and imported goods in their purchases; between the two, the Armington condition [54] is met. Specifically, an incomplete substitution occurs between imported and domestic products. In terms of price selection, this model assumes that the imported goods' price is exogenously provided, the imported goods' price is determined by the international market price, and the importer is the price taker:

$$QC(CC) = AA(CC) \cdot \left[\delta(CC) \cdot QD(CC)^{sa(CC)} + \delta\prime(CC) \cdot IMP(CC)^{sa(CC)}\right]^{\frac{1}{sa(CC)}} \quad (7)$$

where AA(CC) is the scale coefficient; δ(CC) and δ′(CC) represent the shared parameters; sa(CC) represents the conversion elasticity; QD(CC) and IMP(CC) represent domestically produced self-sale and imported goods, respectively; and QC(CC) represents domestic goods. Similarly, assuming that all countries in an international market sector are a smaller size, export commodities are also determined exogenously by international market price. The total domestic output is sold at home and abroad in accordance with the principle of profit maximization. Manufacturers will optimize the combination between domestic sales and exports, and this combined relationship is allocated through the fixed conversion-elastic (CET) function.

$$Q(CC) = AT(CC) \cdot \left[\epsilon(CC) \cdot QD(CC)^{st(CC)} + \epsilon'(CC) \cdot EXP(CC)^{st(CC)}\right]^{\frac{1}{st(CC)}} \quad (8)$$

In Formula (8), AT(CC) is the scale coefficient; ε(CC) and ε′(CC) denote the shared parameters; st(CC) is the conversion elasticity; QD(CC) and EXP(CC) are the domestically produced self-sale and exported goods, respectively; and Q(CC) represents domestically produced goods.

3.1.4. Dynamic Module

The variables in dynamic equations can be roughly divided into two categories. The first category is exogenous growth, which is represented by changes in the labor supply. This model uses a recursive dynamic mechanism, that is, through the dynamic changes in labor force growth LST(TH) and capital accumulation KST(TH) to modify the model. In the long-term, the labor force and population maintain the same proportion of growth and decline, and the population is hardly affected by economic policies. Therefore, such variables in the dynamic CGE model are generally exogenous, with the core equation is as follows:

$$LST(TH + 1) = [1 + gpop(TH)] \cdot LST(TH) \quad (9)$$

where LST(TH + 1) and LST(TH) represent the labor supply in period $t + 1$ and period t, respectively; and gpop(TH) represents the population growth rate in period t. The data comes from *World Population Prospects 2017* released by the Population of the UN Department of Economic and Social Affairs. The other category is capital control variables. The capital stock growth rate is driven by investments, and the size of these investments is affected by the rate of return. Among them, each department's labor force in the base year is given exogenously. The current capital stock is equal to the previous capital stock, plus new capital, and minus depreciation. The distribution of new capital among different sectors uses the CET function to maximize capital gains:

$$KST(TH + 1) = [1 + gk(TH)] \cdot [(1 - dep) \cdot KST(TH) + INVPS(TH)] \quad (10)$$

where KST(TH) and KST(TH + 1) represent the capital stock in period t and period $t + 1$, respectively; INVPS(TH) denotes the total investment in period t; dep is the depreciation rate of macroeconomic capital; and gk(TH) represents the growth rate of capital in period t. This model is calculated based on the GDP forecast data from the Organization for Economic Co-operation and Development (OECD).

3.2. Carbon Module

3.2.1. Carbon Tax Module

In this model, the carbon tax—which involves collecting carbon emissions in the production-end sector and energy products in the consumption end—is exogenous. Carbon tax revenue is calculated by multiplying carbon tax by carbon emissions, namely:

$$\text{TXCO2(TH)} = \text{TRCO2(TH)} * [\text{TCO2(TH)} + \text{FDCO2(TH)}] \tag{11}$$

where TRCO2(TH) represents the carbon tax price; TXCO2(TH) is the total carbon tax revenue; and TCO2(TH) and FDCO2(TH) represent the total carbon dioxide emissions at the production and consumption end, respectively. As a new cost, carbon tax is included in the pricing formula for energy products at the production and consumption end, respectively:

$$\text{PQXEN(TH)} = \text{PQXEN0(TH)} + \text{TRCO2EN(TH)} \tag{12}$$

$$\text{PCQXN(TH)} = \text{PCXEN0(TH)} + \text{TRCO2EN(TH)} \tag{13}$$

where TRCO2EN(TH) represents the carbon tax price per unit of energy, PQXEN(TH) and PCQXN(TH) represent the prices of energy products with carbon prices at the production and consumption end of period t when carbon tax is imposed, respectively. Further, PQXEN0(TH) and PCXEN0(TH) represent the production and consumption prices of energy products without carbon costs, respectively.

3.2.2. Carbon Trading Module

The carbon emissions trading plan is the same as the market plan for other commodities, as carbon emission credits are regarded as commodities. However, the government controls the number of carbon emissions rights; in the carbon-trading market, the total supply of carbon emission rights will be determined according to the government's emissions reduction targets. The setting of the carbon cap primarily depends on the corresponding emissions reduction target, with the specific formula set as follows:

$$\text{Carbon(TH)} = \sum_{\text{PS}} [1 - \text{tcer(PS, TH)}] \cdot \text{CO2ref(PS, TH)} \tag{14}$$

where Carbon(TH) represents the carbon allowance, or the total carbon emissions supply in the carbon-trading market at time t; CO2ref(PS, TH) represents various industries' benchmark carbon emissions during period t; and tcer(PS, TH) is the sectoral carbon emission reduction rate set by the government according to its emissions reduction target. The formula for total carbon emissions is as follows:

$$\text{TCO2(TH)} = \sum_{\text{PS}} \text{CO2(PS, TH)} \tag{15}$$

where TCO2(TH) represents the total carbon emissions of all industries in period t; and CO2(PS, TH) represents different industries' carbon emissions, with prices in the carbon-trading market determined by supply and demand. From production cost perspective, an increase in carbon costs in the energy sector will increase energy prices, and using energy as an intermediate input will increase the cost of using carbon-containing energy products. As an alternative, the cost of using low- and non-carbon energy products is relatively lower, with the following carbon cost formula:

$$\text{PQXEN(TH)} = \text{PQXEN0(TH)} + \text{PCO2EN(TH)} \tag{16}$$

where PQXEN(TH) in carbon trading represents the price of energy products with carbon prices in period t; PQXEN0(TH) represents the production price without carbon costs; and PCO2EN(TH) represents the carbon price per unit of energy. When allocating carbon trading, the government will also give enterprises a partial emissions exemption, as businesses with such an exemption will not need to pay these costs. The formula for this exemption is calculated as follows:

$$\mathrm{TFP(TH)} = \sum_{\mathrm{PS}} \mathrm{fp(PS)} \cdot \mathrm{CO2ref(PS,TH)} = \sum_{\mathrm{PS}} \mathrm{FP(PS,TH)} \tag{17}$$

$$\mathrm{ET(PS,TH)} = \mathrm{TCO2(PS,TH)} - \mathrm{FP(PS,TH)} \tag{18}$$

where TFP(TH) represents the total free allocation quota in period t; FP(PS,TH) represents various industries' free emissions in period t, and ET(PS,TH) represents the carbon quota that can be used for inter-sectoral transactions. According to this formula, when the free allocation ratio fp(PS) in the total quota decreases, carbon-intensive industries must purchase more carbon emissions to ensure production and operation. Carbon revenue equals the carbon price per unit of CO2 multiplied by the carbon emissions excluding the exemption, or:

$$\mathrm{TXCO2(TH)} = \mathrm{PCO2(TH)} \cdot [\mathrm{TCO2(TH)} - \mathrm{TFP(TH)}] \tag{19}$$

where PCO2(TH) represents the carbon price per unit of CO2.

3.3. Technological Progress Module

Climate-friendly technologies include not only technical means for the energy sector to improve energy efficiency, but also technical means for the end consumer and other industrial sectors to reduce carbon dioxide emissions. By introducing the non-carbon energy investment share NClindex (TH), this module creates an endogenous logic curve describing the state of climate-friendly technological progress in the entire system. According to Pan and Kohler [50], the specific formula is as follows:

$$\mathrm{SNCT(TH)} = \mathrm{LGCd} + \frac{\mathrm{LGCa}}{\left\{1 + \mathrm{LGCc} \cdot \mathrm{EXP}\left[-\mathrm{LGCb} \cdot \left(\mathrm{NUMYEAR(TH)} - \frac{\mathrm{LGCm}}{\mathrm{NClindex(TH)}}\right)\right]\right\}^{\frac{1}{\mathrm{LGCc}}}} \tag{20}$$

where SNCT(TH) denotes the share of climate-friendly technology, which is calculated from the following parameters: LGCa, or saturation; LGCb, or the average growth rate; LGCc, or the acceleration in growth; LGCd, or the initial level; LGCm, or the time to maximum growth; NUMYEAR(TH), or the year; and NClindex(TH), or the non-carbon energy investment ratio. where SWSNCT denotes the change in climate-friendly technology, indicated as either zero or one; when SWSNCT equals one, Formulas (21) is run. Changes in the share of climate-friendly technologies will change the carbon emissions coefficient, which will affect the entire society's carbon dioxide emissions. As noted in Formulas (21), the changes in VcofCO2CC(CC,PS,TH) is calculated as follows:

$$\mathrm{VcofCO2CC(CC,PS,TH)} = \begin{cases} \mathrm{cofCO2CC(CC,PS,TH)} & \mathrm{TH} = 1 \\ \mathrm{VcofCO2CC(CC,PS,TH-1)} \cdot [1 - \mathrm{SNCT(TH-1)} \cdot \mathrm{SWSNCT(TH-1)}] & \mathrm{TH} > 1 \end{cases} \tag{21}$$

where cofCO2CC(CC,PS,TH) denotes the carbon emissions coefficient.

3.4. Closing Module

To retain the model with a unique solution, the CGE model must set micro- and macro-closures to ensure that the constraint conditions are consistent with the number of endogenous variables. The economy's market equilibrium solves the equilibrium price, and the equilibrium price is determined by solving the nonlinear equation system. This includes the intermediate and final demand equations as well as the calculation of residential and government income, savings and investments, and the trade balance. This model uses neoclassical closures, or by setting closures in commodity and factor markets, institutional revenues and expenditures, foreign markets, and investment savings. It specifically includes: (1) the commodity and factor market equilibrium; (2) residents' total consumption, which equals their disposable income minus savings; and total government savings, which equals the government's income minus consumption and transfer payments to residents; (3) total investments equal total savings; and (4) the difference between imports and exports equals any foreign investments.

3.5. Model Data

The input-output analysis method created by Leontief [55] provides the possibility for modern economics to move towards quantitative analyses. In a checkerboard-like table, the relationships between production, the factor input, consumption, investments, and trade are clearly expressed as quantities, and the complete, overall picture and structure of the flow of products in the economic system is revealed. Part of the relationship between industrial sectors constitutes the core of the input-output table, reflecting the mutual influence and interdependence among various industrial sectors. Stone [56] extended the input-output analysis to institutional departments and established income accounts to record the product and income flows as well as transfers of income between institutions. The subsequent social accounting matrix system has become the necessary database to model the current computable general equilibrium model and other large-scale quantitative economic structural models.

As the input-output table for 2017 depicts, in the latest data released by the National Bureau of Statistics at the beginning of this research, 2017 is selected as this study's base year. The model's main data includes the following three types: First, the China Social Accounting Matrix (SAM) provides a foundation for the CGE model (Table 1). According to the Input-Output Table of China's 149 Sectors in 2017, we merged then split the energy input-output tables of 29 sectors, including coal, oil, natural gas, thermal, and new energy power. The fiscal and taxation data in the social accounting matrix comes from the *2018 Tax Yearbook* and *2018 Fiscal Yearbook*. Table 1 also displays the resulting macro-social accounting matrix. Second, the exogenous elasticity of substitution includes that between inputs in the production function, the substitution elasticity between imported and domestic products from the CES function within the foreign trade module, and the substitution elasticity between exports and domestic products in the CET function. Our data is derived from the Global Trade Analysis Project database. Third, we calculate the carbon dioxide emissions coefficient per sector by calculating the sectoral carbon dioxide emission and energy product consumption. The carbon dioxide emissions data for this calculation comes from China's Carbon Emission Accounts and Datasets (CEADs) database.

Table 1. Macro SAM table.

Unit: 100 Million Chinese Yuan (Calculated Based on the Producer Price of the Year)

Income \ Expenditure	Production Activities	Product	Labor	Capital	Household	Government	Tax	ROW	Investment	SUM
Production Activities		2,257,733								2,257,734
Product	1,434,518				324,546	125,341		163,847	369,146	2,417,397
Labor	423,268									423,268
Capital	304,969									304,969
Household			423,268	304,969		45,615				773,852
Government							133,372			133,372
Tax	94,978	26,431			11,961					133,372
ROW		133,232							30,615	163,847
Investment					437,345	−37,584			5377	405,138
SUM	2,257,733	2,417,397	423,268	304,969	773,852	133372	133,372	163,847	405,138	7,012,948

4. Constraints and Scenario Settings

4.1. Discussion on Peak Carbon Emissions

China's "30–60" commitment—to peak carbon by 2030 and carbon-neutral production by 2060—adds new constraints to China over the next 40 years, with 2 key factors. One is the apex of peak carbon in 2030, and the other is the GDP growth path until 2060.

Regarding the former, we must first solve the problem of basic carbon emissions data, as the existing carbon emissions database primarily includes data from the CEADs database, British Petroleum (BP), and the Ministry of Ecology and Environment (Figure 3). As the Ministry's data is not sufficiently continuous, the two time-continuous CEADs and BP databases are more suitable for research. Additionally, the carbon emissions in 2005 (not included in the carbon sink) as calculated using the three databases were 5.4, 6.1, and 5.98 billion tons, respectively. In 2014, the 3 databases' carbon emissions were 9.44, 9.24, and 10.28 billion tons, respectively. In 2017, the statistical carbon emissions for CEADs and BP were 9.34 and 9.3 billion tons, respectively. The numerical difference between these databases has significantly narrowed in recent years. Considering that the CEADs contains emissions data for 29 industries—which is convenient for a structural analysis and CGE model simulation—we ultimately chose the CEADs for our analysis. To avoid the uncertainty caused by the significant differences between the early carbon emissions data found in different databases, we selected the carbon emissions from the 2017 CEADs (9.34 billion tons) as our calculation benchmark.

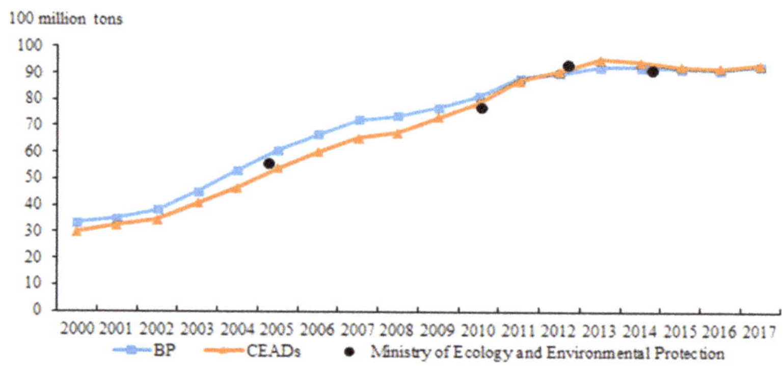

Figure 3. Comparison of China's carbon emission data in various databases.

Regarding China's carbon peak target, President Xi Jinping's critical speech at the General Debate of the 75th United Nations General Assembly on 22 September 2020 indicated that China will enhance its nationally determined contribution. Moreover, the nation would strive to reach peak carbon dioxide emissions by 2030 and carbon neutrality by 2060, or the "30–60 dual carbon target". Specifically, the carbon emission intensity in 2030 would decrease by more than 65% compared with that in 2005. Combined with the declining rate of carbon emissions intensity in that year compared with 2005, as recently announced by the State Council, the 2017 carbon intensity was 46% lower than that in 2005. It is estimated that carbon emissions in 2030 will reach 11.7 billion tons; after subtracting 910 million tons of carbon sinks, we predict that net emissions will reach 10.8 billion tons in 2030, which is an approximate median of the peak data calculated by Tsinghua University, the World Resources Institute, and other institutions.

4.2. Discussion of the GDP Growth Rate

Our research also considers the growth trends of various production factors, such as capital, labor, human capital, and total factor productivity. We refer to relevant research on China's economic growth forecast, both domestically and internationally, and the outline of the 14th 5-Year Plan, to assume that GDP will double from 2020 to 2035, and the GDP growth rate will decline from 6% in 2019 at a uniform rate. It can be concluded that the compound GDP growth rate from 2020 to 2030 is 5%. By 2035, China's total GDP will reach 240 trillion yuan, realizing the nation's long-term goal of doubling its total economic output in 2020. The compound growth rate of the nation's GDP from 2040 to 2060 is estimated at 3%.

In summary, China only has 10 years to achieve its peak carbon goal and 30 years to achieve carbon neutrality, which is a much shorter duration than European countries and the United States. Moreover, China's carbon emissions only have 8% room for improvement, with an average annual growth of 0.77%, and it is highly difficult to maintain a relatively high level of GDP growth. How can China meet this challenging goal? We construct various policy scenarios for a simulation in an attempt to discover the optimal path toward implementation.

4.3. Scenario Setting

To quantitatively analyze the different effects of the two previously mentioned paths in the two dimensions of carbon neutrality and economic development, we focus on the scenarios listed in Table 2.

Table 2. Scenario setting.

Scenario Category	Scenario Code	Setting
BAU	No exogenous intervention	The GDP growth rate drops uniformly from 6% in 2019; the compound GDP growth rate from 2020 to 2030 will be 5%, and the compound annual GDP growth rate from 2040 to 2060 will be 3%.
Carbon price policy	S1.a Carbon tax	Combined with the profitability and model tests of various domestic industries, we set a unified tax rate for the whole society and the tax rate increases year by year. The maximum carbon tax rate will not exceed 1800 CNY/ton. [1]
	S1.b Carbon trading	According to the "the current Guangdong Province's Implementation Plan for the Allocation of Carbon Emission Allowances in 2020", the model sets the industries for carbon trading as: petrochemicals, chemicals, building materials, steel, non-ferrous metals, papermaking, electricity, aviation and their respective free carbon emission allowances.
	S1 Carbon Tax + Carbon Trading	Carbon tax and carbon trading implemented simultaneously
Technological progress	S2.a	Optimistic prospect of technology development
	S2.b	Pessimistic prospect of technology development
	S2	S1+ neutral prospect of technology development Technology curve

[1]: Countries that have levied carbon taxes currently have a carbon tax rate of approximately CNY 80–800 per ton of carbon, but most of them are developed countries.

5. Results and Analysis

5.1. Baseline Scenario Results

In the scenario that only considers the goal of doubling the total economic output or per capita income by 2035, the development trend of China's economy and carbon emissions is that the latter will continue to rise; carbon emissions are expected to reach 41.1 billion tons in 2060 (Figures 4 and 5).

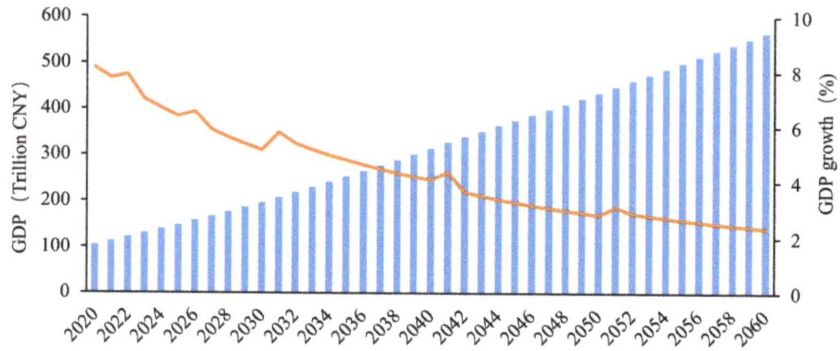

Figure 4. GDP and its annual growth rate in the baseline scenario (BAU) from 2020 to 2060.

Figure 5. Carbon emissions in the baseline scenario (BAU) from 2020 to 2060.

5.2. Results of the Carbon Price Policy Scenario

Most people intuitively feel that continuously increasing the level of carbon tax can fulfill China's "30–60" goal. However, this dual carbon goal requires an incredibly high carbon price. As Figure 6 indicates, the carbon tax rate and carbon trading price would increase annually to 2000 CNY per ton, which would collapse China's economy.

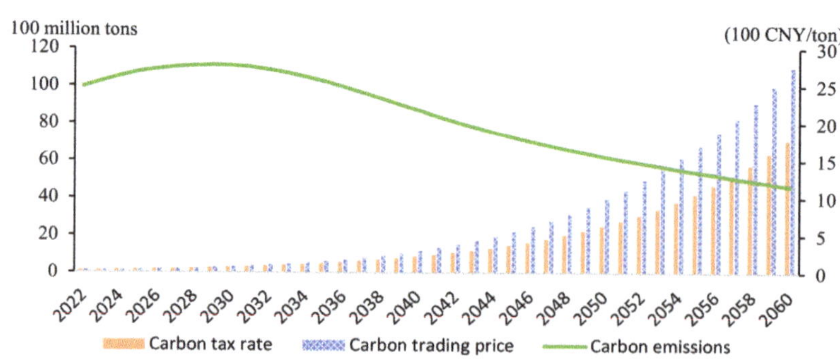

Figure 6. Carbon price and carbon emissions.

In this instance, carbon emissions continue to increase, with no inflection point to achieve any peak and neutralization; this can only shift the carbon emissions curve under the BAU scenario downward (Figure 7).

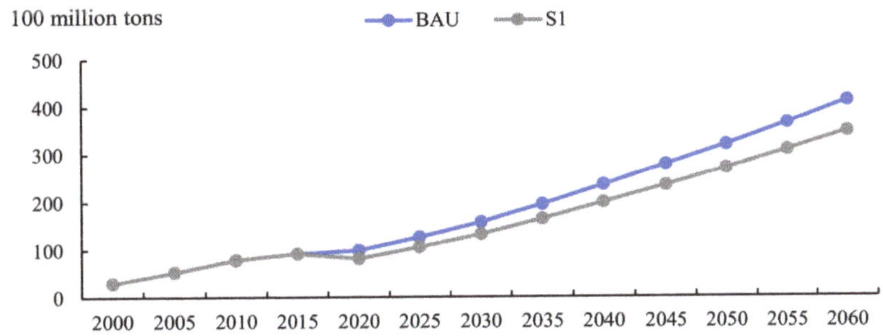

Figure 7. Carbon emissions under carbon price policy S1 from 2020 to 2060.

In summary, it can be observed that the carbon-pricing policy can only shift the carbon emissions curve downward, which can reduce emissions in the short-term, but will cause certain economic losses. The higher the carbon price, the greater the economic loss, consistent with the conclusions of most carbon tax and trading policy studies [23,28,34]. While the carbon price policy positively impacts a reduction in emissions, the fundamental transformation of carbon neutralization, this cannot be achieved solely through a carbon tax and carbon trading.

5.3. Results of Technological Progress Scenario

As previously mentioned, if we only rely on carbon pricing and do not abandon GDP constraints, it will be difficult to achieve carbon neutrality solely through carbon pricing. Therefore, it is necessary to consider technological progress. Developing climate-friendly technologies will transform the production costs of zero-emissions technology and the structure of energy consumption.

Based on the S1 scenario, we levy an CNY 100 carbon tax, implement carbon trading in 8 major industries, and introduce a technology curve. The simulation reveals that the combination of carbon tax, carbon trading, and technological progress should fulfill the dual goals of maintaining growth and carbon neutrality. As Figure 8 indicates, compared with the baseline case, the GDP loss under S1 will be substantial, with a GDP loss of CNY 42 trillion in 2030 and CNY 145 trillion in 2060. In the S2 scenario, the loss of GDP is negligible; even after 2040, the GDP exceeds the baseline, indicating that technological progress has offset or even exceeded carbon pricing's negative impact on economic growth. Carbon emissions peak in 2030 at 10.9 billion tons, then decreases annually to converge to the net-zero goal.

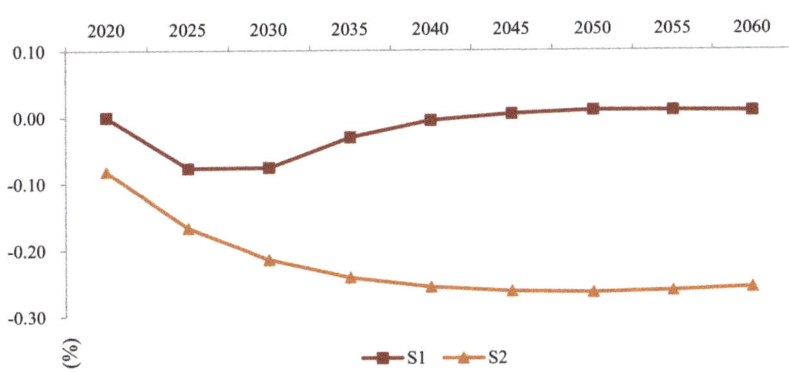

Figure 8. The GDP loss under S1 and S2 scenario compared with the BAU scenario.

The simulation results demonstrate that because the power of technological progress has only had a partial influence in the initial stage, carbon taxes and carbon trading still pressure the economy to increase costs. However, this pressure among various industries is alleviated, compared with the situation without technological progress.

The industrial output significantly changed compared to the benchmark situation. Moreover, the output of non-thermal power generation, such as photovoltaic, hydro-, wind, and nuclear power will significantly expand in 2030, while coal processing and mining will significantly shrink. Further, the proportion of output among the agricultural, forestry, animal husbandry and fishery, public services, and light industries significantly decreased over time. The proportion of output for equipment manufacturing, real estate and leasing, and information and financial services increased annually (Figures 9 and 10).

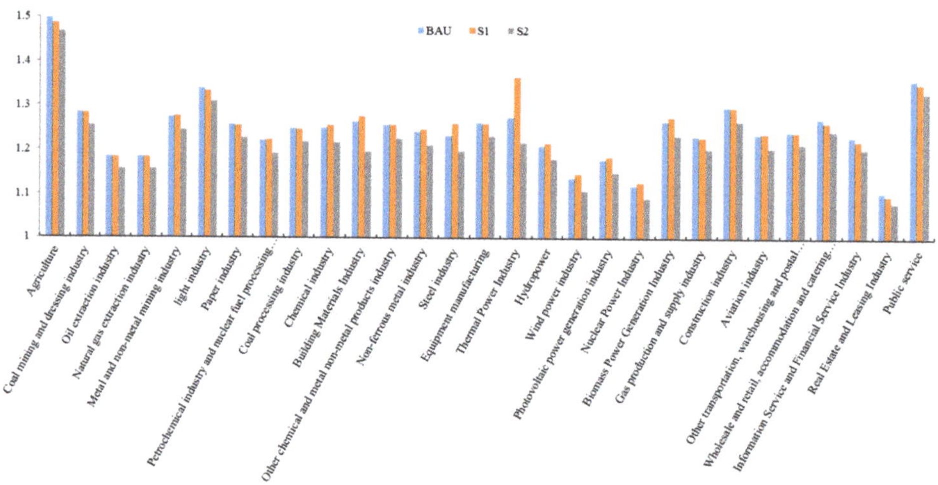

Figure 9. PPI differences in various industries under BAU, S1, and S2 scenarios in 2030.

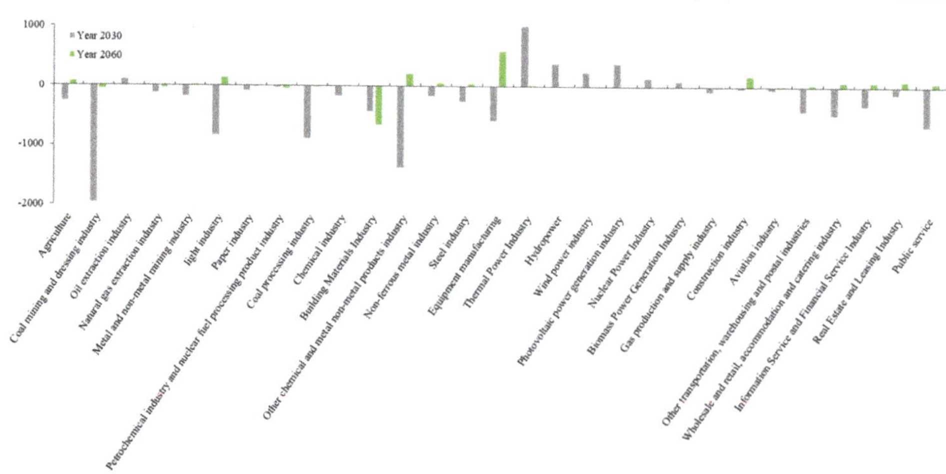

Figure 10. Changes in output of various industries compared with BAU under S2 scenarios in 2030 and 2060.

Compared with the "J"-shaped growth of carbon price under the S1 scenario, the carbon price under the S2 scenario shows an inverted "U"-shaped change with the development of climate-friendly technologies. The carbon price will rise year by year from the initial 10 CNY/ton to 140 CNY/ton in 2030, after which the carbon price will begin to decline steadily. As shown in Figure 11, when carbon prices peak, carbon emissions will also peak at 10.9 billion tons. The climate-friendly technology curve grows in an "S" shape, with a value of SNCT of 0.32 in 2030 and 0.99 in 2060. Investment, as an endogenous variable of the S-shaped technology curve, is the key to promoting technological progress. The model estimates that green technology R&D investment will account for about 2% of GDP, and it will increase year by year, helping R&D technology to cross the laboratory stage of the S-curve and use it for commercial applicato0-ions.

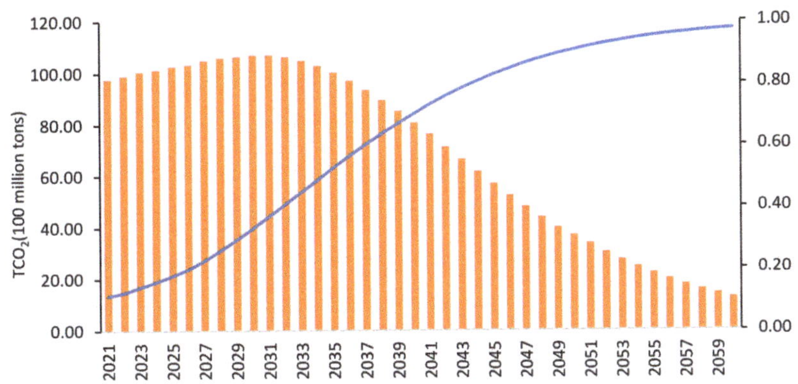

Figure 11. Carbon emissions and technological progress under S2 scenarios. Note: SNCT is a value between 0 and 1.

According to Table 3, the carbon price policy (S1 scenario) has little impact on the energy consumption structure. Fossil energy consumption will still dominate, and new energy consumption will slightly increase. However, the energy structure of the S2 scenario has significantly changed, with fossil energy gradually being withdrawn from the market, and new energy gradually monopolizing the energy consumption market.

Table 3. The proportion of energy consumption under the S1 scenario & S2 scenario.

	S1			S2		
	2020	2030	2060	2020	2030	2060
Coal	0.1%	10%	11.2%	49.3%	40.6%	2.9%
Oil	49.6%	47.7%	45%	8.4%	8.3%	0.8%
Natural Gas	4.4%	4.3%	4.3%	3.1%	2.9%	0.3%
Thermal Power	0.8%	0.9%	1%	23.3%	20.9%	1.5%
New Energy Power	35.1%	37.1%	38.5%	16%	27.4%	94.5%

5.4. Sensitivity Analysis of Technical Curve

It is assumed that the climate-friendly technology curve in the S2 scenario is the benchmark prospect of technological progress, and the pessimistic prospect refers to the situation in which the technological development is less than expected or difficult to commercialize. Intuitively, the technology curve in this scenario is flatter than the benchmark curve: it stays in the laboratory for a longer duration, with high resistance to the large-scale application of technology, or the technology's permeability is low (Figure 12). In reality, it corresponds to advanced and uncertain technologies, such as hydrogen energy and carbon capture, among others, while the optimistic outlook presents the opposite characteristics.

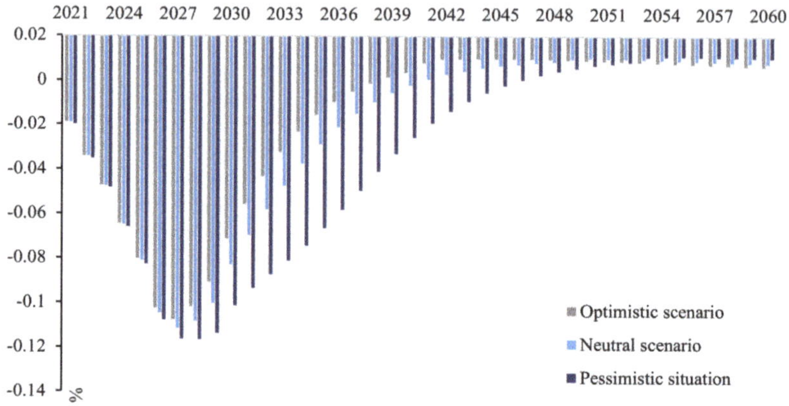

Figure 12. Changes in GDP of different technology curves.

As Figure 13 illustrates, the GDP under the pessimistic technology outlook experiences a relatively large negative impact, and the overall carbon emission curve increases, with a later, higher peak. Optimistic prospects provide the opposite results, but it is difficult to achieve such rapid technological development.

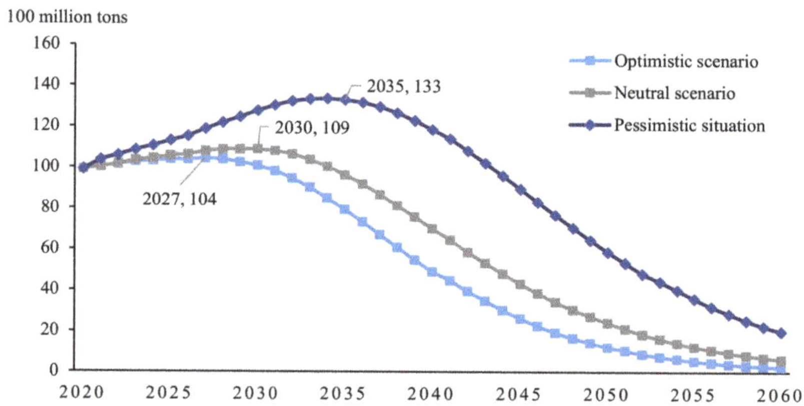

Figure 13. Carbon emission curve of different technologies.

In short, regardless of the speed of technological development, the shape of the carbon emissions curve can change, and inflection points can appear. The only difference is the time at which peak carbon and carbon neutrality occur. Only technological progress can meet the dual constraints of carbon emissions and economic development.

6. Conclusions

Based on the energy-environment-economy triple-coupling (3E-CGE) model, we endogenously generate climate-friendly technologies into the model's analysis framework by depicting the logic curve in the technology's full life cycle and modify the energy and carbon emissions modules within the CGE model. Based on the general equilibrium analysis of this CGE model, we can draw the following conclusions and insights.

First, regarding the simulation of peak carbon and carbon neutralization results, the endogenous CGE model significantly differs from the exogenous CGE model, especially in the long-term. Compared with other results, we note that the endogenous CGE model is more reasonable in optimistic, neutral, and pessimistic technological prospects.

Second, the most appropriate development path for China involves a combination of carbon tax and carbon trading policies while steadily developing climate-friendly technologies. This includes implementing a uniform carbon tax for all industries at 100 CNY per ton, and carbon trading in eight high-emissions industries with increasing carbon prices before the peak occurs. Moreover, climate-friendly technologies, which have begun to develop steadily since 2017, should mature in 2047. On the optimal path, the country peaks in 2030 at 10.9 billion tons, then decreases annually to reach net-zero carbon emissions in 2060; additionally, the dual effects of economic growth and carbon neutrality are achieved, with a stable economic growth rate of 2.4% in the later years.

Third, climate-friendly technologies play an important role in achieving China's goal of peak carbon in 2030 and carbon neutrality by 2060. The progress of climate-friendly technologies will gradually decrease the total carbon emissions from 2030, and will converge to a net zero in 2060, or approximately 60 million tons. The development of climate-friendly technologies will profoundly impact China's economic structure.

The model's results demonstrate that developing climate-friendly technologies gradually evolves the economic structure, from industry-dominated to emerging industry-dominated. Output from the new energy and service industries has rapidly expanded in the past 40 years. Meanwhile, in creating endogeneity among technologies, we can fully describe the interactions among technologies, investments, and carbon emissions. Although developing climate-friendly technologies requires substantial investments, this also stimulates economic growth and creates a mutually beneficial situation between economic growth and environmental improvements.

Regarding the traditional means of reducing emissions, such as carbon taxes and trading, carbon-pricing policies can quickly reduce the carbon emissions of energy-intensive industries in the short-term but will cause economic losses. Any economic recession will worsen as carbon pricing increases. Therefore, it is only theoretically feasible to use carbon pricing to achieve carbon neutrality. A carbon price of up to thousands of CNY per ton is an unbearable pressure for all industries and will inevitably collapse the economy. We also simulated and compared the benefits of carbon tax and carbon trading. Under the same emissions reduction target, carbon trading alone is better for the economy than only levying carbon taxes, as the former will result in less economic damage.

Collectively, carbon neutrality is not only an environmental governance issue, but also involves all aspects of society. It involves profound changes affecting all of Chinese society, including large-scale arrangements and advance planning. The government can begin with the carbon-trading market and gradually expand the scope to include all industries. Simultaneously, various climate-friendly technologies require more precise calculations in terms of their development level and potential. The government can provide preferential policies for investment in climate-friendly technologies, increase the R&D and promotion of such technologies, guide the upgrading of industrial production technologies, promote the formation of an economically beneficial zero-emissions production capacity, and intensify efforts to phase out existing high-carbon emissions assets. Moreover, a company's improved financial performance is important in a low-carbon economy, with such positive results as increasing companies' return on assets [57].

The endogenous technology advancement CGE model used in this article is a real economic model established based on an input-output table. It only introduces a curve for climate-friendly technology that covers all industries and does not describe the technology in detail. We can combine more knowledge with capital flow statements and data on the levels of climate-friendly technological development, capital growth rate, and depreciation rate among various industries. In doing so, we can further optimize the model and introduce a financial module to integrate both real and virtual economies. We can conduct more simulations on green credit and examine its effects on technological progress, dual-carbon goals, and economic growth in order to bring the model's results closer to reality.

Author Contributions: Conceptualization and methodology, H.P.; model construction, H.P. and S.L.; simulation analysis, X.L. (Xinyue Lin); data curation, X.L. (Xinyue Lin); writing—original draft preparation, S.L.; writing—review and editing, visualization, modification, project administration, supervision, funding acquisition, X.L. (Xiaoxue Liu). All authors have read and agreed to the published version of the manuscript.

Funding: Supported by the Major Project of the National Social Science Fund of China (21&ZD151), the earmarked fund for China Agriculture Research System (CARS-170601), Beijing Philosophy and Social Sciences capital circulation research base (JD-ZD-2021-021).

Data Availability Statement: The data used to support the findings of this study are available from the corresponding author upon request.

Acknowledgments: We thank all the participants for their collaboration.

Conflicts of Interest: The authors declare no conflict of interest.

References

1. Bi, H.M.; Xiao, H.; Sun, K.J. The Impact of Carbon Market and Carbon Tax on Green Growth Pathway in China: A Dynamic CGE Model Approach. *Emerg. Mark. Financ. Trade* **2019**, *55*, 1312–1325. [CrossRef]
2. Nestor, D.V.; Pasurka, C.A. CGE model of pollution abatement processes for assessing the economic effects of environmental policy. *Econ. Model.* **1995**, *12*, 53–59. [CrossRef]
3. Jorgenson, D.W.; Wilcoxen, P.J. Intertemporal general equilibrium modeling of U.S. environmental regulation. *J. Policy Modeling.* **1990**, *12*, 715–744. [CrossRef]
4. Wang, K.; Liu, F.M.; Yin, M.J.; Liu, J.L. Research on China's carbon emissions pathway under the 1.5 °C target. *Prog. Clim. Change Res.* **2021**, *17*, 7–17. (In Chinese) [CrossRef]
5. IPCC; Allen, M.; Babiker, M.; Chen, Y.; Zickfeld, K. Summary for Policymakers. In *Global Warming of 1.5 °C*; An IPCC Special Report; IPCC: Geneva, Switzerland, 2018.
6. Dowlatabadi, H. Integrated assessment models of climate change: An incomplete overview. *Energy Policy* **1995**, *23*, 289–296. [CrossRef]
7. Nakata, T. Energy-economic models and the environment. *Prog. Energy Combust. Sci.* **2004**, *30*, 417–475. [CrossRef]
8. Dong, K.; Sun, R.; Wu, J.; Hochman, G. The growth and development of natural gas supply chains: The case of China and the US. *Energy Policy* **2018**, *123*, 64–71. [CrossRef]
9. Kohler, J.H. Long Run Technical Change in an Energy-Environment-Economy (E3) Model for an IA System: A Model of Kondratiev Waves. *Integr. Assess.* **2003**, *4*, 17–55. [CrossRef]
10. Zhao, F.; Zhang, D.N.; Zhang, J. Research on Impulse Response of Provincial Underground Economy to Environmental Pollution in China? Based on a Spatial Econometric Model. *Int. J. Econ. Manag. Sci.* **2021**, *10*, 601.
11. Jiang, C.; Xue, Y.; Huang, J.; Xue, F.; Wen, F.; Li, K. Aggregated impact of allowance allocation and power dispatching on emission reduction. *J. Mod. Power Syst. Clean Energy* **2017**, *5*, 936–946. [CrossRef]
12. Jia, Z.J.; Lin, B.Q. Rethinking the choice of carbon tax and carbon trading in China. *Technol. Forecast. Soc. Change* **2020**, *159*, 120187. [CrossRef]
13. Mardones, C.; García, C. Effectiveness of CO_2 taxes on thermoelectric power plants and industrial plants. *Energy* **2020**, *206. prepublish*. [CrossRef]
14. Ntombela, S.M.; Bohlmann, H.R.; Kalaba, M.W.; Bateman, I.J. Greening the South Africa's Economy Could Benefit the Food Sector: Evidence from a Carbon Tax Policy Assessment. *Environ. Resour. Econ.* **2019**, *3*, 891–910. [CrossRef]
15. The Institute of Climate Change and Sustainable Development of Tsinghua University. Comprehensive report on China's Long-Term Low-Carbon Development Strategies and Pathways. *Chin. J. Popul. Resour. Environ.* **2020**, *18*, 1–25. (In Chinese) [CrossRef]
16. Ajay, M.; Adair, T.; Noëmie, L.R. *Green, Reliable and Viable: Perspectives on India's Shift Towards Low-Carbon Energy: Perspectives on India's Shift Towards Low-Carbon Energy*, 1st ed.; CRC Press: London, UK, 2019.
17. Galeotti, M.; Lanza, A. Desperately seeking environmental Kuznets. *Environ. Model. Softw.* **2004**, *20*, 1379–1388. [CrossRef]
18. Beyene, S.D.; Kotosz, B. Testing the environmental Kuznets curve hypothesis: An empirical study for East African countries. *Int. J. Environ. Stud.* **2020**, *77*, 636–654. [CrossRef]
19. Hu, J.Q.; Hernandez del Valle, A.; Martinez Garcia, M.A. Environmental Pollution and Economic Growth in China: A Test of the Environmental Kuznets Curve. *J. Geosci. Environ. Prot.* **2017**, *5*, 92–100. [CrossRef]
20. Bergman, L. General equilibrium effects of environmental policy: A CGE-modeling approach. *Environ. Resour. Econ.* **1991**, *1*, 43–61. [CrossRef]
21. Bergman, L. Energy and environmental constraints on growth: A CGE modeling approach. *J. Policy Modeling* **1990**, *12*, 671–691. [CrossRef]
22. Liu, W.; Cai, Z.Z. Technological Progress, Industrial Restructuring and Improvement of National Intermediate Consumption. *Econ. Res.* **2008**, *4*, 4–14. (In Chinese)

23. Zhou, J.F.; Wu, D.; Chen, W. Cap and Trade Versus Carbon Tax: An Analysis Based on a CGE Model. *Comput. Econ.* **2021**. *prepublish*. [CrossRef]
24. Zhang, W.W.; Zhao, B.; Gu, Y.; Sharp, B.; Xu, S.C.; Liou, K.N. Environmental impact of national and subnational carbon policies in China based on a multi-regional dynamic CGE model. *J. Environ. Manag.* **2020**, *270*, 110901. [CrossRef]
25. Cao, J.; Ho, M.S.; Jorgenson, D.W.; Nielsen, C.P. China's emissions trading system and an ETS-carbon tax hybrid. *Energy Econ.* **2019**, *81*, 741–753. [CrossRef]
26. Chai, Q.; Xiao, Z.; Lai, K.-H.; Zhou, G. Can carbon cap and trade mechanism be beneficial for remanufacturing? *Int. J. Prod. Econ.* **2018**, *203*, 311–321. [CrossRef]
27. Tang, L.; Zhang, L.; Yu, L.A. The Impact of Carbon Emission Trading Scheme on Emission Abatement Cost Based on Dynamic CGE Model. *J. Syst. Sci. Math. Sci.* **2019**, *1*, 51–64. (In Chinese) [CrossRef]
28. Tang, L.; Wu, J.; Yu, L.; Bao, Q. Carbon emissions trading scheme exploration in China: A multi-agent-based model. *Energy Policy* **2015**, *81*, 152–169. [CrossRef]
29. Liu, Y.; Lu, Y. The Economic impact of different carbon tax revenue recycling schemes in China: A model-based scenario analysis. *Appl. Energy* **2015**, *141*, 96–105. [CrossRef]
30. Goulder, L.H.; Schein, A.R. Carbon Taxes Versus Cap and Trade: A Critical Review. *Clim. Change Econ.* **2013**, *4*, 1–28. [CrossRef]
31. Lin, B.; Li, X. The effect of carbon tax on per capita CO_2 emissions. *Energy Policy* **2011**, *39*, 5137–5146. [CrossRef]
32. Wittneben, B.B. Exxon is right: Let us re-examine our choice for a cap-and-trade system over a carbon tax. *Energy Policy* **2009**, *37*, 2462–2464. [CrossRef]
33. Yang, L.; Li, F.; Zhang, X. Chinese companies' awareness and perceptions of the Emissions Trading Scheme (ETS): Evidence from a national survey in China. *Energy Policy* **2016**, *98*, 254–265. [CrossRef]
34. Lin, B.; Jia, Z. The energy, environmental and economic impacts of carbon tax rate and taxation industry: A CGE based study in China. *Energy* **2018**, *159*, 558–568. [CrossRef]
35. Nong, D.; Nguyen, T.H.; Wang, C.; Van Khuc, Q. The environmental and economic impact of the emissions trading scheme (ETS) in Vietnam. *Energy Policy* **2020**, *140*, 111362. [CrossRef]
36. Choi, Y.; Liu, Y.; Lee, H. The economy impacts of Korean ETS with an emphasis on sectoral coverage based on a CGE approach. *Energy Policy* **2017**, *109*, 835–844. [CrossRef]
37. Zhou, Y.; Fang, W.; Li, M.; Liu, W. Exploring the impacts of a low-carbon policy instrument: A case of carbon tax on transportation in China. *Resour. Conserv. Recycl.* **2018**, *139*, 307–314. [CrossRef]
38. Shi, J.R.; Cai, H.L.; Tang, L.; Yu, L.A. Impacts of Carbon Emission Trading on China: Based on CGE Model. *Chin. J. Manag. Sci.* **2015**, *11*, 801–806. (In Chinese)
39. Vera, S.; Sauma, E. Does a carbon tax make sense in countries with still a high potential for energy efficiency? Comparison between the reducing-emissions effects of carbon tax and energy efficiency measures in the Chilean case. *Energy* **2015**, *88*, 478–488. [CrossRef]
40. Okushima, S.; Tamura, M. What causes the change in energy demand in the economy? *Energy Econ.* **2009**, *32*, S41–S46. [CrossRef]
41. Jaffe, A.B.; Newell, R.; Stavins, R. Chapter 11 Technological Change and the Environment. In *Handbook of Environmental Economics*, 1st ed.; American Cancer Society: Boston, MA, USA, 2003.
42. Grubb, M.; Köhler, J.; Anderson, D. Induced Technical Change in Energy and Environmental Modeling: Analytic Approaches and Policy Implications. *Annu. Rev. Energy Environ.* **2002**, *27*, 271–308. [CrossRef]
43. Löschel, A. Technological change in economic models of environmental policy: A survey. *Ecol. Econ.* **2002**, *43*, 105–126. [CrossRef]
44. Griliches, Z. Hybrid Corn: An Exploration in the Economics of Technological Change. *Econom. J. Econom. Soc.* **1957**, *25*, 501. [CrossRef]
45. Solow, R.M. A Contribution to the Theory of Economic Growth. *Q. J. Econ.* **1956**, *70*, 65–94. [CrossRef]
46. Romer, P.M. Increasing Returns and Long-Run Growth. *J. Political Econ.* **1986**, *94*, 1002–1037. [CrossRef]
47. Arrow, K.J. The Economic Implications of Learning by Doing. *Rev. Econ. Stud.* **1962**, *29*, 155–173. [CrossRef]
48. Köhler, J.; Grubb, M.; Popp, D.; Edenhofer, O. The Transition to Endogenous Technical Change in Climate-Economy Models: A Technical Overviewto the Innovation Modeling Comparison Project. *Energy J.* **2006**, *27*, 17–55. [CrossRef]
49. Pan, H. Dynamic and endogenous change of input–output structure with specific layers of technology. *Struct. Change Econ. Dyn.* **2006**, *17*, 200–223. [CrossRef]
50. Pan, H.; Köhler, J. Technological change in energy systems: Learning curves, logistic curves and input–output coefficients. *Ecol. Econ.* **2007**, *63*, 749–758. [CrossRef]
51. Wang, K.; Wang, C.; Chen, J. Analysis of the economic impact of different Chinese climate policy options based on a CGE model incorporating endogenous technological change. *Energy Policy* **2009**, *37*, 2930–2940. [CrossRef]
52. Kristkova, Z.S.; Van Dijk, M.; Van Meijl, H. Projections of long-term food security with R&D driven technical change—A CGE analysis. *NJAS–Wagening. J. Life Sci.* **2016**, *77*, 39–51. [CrossRef]
53. Hazilla, M.; Kopp, R.J. Social Cost of Environmental Quality Regulations: A General Equilibrium Analysis. *J. Political Econ.* **1990**, *98*, 853–873. [CrossRef]
54. Armington, P.S. A Theory of Demand for Products Distinguished by Place of Production. *Staff. Pap. Int. Monet. Fund* **1969**, *16*, 159–178. [CrossRef]
55. Leontief, W. *The Structure of American Economy 1919–1939*, 1st ed.; Oxford University Press: London, UK, 1941.

56. Stone, R. *System of National Accounts*, 1st ed.; UN: Geneva, Switzerland, 1947.
57. Secinaro, S.; Brescia, V.; Calandra, D.; Saiti, B. Impact of climate change mitigation policies on corporate financial performance: Evidence-based on European publicly listed firms. *Corp. Soc. Responsib. Environ. Manag.* **2020**, *27*, 2491–2501. [CrossRef]

International Journal of
Environmental Research and Public Health

MDPI

Article

Impact on Carbon Intensity of Carbon Emission Trading—Evidence from a Pilot Program in 281 Cities in China

Wanlin Yu and **Jinlong Luo** *

School of Economics, Shandong University of Technology, Zibo 255000, China
* Correspondence: qqn123n@163.com; Tel.: +86-132-8069-0015

Abstract: China's carbon emissions trading scheme (ETS) is an institutional arrangement that China intends to explore as a means of energy conservation and emission reduction. It is the core of China's goal of achieving carbon peaking and carbon neutrality. This paper regards the introduction of pilot carbon emission trading policies as a quasi-natural experiment. Propensity Score Matching (PSM), Differences-in-Differences (DID), and spatial Durbin methods were used to evaluate the policy effects of pilot carbon emission trading policies on the carbon intensity of Chinese cities. We empirically tested the impact mechanism using the panel data of 281 cities at the prefecture level and above in China from 2006 to 2019. The results show that (1) the pilot policy of carbon emission trading has significantly reduced the carbon intensity of Chinese cities and shows characteristics of heterogeneity; (2) the dynamic effect test shows that the mitigation effect of the pilot carbon emission trading policy has increased gradually with time; (3) the mediation effect shows that the pilot carbon emission trading policy alleviates urban pollution in the region by improving the level of environmental governance and jointly reduces urban carbon intensity by increasing the level of green technology innovation; (4) the Durbin test suggests that pilot carbon emissions trading policy enforcement can significantly improve the carbon intensity of the area surrounding the city. In summary, the national carbon emissions trading market appears to be a successful experiment that also can contribute to China's sustainable development. Its promise in achieving the "double carbon" target provides important policy implications.

Keywords: carbon emission trading pilot; carbon intensity; green technology innovation; environmental governance level

Citation: Yu, W.; Luo, J. Impact on Carbon Intensity of Carbon Emission Trading—Evidence from a Pilot Program in 281 Cities in China. *Int. J. Environ. Res. Public Health* **2022**, *19*, 12483. https://doi.org/10.3390/ijerph191912483

Academic Editors: Paul B. Tchounwou, Taoyuan Wei and Qin Zhu

Received: 4 August 2022
Accepted: 27 September 2022
Published: 30 September 2022

1. Introduction

In the context of economic globalization, climate change is a major challenge for the survival and development of mankind in the 21st century, while the economic development of countries around the world always comes at the cost of energy consumption [1]. The "Statistical Review of World Energy" released by BP shows that global energy demand grew 2.9 percent in 2018, while carbon emissions rose 2.0 percent to reach their highest point in the 21st century. Global primary energy consumption grew 2.9 percent, almost double the average growth rate of 1.5 percent over the past decade [2]. At the same time, carbon emissions from energy consumption grew by 2%, also the highest in years. The new carbon emissions amounted to 600 million tons, which is equivalent to adding a third of the emissions produced by the planet's passenger cars. Therefore, it is of great significance to implement effective means to achieve rapid carbon peaking and net zero emissions [3].

As the world's largest developing country, China has become the world's largest carbon emitter. China's carbon dioxide emissions reached 11.3 billion tons in 2021, accounting for 33 percent of the global total [4]. The Chinese government has announced its intentions to undertake increasingly forceful measures with the goal of achieving a carbon peak before 2030 and carbon neutrality by 2060 [5]. This demonstrates China's determination to achieve its "dual carbon" goal of carbon emissions and carbon neutrality and to actively undertake the corresponding obligations of its international treaty obligations.

Carbon taxes and emissions trading systems are internationally recognized as effective tools to reduce carbon emissions. According to China's current situation, in order to ensure people's livelihoods, China temporarily does not tax carbon dioxide emitted by coal and natural gas used by individuals. For China, the ETS has become the main tool to reduce carbon emissions.

In December 1997, the Kyoto Protocol was adopted as the first additional agreement to the United Nations Framework Convention on Climate Change (UNFCCC). As part of that agreement, market mechanisms were recognized as a new path to reduce greenhouse gas emission—that is, the right to emit carbon dioxide became regarded as a commodity, thus forming the basis of carbon trading systems [6]. The European Emissions Trading System (EUETS), the world's largest carbon market, came into operation in 2005. The scheme imposes emission limits on member countries; the sum of national emission allowances does not exceed the emissions allowed under the Protocol. The allocation of emission allowances takes into account factors such as historical emissions, projected emissions, and emission standards of member countries [7].

The EU emissions trading system uses "Cap-and-Trade" rules. In order to limit the total amount of greenhouse gas emissions, administrative permits for emissions are bought and sold. The three major principles are the total trade principle, decentralized governance mode, and development characteristics. Under the EUETS, EU member state governments must agree to national emission caps set by the EUETS. Within this cap, companies can sell or buy additional credits in addition to their allocated emissions, provided that overall emissions fall within a specific quota. Firms that emit excess emissions beyond their allocated or purchased allotment are penalized, while those with surplus allowances can keep the emissions for future use or sell them to other firms. The EUETS has played an exemplary role in the world's development of carbon trading markets.

China's carbon market construction started with local pilots [8,9] based on the EUETS. In 2011, the Chinese government listed seven provinces and cities, including Beijing, as pilot areas of the ETS. In 2013, these pilot carbon markets began online transactions. The aim of the program is to cost-effectively reduce greenhouse gas emissions of enterprises in the pilot provinces and cities. The goals include training talent and accumulating experience to lay the foundation for a national carbon market [10]. At present, a national carbon market has started with the power generation industry (2225 enterprises). Eight industries with high energy consumption, including power, petrochemical, chemical, building materials, steel, non-ferrous, paper-making, and civil aviation, will be included in the national carbon market. It is expected to gradually include another seven industries over the 14th Five-Year Plan period.

The carbon emission trading scheme (ETS), regarded as a vital market-driven carbon mitigation instrument, could trigger technology innovation and accelerate a green economic transition [11]. In 2015, China's CO_2 emission from fossil fuel consumption was about 9 billion tons. During the 14th Five-Year Plan period, overall carbon intensity is expected to decrease by 18% and energy consumption per unit of GDP will be reduced by 13.5 percent. Now, the ETS has introduced a system innovation. How to reduce the carbon intensity of cities? What are the pathways that affect carbon intensity? This study will evaluate the ETS policy from the perspective of regional carbon emissions. A thorough review of the pilot policy's impact on carbon emissions, and its relationship to China's overall development, will provide valuable experience for China's efforts to deepen the reform and transformation of its pattern of economic development.

The rest of this study will be divided into the following parts. Part 2 is a literature review. Part 3 is a theoretical hypothesis. Part 4 is the data and empirical framework. Part 5 is the regression analysis. Part 6 further analyzes the mediating effect and spillover effect. Part 7 concludes and makes policy recommendations.

2. Literature Review

Because carbon emissions cause negative externalities [12], the arguments of Pigou [13] suggest government intervention through means such as taxation. Coase [14] held the opposite opinion, believing that the government should regulate property rights and allow the market to respond to externalities. In both cases, the instruments of the market are used to address externalities. Dales [15] proposed commercializing pollution on the basis of Coase, arguing that the pollution caused by companies is the property of the government, and that businesses should be able to buy and sell freely in the market. This was the embryonic form of the modern emissions trading system.

As mentioned above, although China's ETS has borrowed some practices from the EUETS, it is different. First, the EUETS consists of a "three-pillar" system of "carbon trading", "carbon tax", and "carbon border tax". This is slightly different from a carbon emission quota, which is the basis of carbon emission trading in China. Second, the EU emissions trading scheme adopted a cap-and-trade principle. That is, on the premise that the total amount of emissions does not exceed the allowable upper limit, each emission source can adjust its emissions through exchange of permits. The upper limit will be reduced year by year. By contrast, carbon trading in China is divided into a primary market and a secondary market. The primary market is mainly for "quota creation", which is managed by national authorities and entrusted to agencies to create and distribute carbon emission rights quotas. The participants in the secondary market are mainly enterprises and financial institutions. Third, the trading rules published by the Shanghai Ring Exchange have price fluctuation limits within a daily limit. The EU carbon price, on the other hand, has no price limit. Carbon prices in the European Union have risen rapidly in recent years, more than doubling from pre-pandemic levels. Fourth, the industry coverage of the EU carbon trading system, which started with the power industry and energy-intensive industries, gradually expanded to the transportation sector and the production of specific products such as steel and cement. At present, China's carbon emission trading market is focused on the electric power industry [16,17].

Existing research on emissions trading can be broadly divided into two categories. The first category focuses on assessing the efficiency of the ETS design, including the effectiveness of a carbon price in reducing emissions [18,19], the controllability of transaction costs [20,21], and the rationality of quota allocation [22,23]. The second category focuses on how the ETS affects macroeconomic variables. This study is in the second category.

From the perspective of energy conservation and emission reduction, earlier studies mostly used scenario simulation to evaluate carbon emission trading. In terms of energy saving, most scholars have used data simulation analysis. It has been found that ETS can effectively reduce the consumption of non-renewable energy [24,25]. In terms of emission reduction, Zhang et al. [26] simulated ETS implementation in China and found that interregional commodity exchanges can alleviate carbon emissions, based on China's provincial panel data [27]. The simulations were analyzed in the case of both unconstrained and constrained countries to assess the potential effectiveness of ETS in China. The study found that ETS had the potential to reduce carbon intensity by 20.06% without having a negative effect on GDP.

The development of the ETS systems in Europe and China provides the opportunity to turn the simulation into reality. Most studies have found that ETS has reduced carbon in pilot areas in China. Computable General Equilibrium (CGE) and Difference in Difference (DID) models have been the main empirical evaluation methods used in recent years. Liu et al. [1], using a regional CGE model, found that the Hubei province pilot ETS reduced carbon emissions by about 1% in 2014. In an empirical study, Yucai et al. [28] used DID to model the effect of the pilot ETS on energy conservation and emissions reduction; the results showed that regulated industry energy consumption in the ETS pilot areas decreased by 22.8% and carbon emissions by 15.5%.

Some scholars also have studied the possible economic losses caused by the implementation of ETS. Most scholars have found that EUETS has had no adverse effect on corporate profits and social welfare [29]. For China's carbon trading market, however, Wang and Pan [30] found that the implementation of ETS has led to a 0.28% decline in GDP. This is because China's economic development has been dependent on natural resources. Hubler et al. [31] found that the economic losses of ETS in China may be around 1%.

In conclusion, the existing papers mainly study the impact of ETS on energy saving, emission reduction, and economic loss. However, there are few studies on comprehensive macro indicators, such as urban carbon emission intensity. Urban carbon emission intensity is defined as the ratio of CO_2 emissions to GDP in a city within a year. This indicator has been widely used to evaluate China's "double carbon" target [32].

In China, most studies on ETS use a CGE model or a DID model. There are a number of limitations with these studies. CGE modeling is subject to defects such as difficulty in meeting the assumptions on which it is premised, strong subjectivity of parameter setting, and difficulty in determining whether its feedback mechanism measures real effects. The DID model requires homogeneity of the sample, whereas in reality, there is heterogeneity in relevant characteristics between the treated and control localities. In addition, most of the relevant studies start from the provincial level, while implementing carbon emission trading policies depends more on whether urban units can strictly implement the orders of their superiors. Further, earlier studies have ignored the influence of spatial factors on carbon intensity, although spatial factors have an important impact on carbon intensity and neglecting spatial factors may lead to bias in simulation results.

Against this background, this study makes the following contributions. First, the introduction of pilot carbon-emission trading policies is regarded as a quasi-natural experiment. This allows the use of a PSM-DID model estimation method to assess the impact of ETS on urban carbon intensity. The quasi-natural experiment not only meets the requirements of a DID model, but also ensures optimal matching because of the large samples. This gives more credibility to the research conclusions. Second, this study focuses on carbon intensity at the city level. Considering that cities are an important part of local government institutions in China, this makes the policy effect more plausible. Third, this study uses spatial Durbin to test the spillover effect of ETS on surrounding areas, thus going beyond the previous focus on the local area, which has ignored the surrounding area. This provides a more complete picture of the impact of emissions trading policies.

3. Theoretical Background

The carbon emission trading system is mainly an exercise of the "Porter hypothesis," which holds that appropriate environmental regulation can encourage enterprises to carry out more innovative activities [2]. These innovations will increase the productivity of firms, thereby offsetting the costs of environmental protection and reducing total carbon emissions at the societal level. Theoretically, the system is dominated by the government, which uses market mechanisms to promote energy conservation and emission reduction [33]. First, the ETS sets a relatively strict carbon allowance for each company. Within this limit, companies can carry out free carbon emissions. The excess needs to be purchased from the carbon emissions retained by other companies. In essence, carbon permits have become a commodity [34]. Because firms aim at profit maximization, companies make good use of free credits while trying to avoid exceeding that limit; otherwise, high production costs will be incurred. Second, ETS can promote corporate emission reduction because firms will sell unused emissions credits if the carbon price is higher than the firm's marginal cost of emission reduction [35]. Therefore, a market-based trading system can be effective in mitigating carbon emissions.

Establishing a carbon emission trading system can force enterprises to innovate, and technological progress is one of the three major factors affecting the environment [36]. Green technology innovation depends on increasing investment in such innovation. The emissions trading system encourages companies to actively develop and apply green technologies [2]. Companies that invest more resources in reducing carbon emissions can sell surplus carbon emission credits to high-carbon emission enterprises and obtain high profits [37]. Firms will tend to accelerate the process of green technology development in order to achieve higher profits. This is the incentive effect. Conversely, for high carbon emission enterprises, it is necessary to buy carbon emission credits from sellers, which will increase production costs, compress profit margins, and reduce these firms' competitiveness. Under this pressure, enterprises have to carry out technological innovation [2]. This is the punishment effect.

The incentive effect and punishment effect of market-based environmental regulation such as ETS give the government more tools for environmental governance. Because carbon dioxide does not harm health or production in the short run, and because it is costly to enforce non-market forms of governance, it has been difficult for the focus of environmental governance to shift quickly in the direction of reducing carbon emissions. By encouraging innovation and providing opportunities for profit, ETS has effectively improved the environmental governance level while ensuring that normal activities and production can continue.

Improved environmental governance can promote a change of regional energy structure. In particular, ETS has the potential to reduce coal consumption [28]. This paper applies the new economic geography to evaluate such changes. Firms will always look for the optimal location in order to maximize profits [38]. Theoretically, a carbon emission trading system should have policy spillover effects [39], including alleviating regional carbon emissions. It can also encourage high-tech enterprises to continue to innovate through its incentive mechanism. However, it will also cause a large number of polluting enterprises to incur high production costs due to its punishment mechanism. This is because polluting enterprises in the region face increased production costs due to the need to buy carbon emission rights, which reduces their profits. If there is no ETS policy in the surrounding areas, polluting enterprises are expected to migrate to the surrounding areas. Conversely, high-tech firms from surrounding areas are expected to migrate to the ETS area in order to increase their profits by selling carbon credits. The transfer behavior of the two types of enterprises can reduce carbon emissions in the ETS region while increasing emissions in the area around the ETS [40]. The theoretical background of this study is shown in Figure 1. Accordingly, the following hypothesis is proposed:

Hypothesis 1 (H1). *The carbon emissions trading system has reduced the carbon intensity of the pilot cities in China.*

Hypothesis 2 (H2). *Green technology innovation is one mechanism through which the carbon emissions trading system reduces regional carbon intensity.*

Hypothesis 3 (H3). *Improving urban environmental governance is another mechanism through which the carbon emission trading system alleviates regional carbon intensity.*

Hypothesis 4 (H4). *The carbon emission trading system has increased the carbon intensity of the areas surrounding the pilot cities.*

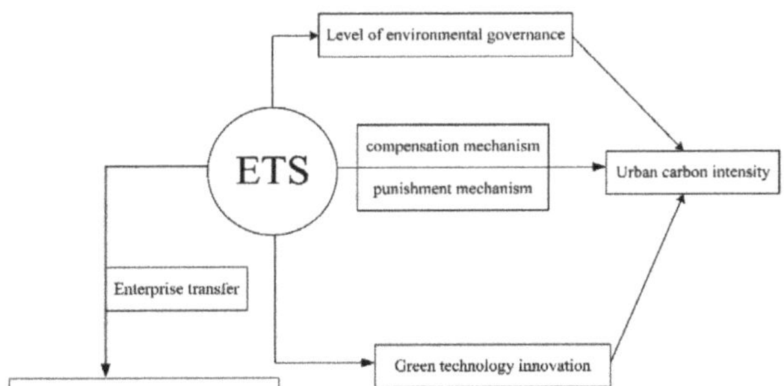

Figure 1. Theoretical background of the study.

4. Data and Methodology

4.1. Data Sources

Through screening and matching, this paper selected panel data of 281 cities in China from 2006 to 2019 as the research object. A total of 37 cities at the prefecture level and above were designated as pilot carbon emission trading cities. These 37 cities constitute the experimental group, and the remaining cities were analyzed as the control group. Most of the data in this study come from data already publicly available in China, including the National Bureau of Statistics (https://data.stats.gov.cn/, accessed on 15 August 2021), the China City Statistical Yearbook, the China Energy Statistical Yearbook, and the Statistical Yearbooks of 281 cities. The data were drawn from reports on social development, including statistics on the main energy consumption of local industries above city size, total industrial output value, urbanization rate, etc. Patent data were derived from the State Intellectual Property Office (https://www.cnipa.gov.cn/, accessed on 20 August 2021).

4.2. Variable Selection

Urban carbon intensity is based on the total amount of carbon emissions to be measured. In this study, a material balance algorithm was used to calculate the total carbon emissions. Carbon emissions are estimated using the chemistry of carbon dioxide produced during energy consumption.

$$Carbon_{it} = \sum_{v=1}^{n} Q_{vt} \times W_v \times M_v \times R_v \times 44/12 \quad (1)$$

Q_{vt} is the annual actual consumption of the V type of energy in the city in year t. According to the 26 fossil fuels listed in the China Energy Statistical Yearbook, they are combined into nine final energy consumption types: coal, coke, crude oil, gasoline, kerosene, diesel, fuel oil, natural gas, and electricity. Because electricity is not a direct energy source, the concept of secondary energy reflects the fact that electricity is produced by consuming other energy; therefore, this study will not measure electricity separately. W_v, M_v, R_v are the energy calorific value conversion coefficient, carbon emission coefficient, and carbon oxidation factor, respectively. The data come from the average low calorific value of the China Energy Statistical Yearbook and IPCC (2006). As (44/12) is known to be the ratio of carbon dioxide to carbon molecular weight, the carbon dioxide emissions of 281 cities in China from 2006 to 2019 can be calculated. Since this study uses historical CO_2 emission data, and is not based on carbon trading schemes, it should be considered post hoc analysis, and therefore calculation errors caused by different ways of allocating emission reduction targets can be avoided.

By referring to relevant literature and considering the actual situation [34], the following control variables were selected to conduct propensity matching scores: regional economic development level (PGDP), industrial structure (IND), urban population (PP), degree of openness to the outside world (OPEN), efficiency of financial development (FS), scale of financial development (FD), and government environmental intervention (WODK). The specific calculation methods are shown in Table 1.

Table 1. Description of variables.

Index	Measure
PGDP	Real GDP per capita in cities is measured in logarithms (Yuan per person)
IND	Ratio of the added value of secondary production to the gross city product (%)
PP	The logarithm of the resident population at the end of the year (million)
OPEN	Ratio of foreign investment to gross city product (%)
FS	The ratio of total social loans to gross urban product (%)
FD	The ratio of total social savings to gross urban product (%)
WODK	The proportion of the use of environmental words in the total words of the government work report. (e.g., environmental protection, green, low-carbon, energy-saving and emission reduction, etc.) (%)

4.3. Model Setting

The reference measurement model of this paper is set as follows:

$$Carbon_{it} = \alpha_0 + \alpha_1 treated_{it} * time_{it} + \sum_{i=1}^{N} \beta_j control_{it} + \mu_i + \gamma_t + \varepsilon_{it} \quad (2)$$

where i represents the individual city and t represents the year. $Carbon_{it}$ is the carbon emission of city i in year t. The year dummy variable $time_{it}$ takes a value of 0 before the introduction of the carbon emission trading pilot policy (the policy impact point is set as 2013) and 1 after the establishment, and $treated_{it}$ is a group dummy variable. The ETS pilot cities are assigned a value of 1, non-ETS pilot cities are assigned a value of 0, and $treated_{it} * time_{it}$ is the interaction term of the two and takes the value of 0 or 1. Here, 1 represents the pilot cities after 2013, and 0 represents non-pilot cities and the pilot cities before 2013. The coefficient α_1 before the interaction term of $treated_{it} * time_{it}$ is an important explanatory variable that represents the policy effect of emissions trading on urban carbon intensity. This paper will introduce various control variables affecting urban carbon intensity into the multi-stage DID regression. The bidirectional fixed effect of city and year will be introduced for further analysis.

4.4. Propensity Score Matching Results and Descriptive Statistics

4.4.1. Counterfactual Matches with the Equation Estimates

Rosembaum and Rubin proposed the Propensity Score Method [41]. Simulation experiments show that the ATT can obtain unbiased estimation results under a series of assumptions. It can be defined as "an algorithm that matches the treatment group and the control group based on the conditional probability of participants, namely the propensity score, under the condition of given observable characteristics". The propensity score is defined as:

$$P(X_i) = Pr\{exp_i = 1|X_i\} \quad (3)$$

According to Equation (3), the propensity score similarity between the treatment group and the control group is matched, and its effectiveness depends on two preconditions. The first is conditional independence. The second is that the conditions for common support are met. The independence condition means that ETS pilot cities or non-pilot cities are independent of carbon intensity after controlling the common influencing factor X, and the common support condition ensures that cities in each treatment group can match cities in the control group through propensity score matching. The average treatment effect ATE of city i can be expressed as

$$E[\triangle_i] = E\left[lny_i^1\left(ny_i^1, fy_i^1\right)\middle|exp_i = 1, P(X_i)\right] - E\left[lny_i^0\left(ny_i^0, fy_i^0\right)\middle|exp_i = 1, P(X_i)\right] \quad (4)$$

To estimate *P(X)* is to estimate the probability that the city is or is not an ETS pilot. A Probit or Logit binary choice model is most commonly used. In this paper, a Probit model is used to obtain the predicted probability value P_i of city i in the treatment group and P_j of city j in the control group. The average treatment effect (ATT) of ETS on carbon intensity is as follows:

$$\beta = \frac{1}{M}\sum_{i\in(exp=1)}\left[lny_i(ny_i, fy_i) - \sum_{i\in(exp=0)} Y(NY, FY)(p_i, p_j) lny_j(ny_j, fy_j)\right] \quad (5)$$

where M is the number of cities in which ETS was piloted. $Y(NY, FY)(p_i, p_j)$ represents the case when $lny_i^0(ny_j^0, fy_j^0)$ of city j is replaced by $lny_i^0(ny_i^0, fy_i^0)$ of city i. This represents the weight assigned to $lny_i^0(ny_j^0, fy_j^0)$ of city j. When the corresponding assumptions are met, especially when the mean values of variables in the treatment group and the control group are not different, the propensity score matching method can obtain the ATT, and a "clean" policy treatment effect can be obtained. Of course, being able to eliminate this noise completely requires being able to control for all variables that may have an impact on choice and outcome when matched. According to the matching method (radius matching, caliper matching, local linear regression matching, etc.), the weight function selection is also different. This study first selects the nearest neighbor matching with k = 4, and then selects other matching methods in the robustness test.

4.4.2. Plot of Propensity Score Matching Kernel Density Function

The quality of propensity score matching can be examined by a plot of kernel density functions. If there is more overlap between the treatment group and the control group in the figure, this indicates that the test propensity match score is better. Figure 2 shows the kernel density function of the two groups of cities before and after propensity score matching. The solid line represents the cities in the processing group, and the dashed line represents the cities in the control group. As shown in Figure 2, prior to PSM, the two groups showed large differences in both skewness and kurtosis. After PSM, the change trend of the two groups is consistent, and there is a high degree of line segment coincidence. This indicates that the propensity score matching has a significant effect. This provides a good data basis for the use of the DID method in the empirical part of this paper.

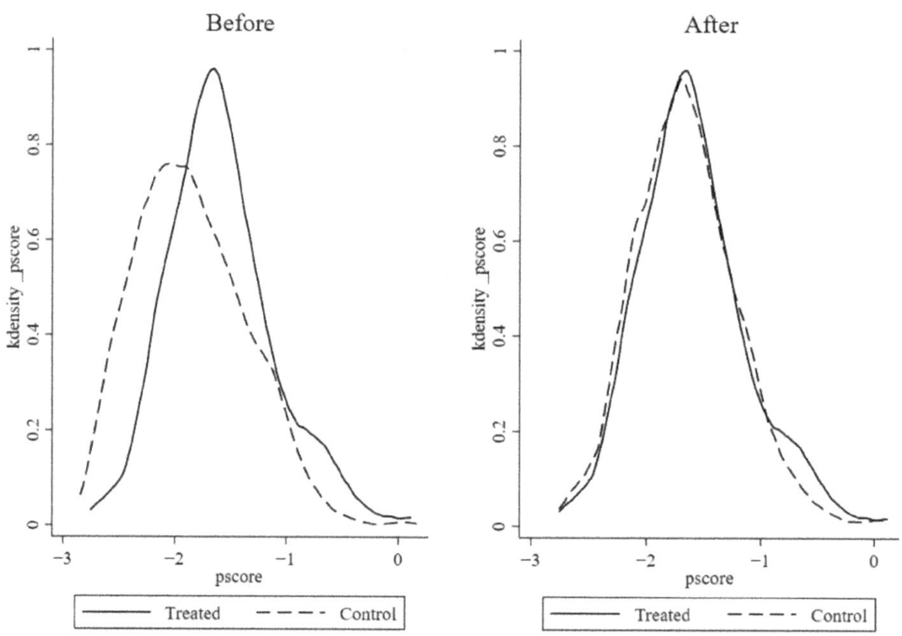

Figure 2. Kernel density function before and after PSM in treatment group and control group.

4.4.3. Balance Test of Propensity Score Matching and Variables of Descriptive Statistics

In order to make the results of PSM more robust, the results should satisfy the two groups of cities, and there is no obvious difference in each matching variable. The method to judge whether PSM is effective generally carries out the balance test of propensity score matching. Note the absolute value of the standard deviation of the matching variable. If the absolute value of the standard deviation is smaller, it indicates that the matching effect is better. Table 2 results show that most of the matching variables decrease significantly in the absolute value of the PSM standard deviation. The t-test value also changed from significant to insignificant. This indicates that the null hypothesis that the mean of each variable is consistent after matching is accepted. Propensity matching scores are valid. Table 3 shows the descriptive statistics of variables after PSM.

Table 2. Balance test of propensity score matching.

Variable	Sample Match	The Mean		Standard Deviation (%)		t Test $p > \|t\|$
		Treat	Control	Deviation	To Reduce	
PGDP	Before	10.576	10.432	20	99.3	0.000
	After	10.576	10.575	0.1		0.982
IND	Before	3.818	3.844	−11.2	99.3	0.026
	After	3.818	3.836	7.8		0.203
PP	Before	6.084	5.962	19	97	0.000
	After	6.084	6.088	−0.6		0.926
OPEN	Before	0.003	0.003	23.7	95.6	0.000
	After	0.003	0.003	1		0.877
FS	Before	0.725	0.727	−1.2	−550.3	0.824
	After	0.725	0.706	7.8		0.182
FD	Before	0.820	0.812	1.8	−236.8	0.694
	After	0.820	0.794	5.9		0.323
WODK	Before	0.003	0.003	11.5	78.1	0.017
	After	0.003	0.004	-2.5		0.692

Table 3. Descriptive statistics.

Variable	Size	Means	Std. Dev.	Min.	Max.
CARBON	3432	−2.876	0.650	−5.113	−0.355
PGDP	3432	10.453	0.691	7.926	13.056
IND	3432	3.840	0.242	2.460	4.450
PP	3432	5.980	0.611	3.959	8.136
OPEN	3432	0.003	0.003	0	0.019
FS	3432	0.727	0.263	0.083	2.547
FD	3432	0.814	0.411	0.112	2.683
WODK	3432	0.003	0.001	0	0.012

5. Empirical Analysis

5.1. Results of Dual Difference Regression

In order to more clearly identify the causal impact of ETS on urban carbon intensity, the above control variables will be introduced in this section. The model combining city individual fixed effects (Id) and year fixed effects (Year) is used for further analysis, and the results are shown in Table 4. The results in column (1) show that, without adding any control variables, the coefficient of Treated*time is significantly negative at the level of 1%. The results in columns (2)–(4) show that, after the introduction of other control variables, the coefficient of Treated ∗ time is significantly negative at the 1% level. This indicates that ETS can effectively reduce urban carbon intensity. In order to make the results more reliable, Column (5) shows the test results of the generalized method of moments estimation for dynamic instrumental variables. The coefficient of Treated ∗ time is still significantly negative at the level of 1%, which further verifies the conclusion of this paper. This empirical study also preliminarily shows that the introduction of pilot carbon emission trading policies can effectively reduce urban carbon intensity, and Hypothesis 1 is established.

Table 4. Dual difference regression.

Variable	(1)	(2)	(3)	(4)	(5)
Treated ∗ time	−0.625 *** (0.264)	−0.173 *** (0.026)	−0.150 *** (0.024)	−0.142 *** (0.024)	−0.316 *** (0.027)
PGDP		−0.643 *** (0.009)	−0.618 *** (0.010)	−0.636 *** (0.011)	−0.288 *** (0.024)
IND		0.178 *** (0.034)	0.155 *** (0.031)	0.248 *** (0.038)	0.116 ** (0.051)
PP			−0.616 *** (0.075)	−0.576 *** (0.083)	−0.197 *** (0.013)
OPEN			1.394 (1.962)	1.759 (1.836)	−3.730 (2.606)
FS				0.127 *** (0.037)	0.501 *** (0.043)
FD				0.010 (0.034)	−0.109 *** (0.027)
WODK				3.211 (1.96)	9.683 * (5.746)
Constant	−4.061 *** (0.001)	1.952 *** (0.156)	5.448 *** (0.427)	4.932 *** (0.488)	−0.911 *** (0.298)
Id	YES	YES	YES	YES	NO
Year	YES	YES	YES	YES	YES
R_2	0.153	0.894	0.903	0.905	0.451
Sample size	3432	3432	3432	3432	2839

Note: *, **, and *** represent the significance levels of 10%, 5%, and 1% respectively. The clustering standard error is shown in brackets.

5.2. Heterogeneity Analysis

5.2.1. Regional Heterogeneity Test

On the whole, ETS can effectively reduce the carbon intensity of cities. However, different cities in China are located in different external environments. This leads to obvious differences across urban regions. In particular, the pilot policy of carbon emission trading has great relevance to the energy environment. The economically developed eastern region and the economically less-developed central and western regions have obvious differences in infrastructure and other conditions. Table 5 shows the regional heterogeneity results. The results in columns (1)–(3) show that the eastern region is inferior to the central and western regions in terms of coefficient and significance level. This indicates that the ETS policy in the eastern region is less effective in reducing carbon intensity. Relative to the central and western regions, the eastern region has a large population and more developed economy, with a concentration of various industries. As a result, the consumption of electric energy and heat energy caused by industrial electricity and residential electricity is large. Due to the normal economic and social activities in the eastern region, ETS cannot reduce the carbon intensity of the city in a short time. However, in China's central and western areas, the population is lower and the economic development level is weaker. In those regions, the establishment of carbon emission trading pilots can effectively reduce the carbon intensity.

Table 5. Regional heterogeneity test.

Variable	The Eastern Region	The Central Region	In the Western Region
Treated * time	−0.067 * (0.035)	−0.207 *** (0.020)	−0.239 *** (0.021)
Constant	6.132 *** (1.194)	4.524 *** (0.694)	4.777 ** (1.954)
Control	YES	YES	YES
Id	YES	YES	YES
Year	YES	YES	YES
R_2	0.909	0.916	0.900
Sample size	1324	1405	703

Note: *, **, and *** represent the significance levels of 10%, 5%, and 1% respectively. The clustering standard error is shown in brackets.

5.2.2. Quantile Regression Test

The regional heterogeneity of ETS on urban carbon intensity has been analyzed above. This part will analyze the quantile heterogeneity of ETS on carbon intensity—that is, the policy effect of ETS on high and low carbon intensity. It can be seen from Table 6 that, regardless of the value of M (M is the quantile), ETS always has a dampening effect on carbon intensity. Moreover, the impact of ETS on carbon intensity at different quantiles also changes significantly. Specifically, the emission reduction effect of ETS on cities with higher carbon intensity is more obvious. Figure 3 shows the trend of ETS regression on urban carbon intensity quantiles. The horizontal axis in the figure shows the different quantile decimal points of the ETS on urban carbon intensity. The vertical axis shows the regression coefficients of each variable. The dashed lines of the line segments represent the OLS regression estimates of the corresponding explanatory variables. The region between the two dotted lines represents the confidence interval of the OLS regression value (confidence 0.95). The solid lines are the quantile regression estimation results of each explanatory variable. The shaded part is the confidence interval (confidence 0.95) of the quantile regression estimate. Figure 3 further shows that the emission reduction effect of ETS on cities with higher carbon intensity is more obvious.

Table 6. Quantile regression.

Variable	M = 0.1	M = 0.3	M = 0.5	M = 0.7	M = 0.9
Treated * time	−0.203 *** (0.055)	−0.220 *** (0.033)	−0.328 *** (0.021)	−0.435 *** (0.028)	−0.520 *** (0.045)
Constant	0.049 (0.355)	0.041 (0.232)	0.255 (0.345)	0.459 (0.283)	1.513 *** (0.359)
Control	YES	YES	YES	YES	YES
Id	YES	YES	YES	YES	YES
Year	YES	YES	YES	YES	YES
R_2	0.283	0.286	0.289	0.283	0.264
Sample size	3432	3432	3432	3432	3432

Note: *** represent the significance levels of 1%. The clustering standard error is shown in brackets

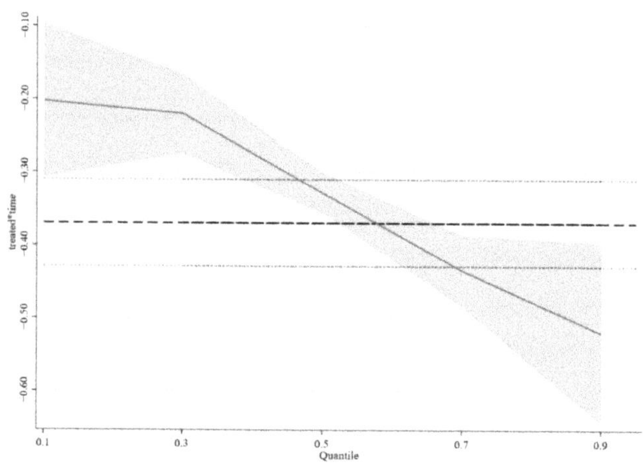

Figure 3. Quantile regression trend of carbon intensity of cities under ETS.

5.3. Robustness Test

5.3.1. Parallel Trend Test

This section presents the results of parallel trend tests. The specific test formula is set as:

$$Carbon_{it} = \alpha_0 + \omega_d \sum_{d=-5}^{d=5} treated_{it} * time_{it} + \sum_{i=1}^{N} \beta_j control_{it} + \mu_i + \gamma_t + \varepsilon_{it} \quad (6)$$

The main variables in the above formula have the same meaning as in Formula (1), where d_5 represents 5 years before the introduction of ETS policy, and d5 represents the 5th year after the introduction of ETS policy. The coefficient ω_d is the focus of this paper's test. If the coefficient estimate is insignificant before ETS, significantly negative after ETS, and shows a difference in marginal effects, then the parallel trend assumption is satisfied. As shown in Figure 4, before the ETS, the effect on carbon intensity is not significant. After the establishment of the ETS, the coefficient is significantly negative. The marginal effect of ETS on carbon intensity mitigation is strengthened over time, showing a long-term emission reduction effect. This proves the rationality of using the PSM-DID method in this paper.

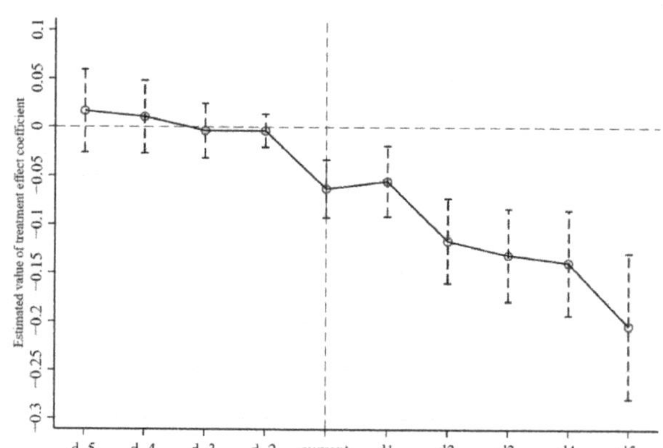

Figure 4. Parallel trend test.

5.3.2. Change the Sample-Matching Method

The nearest neighbor matching method with K = 4 was selected above for data matching and processing. In order to make the above conclusion more robust, this part re-selects the matching party for data matching. In this part, the methods of Mahalanobis distance matching, caliper matching, radius matching, and kernel matching were used to re-match the data. Table 7 shows the results of difference-in-differences estimation by various matching methods. After changing the propensity matching scoring method, the estimated results are close to the regression results above. This indicates that the above regression results are reliable, verifying that that ETS can effectively reduce urban carbon intensity.

Table 7. Results of replacing the matched DID.

Variable	Mahalanobis Distance Matches	Caliper Match	Radius of a Match	Nuclear Match
Treated * time	−0.168 *** (0.026)	−0.141 *** (0.024)	−0.168 *** (0.026)	−0.141 *** (0.024)
Constant	4.530 *** (0.469)	4.932 *** (0.488)	4.530 *** (0.469)	4.861 *** (0.493)
Control	YES	YES	YES	YES
Id	YES	YES	YES	YES
Year	YES	YES	YES	YES
R_2	0.872	0.905	0.872	0.904
Sample size	3934	3432	3934	3434

Note: *** represent the significance levels of 1%. The clustering standard error is shown in brackets.

5.3.3. Placebo Test

The cities in which ETS was piloted may have been chosen as pilots due to their relatively complete infrastructure and high economic development potential. Therefore, in order to eliminate the interference of other unobservable factors with the conclusions of this paper, a placebo test was used to further prove the reliability of the previous conclusions. In this part, the interaction terms are randomly selected 1000 times to check whether the coefficients are significantly different from the benchmark estimation results. The results are shown in Figure 5. The dashed line indicates that the actual estimated coefficient obtained by PSM-DID is −0.142. The coefficient estimate is lower than 1000 random draws. This indicates that the placebo test in this part is valid. Thus, the reliability of the conclusions of this paper is proven.

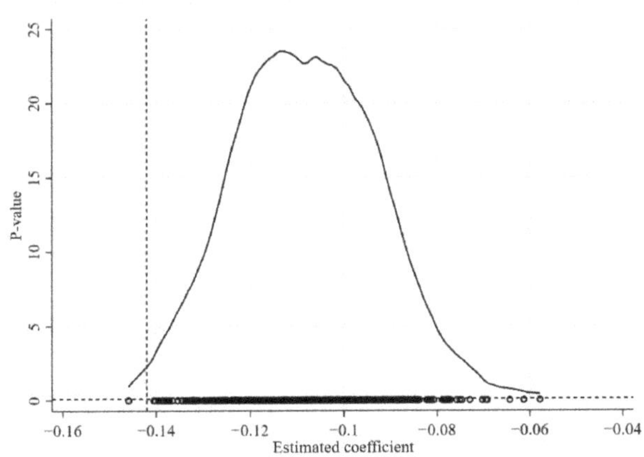

Figure 5. Placebo test.

6. Further Analysis

6.1. The Mediation Effect Test

The above empirical results show that introducing the carbon emission trading pilot policy has alleviated the carbon intensity of cities. Then, how does ETS affect the carbon intensity, and what is the specific mechanism? According to the above theoretical analysis, this paper argues that the pilot carbon emissions trading policy acts through green technology innovation and environmental governance. Therefore, this paper will examine the intermediary mechanism from the two channels of green technology innovation and environmental governance.

$$\mathrm{M}_{it} = \beta_0 + \delta_2 treated_{it} * time_{it} + \sum_{i=1}^{N} w_j control_{it} + \mu_i + \gamma_t + \varepsilon_{1it} \tag{7}$$

$$Carbon_{it} = \theta_1 + \delta_3 treated_{it} * time_{it} + \delta_4 M_{it} + \sum_{i=1}^{N} e_j control_{it} + \mu_i + \gamma_t + \varepsilon_{2it} \tag{8}$$

In the above equation, M represents the mediating variables, which are green technological innovation (Inno) and environmental governance (Trash), respectively. Among them, green technological innovation is represented by the number of green invention patents and green utility model patents granted per capita in cities [42]. A larger value indicates a higher level of green technology innovation. The calculation of the environmental governance level index is measured by the sum of waste water, waste gas, and solid waste generated by the city. A smaller value indicates a higher level of environmental governance.

Traditional parameter estimation methods require the assumption of a normal distribution of data. The use of stepwise regression may have some impact on the assessed policy effects. Therefore, the Sobel test and Bootstrap method were used to test the mediating effect in this part. The Bootstrap test uses the mixed effects hypothesis. In this paper, the original sample was randomly sampled repeatedly with $n = 1000$. The asymmetry in the distribution of indicators was corrected. This can significantly improve the accuracy of model testing under a complex mediation structure.

Table 8 shows the mediation test results. When Inno is used as a mediating variable, the coefficient before Treated $*$ time is significantly positive at the 1% level. This indicates that the introduction of the pilot policy of carbon emission trading has promoted the level of urban green technological innovation. The coefficient of urban carbon intensity is significantly negative at the 1% level. This shows that the improvement of green technology innovation alleviates urban carbon intensity, and the path of "carbon emission trading pilot policy-green technology innovation-urban carbon intensity" is established. This proves Hypothesis 2.

When the level of environmental governance is used as a mediating variable, the coefficient before Treated $*$ time is significantly negative at the 1% level. This shows that the introduction of the pilot policy of carbon emission trading has improved the level of urban environmental governance. The coefficient of urban carbon intensity is significantly positive at the 1% level. This indicates that the improvement of environmental governance will alleviate urban carbon emissions. The path of "Carbon emission trading pilot policy—environmental governance level—urban carbon intensity" is established. This proves Hypothesis 3.

Table 8. Results of mediating effect test.

Variable	Green Technology Innovation		Environmental Governance	
	(1)	(2)	(3)	(4)
	Inno	Carbon	Trash	Carbon
Treated $*$ time	0.931 *** (0.140)	−0.055 *** (0.008)	−0.119 *** (0.039)	−0.079 *** (0.008)
Inno		−0.016 *** (0.001)		

Table 8. *Cont.*

Variable	Green Technology Innovation		Environmental Governance	
	(1)	(2)	(3)	(4)
	Inno	Carbon	Trash	Carbon
Trash				0.013 *** (0.003)
Constant	−30.76 *** (5.346)	1.884 *** (0.317)	3.910 ** (1.515)	1.169 *** (0.305)
Control	YES	YES	YES	YES
Id	YES	YES	YES	YES
Year	YES	YES	YES	YES
Sobel test	Z = −6.094 ***		Z = −2.333 **	
The Bootstrap test	[−0.023, −0.007] (BC)		[−0.004, −0.0003] (BC)	
R^2	0.765	0.979	0.851	0.985
Sample size	3432	3432	3432	3432

Note: ** and *** represent the significance levels of 5%, and 1% respectively. The clustering standard error is shown in brackets.

6.2. Spatial Spillover Effect Test

6.2.1. Model Set and Related Analysis

According to the above analysis, the impact of the pilot carbon emission trading policy on carbon intensity may have a spatial spillover effect, which needs further analysis. Therefore, this paper establishes a spatial econometric model based on the equation:

$$Carbon_{it} = \beta_0 + \beta_1 W \times Carbon_{it} + \beta_2 CD_{it} + \beta_3 W \times CD_{it} + \sum_{i-1}^{N} q_j control_{it} + \mu_i + \gamma_t + \varepsilon_{it} \quad (9)$$

where W is the spatial weight matrix. Equation (7) adds the spatial interaction term ($W \times CD$) of the core explanatory variable (CD) to the equation. The model estimates the spatial spillover effects of the explained and core explanatory variables. Regarding the selection of the spatial weight matrix, this paper chooses the geographical inverse distance matrix to study the possible spatial spillover effect.

Before the spatial econometric analysis, it is necessary to determine whether there is a spatial correlation of urban carbon intensity. In this paper, the global Moran's I index is used to test the spatial correlation of carbon emissions. Table 9 reports the regression results of each year. For 2006~2019, Moran's I index shows significance under the 1% level, which shows a spatial correlation in urban carbon intensity.

Table 9. Results of spatial correlation test.

Year	Moran's I	Z Value	Year	Moran's I	Z Value	Year	Moran's I	Z Value
2006	0.141 ***	27.849	2011	0.128 ***	25.399	2016	0.172 ***	33.931
2007	0.136 ***	26.840	2012	0.128 ***	25.332	2017	0.173 ***	34.062
2008	0.131 ***	25.889	2013	0.133 ***	26.313	2018	0.168 ***	33.182
2009	0.130 ***	25.670	2014	0.143 ***	28.317	2019	0.183 ***	36.064
2010	0.133 ***	26.251	2015	0.156 ***	30.766			

Note: *** represent the significance levels of 1%. The clustering standard error is shown in brackets.

6.2.2. Analysis of Regression Results

Table 10 shows the regression results of the spatial Durbin model with double fixed effects. Columns (1)–(3) represent the direct effect, indirect effect, and total effect after coefficient decomposition respectively. From R^2 and the $Sigma^2$ and log-likelihood statistics, the fit of the model is better and the overall regression reliability is higher. As column

(1) shows, the Treated ∗ time coefficient is −0.141, and is significant at the 1% level. This means that the establishment of carbon emissions trading pilot cities can alleviate local urban carbon intensity, which is consistent with the results of the benchmark in front of the regression. Column (2) shows that the Treated ∗ time coefficient is 0.168 and is significant at the 1% level. This means that the establishment of pilot emissions trading can increase the carbon intensity in areas surrounding the region. W ∗ Treated ∗ time before the time coefficient is 0.399 and is significant at the 1% level. This means that, when pilot emissions trading was set up in this region, the ETS produced a spatial spillover effect, increasing the carbon intensity of the surrounding area. Because of the region's strict carbon trading controls, polluting companies cannot afford the high prices of carbon credits and move to surrounding areas. As the above result shows, the establishment of a pilot emissions trading city not only can reduce the carbon intensity of the city, but also can affect surrounding cities. The environmental regulation in the region, through the strict design of the carbon trading system, relies on the power of the government. The expansion of the implementation of tertiary industries such as service, acceleration of the upgrading of industrial structure, and finally, the improved efficiency of energy utilization can alleviate the carbon intensity of the region, but may cause enterprises to transfer, which can increase the carbon intensity in the surrounding areas. This proves hypothesis 4.

Table 10. Regression results of spatial Durbin model.

Variable	(1)	(2)	(3)
Treated ∗ time	−0.141 *** (0.024)	0.168 *** (0.026)	−0.141 *** (0.024)
W ∗ Treated ∗ time		0.399 *** (0.063)	
Log-likelihood		4258.507	
$sigma^2$		0.006 *** (0.001)	
Control		YES	
Id		YES	
Year		YES	
R^2		0.316	
Sample size		3 934	

Note: *** represent the significance levels of 1%. The clustering standard error is shown in brackets.

7. Conclusions and Recommendations

7.1. Conclusions

This paper regards the carbon emission trading system as a quasi-natural experiment. Using the panel data of 281 cities in China from 2006 to 2019, this paper empirically examined the policy effect and spatial spillover effect of ETS on urban carbon intensity in China by using PSM-DID and spatial Durbin models and analyzed it from multiple perspectives.

First, ETS helps mitigate urban carbon intensity. However, this effect has heterogeneous characteristics. The mitigation effect of the carbon emission trading system on the carbon intensity in the eastern region is not significant. By contrast, the mitigation effect on carbon intensity in the central and western regions is very significant. Compared with the central and western regions, the eastern region has a large population, developed economy and various industries. Industrial and residential consumption of electricity and heat energy is huge. Setting up pilot carbon emission trading in the eastern region, while also promoting economic activities in that region, cannot significantly reduce the carbon intensity of cities in a short period of time. In the central and western regions of China, the population is small, and the level of economic development is weak. Setting up carbon emission trading pilots in those regions can effectively reduce the carbon intensity of the regions. The results of the quantile test show that the emission reduction effect of ETS is more obvious for cities with higher carbon intensity, and the marginal effect of emission reduction is larger for cities with higher carbon intensity, so there is more room for emission reduction.

Second, the parallel trend test shows that the longer the carbon emission trading system is established, the more obvious is the mitigation effect on urban carbon intensity. The longer the carbon emission trading system is established, the more time the pilot enterprises have to carry out technological innovation, and the more obvious the effect of mitigating urban carbon intensity measurement will be.

Third, this paper further analyzes the influence mechanism of ETS from the two aspects of green technological innovation and environmental governance. The results show that the carbon emission trading system can encourage enterprises to carry out technological innovation to reduce emissions, thus alleviating urban carbon intensity. By improving the level of environmental governance and reducing the emission of all kinds of pollutants, this also reduces the corresponding carbon emissions, which then alleviates the carbon intensity. This is consistent with most of the literature results.

Fourth, spatial spillovers show that the ETS, although able to mitigate the carbon intensity of the pilot cities, causes the carbon emissions of the surrounding non-pilot cities to rise. This is because the penalty mechanism of ETS leads to high environmental costs that cause enterprises to transfer to surrounding non-pilot areas. As a result, the carbon intensity of surrounding areas increases.

7.2. Recommendations

First, the development of ETS should always adhere to the combination of "market determination" and "government regulation". On the one hand, policy makers should continue to insist on the decisive role of the market in the allocation of carbon emission rights, and should use supply and demand mechanisms, competition mechanisms, price mechanisms, and other means to promote the effective operation of carbon trading markets. This requires constantly adjusting the incentives of enterprises through surplus carbon emission rights and adjusting the penalties imposed on enterprises with insufficient carbon emission rights through market means. Thus, the cost of carbon emissions is internalized into the cost-benefit analysis of the enterprise and becomes an important variable for maximizing corporate profits, thus promoting carbon emission reduction. On the other hand, policy makers should give full play to the regulating and supporting role of the government. The government should formulate laws and regulations suitable for the healthy and effective operation of the market in order to make up for market failures such as monopoly, information asymmetry, and externalities caused by market limitations, thereby constantly improving the market environment.

Second, the carbon reduction effect of ETS is regionally heterogeneous. There are significant differences between different regions due to their local level of economic development, industrial structure, energy structure, and other factors. Therefore, each transaction pilot cannot adopt a "one size fits all" attitude when formulating policies. The construction of carbon trading markets should be carried out according to local conditions. In this way, achieving carbon reduction targets also can promote high-quality economic development.

Third, scientific and technological research and innovation of enterprises is the key to the carbon reduction effect of the ETS. The government, enterprises, and society should pay special attention to the important role of scientific and technological R&D and innovation in carbon trading policies. It is recommended to continuously increase the R&D investment of all enterprises and encourage them to carry out technological innovation in order to constantly update the production process. This will promote the green development of enterprises. The government should also effectively improve its own environmental governance level in order to improve its ability to prevent and control urban pollution. Through the development of a series of laws and regulations to assist the operation of the ETS system, strict penalties should be imposed on enterprises that violate the system.

Fourth, the spatial spillover effect among different cities increases the carbon intensity of surrounding cities. On the one hand, local governments in non-pilot areas are encouraged to actively learn from the experiences of pilot areas in order to reduce the carbon intensity of the region. However, carbon leakage through spatial spillovers undermines the goal of reducing emissions. New mechanisms should be considered to prevent companies from avoiding emission rules. A hybrid mechanism that combines the carbon ETS with other environmental regulation tools is recommended. For example, a carbon tax could be imposed on emissions in non-pilot areas to discourage firms from leaving pilot areas.

The above is the main content of this paper, but the research of this paper still has limitations. This study uses an econometric approach based on historical data from a pilot carbon trading program in China. We also believe that carbon tax is one of the effective ways to reduce the carbon intensity of cities. If carbon tax projects are implemented in these cities, a more reasonable conclusion can be obtained by comparing the effects of carbon trading and carbon tax. Since China has no plan to implement carbon tax at present, such data cannot be obtained to reconstruct the regression model of carbon tax cases. This prevents more reasonable conclusions from being drawn. If China has some concrete practice in carbon tax, the author will study it.

Author Contributions: Writing, J.L.; Methodology, J.L.; Providing case and idea: W.Y.; Revising and editing: J.L. All authors have read and agreed to the published version of the manuscript.

Funding: This research received no external funding.

Institutional Review Board Statement: Not applicable.

Informed Consent Statement: Not applicable.

Data Availability Statement: The data in this paper were collected through publicly available data. Data about cities are based on the related data from the National Bureau of Statistics (https://data.stats.gov.cn/, accessed on 15 August 2021). Patent data is from the State Intellectual Property Office (https://www.cnipa.gov.cn/, accessed on 20 August 2021). The relevant data can be downloaded from the relevant website. Data can be obtained from the corresponding author on request.

Acknowledgments: The author thanks the anonymous reviewer for critical corrections and Cyndi Berck for professional language editing.

Conflicts of Interest: The authors declare no conflict of interest.

References

1. Liu, Y.; Zhou, Y.; Wu, W. Assessing the impact of population, income and technology on energy consumption and industrial pollutant emissions in China. *Appl. Energy* **2015**, *155*, 904–917. [CrossRef]
2. Hong, Q.; Cui, L.; Hong, P. The impact of carbon emissions trading on energy efficiency: Evidence from quasi-experiment in China's carbon emissions trading pilot. *Energy Econ.* **2022**, *110*, 106025. [CrossRef]
3. Xy, A.; Ping, J.A.; Yao, P.B. Does China's carbon emission trading policy have an employment double dividend and a Porter effect? *Energy Policy* **2020**, *142*, 111492. [CrossRef]
4. Chen, H.; Guo, W.; Feng, X.; Wei, W.; Gong, W. The impact of low-carbon city pilot policy on the total factor productivity of listed enterprises in China. *Resour. Conserv. Recycl.* **2021**, *169*, 105457. [CrossRef]
5. Li, C.; Qi, Y.; Liu, S.; Wang, X. Do carbon ETS pilots improve cities' green total factor productivity? Evidence from a quasi-natural experiment in China. *Energy Econ.* **2022**, *108*, 105931. [CrossRef]
6. Holligan, B. Commodity or Propriety? Unauthorised Transfer of Intangible Entitlements in the EU Emissions Trading System. *Mod. Law Rev.* **2020**, *83*, 979–1007. [CrossRef]
7. Carlen, B.; Dahlqvist, A.; Mandell, S.; Marklund, P. EU ETS emissions under the cancellation mechanism—Effects of national measures. *Energy Policy* **2019**, *129*, 816–825. [CrossRef]
8. Ding, S.; Zhang, M.; Song, Y. Exploring China's carbon emissions peak for different carbon tax scenarios. *Energy Policy* **2019**, *129*, 1245–1252. [CrossRef]
9. Perino, G. New EU ETS Phase 4 rules temporarily puncture waterbed. *Nat. Clim. Change* **2018**, *8*, 262–264. [CrossRef]
10. Keohane, N.O. Cap and Trade, Rehabilitated: Using Tradable Permits to Control U.S. Greenhouse Gases. *Rev. Environ. Econ. Policy* **2009**, *3*, 42–62. [CrossRef]
11. Feng, D.; Dai, Y.; Zhang, S.; Zhang, X.; Long, R. Can a carbon emission trading scheme generate the Porter effect? Evidence from pilot areas in China—ScienceDirect. *Sci. Total Environ.* **2019**, *653*, 565–577.

12. Bian, Y.; Yan, S.; Xu, H. Efficiency evaluation for regional urban water use and wastewater decontamination systems in China: A DEA approach. *Resour. Conserv. Recycl.* **2014**, *83*, 15–23. [CrossRef]
13. Pigou, A.C. Real income and economic welfare. *Oxf. Econ. Pap.* **1951**, *3*, 16–20. [CrossRef]
14. Coase, R.H. The Problem of Social Cost: The Citations. *Cklawreview* **1996**, *71*, 809.
15. Dales, J.H. *Pollution, property and prices: An essay in policy-making and economics*; University of Toronto Press: Toronto, ON, Canada, 1968.
16. Tanaka, K. A comparison study of EU and Japan methods to assess CO2 emission reduction and energy saving in the iron and steel industry. *Energy Policy* **2012**, *51*, 578–585. [CrossRef]
17. Lovcha, Y.; Perez-Laborda, A.; Sikora, I. The determinants of CO_2 prices in the EU emission trading system. *Appl. Energy* **2022**, *305*, 117903. [CrossRef]
18. Seifert, J.; Uhrig-Homburg, M.; Wagner, M. Dynamic behavior of CO2 spot prices. *J. Environ. Econ. Manag.* **2007**, *56*, 180–194. [CrossRef]
19. Fang, G.; Tian, L.; Liu, M.; Fu, M.; Sun, M. How to optimize the development of carbon trading in China—Enlightenment from evolution rules of the EU carbon price. *Appl. Energy* **2018**, *211*, 1039–1049. [CrossRef]
20. Jaffe, A.B.; Stavins, R.N. Dynamic Incentives of Environmental Regulations: The Effects of Alternative Policy Instruments on Technology Diffusion—ScienceDirect. *J. Environ. Econ. Manag.* **1995**, *29*, S43–S63. [CrossRef]
21. Bakam, I.; Balana, B.B.; Matthews, R. Cost-effectiveness analysis of policy instruments for greenhouse gas emission mitigation in the agricultural sector. *J. Environ. Manag.* **2012**, *112*, 33–44. [CrossRef]
22. WråKe, M.; Myers, E.; Burtraw, D.; Mandell, S.; Holt, C. Opportunity Cost for Free Allocations of Emissions Permits: An Experimental Analysis. *Environ. Resour. Econ.* **2010**, *46*, 331–336. [CrossRef]
23. Xiong, L.; Shen, B.; Qi, S.; Price, L.; Ye, B. The allowance mechanism of China's carbon trading pilots: A comparative analysis with schemes in EU and California. *Appl. Energy* **2017**, *185*, 1849–1859. [CrossRef]
24. Burniaux, J.M.; Nicoletti, G.; Oliveira-Martins, J. Green: A global model for quantifying the costs of policies to curb CO_2 emissions. *OECD Econ. Stud.* **1992**, *19*, 49–92.
25. Zhang, Y.J.; Wei, Y.M. An overview of current research on EU ETS: Evidence from its operating mechanism and economic effect. *Appl. Energy* **2010**, *87*, 1804–1814. [CrossRef]
26. Zhang, C.; Wang, Q.; Shi, D.; Li, P.; Cai, W. Scenario-based potential effects of carbon trading in China: An integrated approach. *Appl. Energy* **2016**, *182*, 177–190. [CrossRef]
27. Zhang, Y.J.; Peng, Y.L.; Ma, C.Q.; Shen, B. Can environmental innovation facilitate carbon emissions reduction? Evidence from China. *Energy Policy* **2017**, *100*, 18–28. [CrossRef]
28. Hu, Y.; Ren, S.; Wang, Y.; Chen, X. Can carbon emission trading scheme achieve energy conservation and emission reduction? Evidence from the industrial sector in China. *Energy Econ.* **2020**, *85*, 104590. [CrossRef]
29. Anger, N.; Oberndorfer, U. Firm performance and employment in the EU emissions trading scheme: An empirical assessment for Germany. *Energy Policy* **2008**, *36*, 12–22. [CrossRef]
30. Wang, Y.M.; Pan, S.M. Status and trends of China carbon trading market under financial crisis. *Econ. Theory Bus. Manag.* **2009**, *11*, 57–63.
31. Hübler, M.; Lschel, A.; Voigt, S. Designing an emissions trading scheme for China: An up-to-date climate policy assessment. *ZEW Discuss. Pap.* **2014**, *75*, 57–72. [CrossRef]
32. Su, B.; Ang, B.W. Multiplicative decomposition of aggregate carbon intensity change using input–output analysis. *Appl. Energy* **2015**, *154*, 13–20. [CrossRef]
33. Ansari, M.A. Does trade openness affects global carbon dioxide emissions. *Manag. Environ. Qual. Int. J.* **2019**, *31. ahead-of-print.* [CrossRef]
34. Xuan, D.; Ma, X.; Shang, Y. Can China's policy of carbon emission trading promote carbon emission reduction? *J. Clean. Prod.* **2020**, *270*, 122383. [CrossRef]
35. Hxwa, D.; Zrc, B.; Pyn, C. Environmental and economic performance of China's ETS pilots: New evidence from an expanded synthetic control method. *Energy Rep.* **2021**, *7*, 2999–3010.
36. Grossman, G.M.; Helpman, E. Innovation and Growth in the Global Economy. *Mit Press Books* **1993**, *1*, 323–324.
37. Calel, R.; Dechezleprêtre, A. Environmental Policy and Directed Technological Change: Evidence from the European Carbon Market. *Rev. Econ. Stat.* **2016**, *98*, 173–191. [CrossRef]
38. Krugman, P.R. Increasing Return and Economic Geography. *J. Political Econ.* **1991**, *99*, 483–499. [CrossRef]
39. Li, Z.; Wang, J. Spatial spillover effect of carbon emission trading on carbon emission reduction: Empirical data from pilot regions in China. *Energy* **2022**, *251*, 123906. [CrossRef]
40. Sadayuki, T.; Arimura, T.H. Do regional emission trading schemes lead to carbon leakage within firms? Evidence from Japan. *Energy Econ.* **2021**, *104*, 105664. [CrossRef]
41. Rosenbaum, P.R.; Rubin, D.B. The central role of the propensity score in observational studies for causal effects. *Biometrika* **1983**, *70*, 41–55. [CrossRef]
42. Mz, A.; Yg, A.; Ll, A.; Dz, B. Impacts of carbon emission trading schemes on the development of renewable energy in China: Spatial spillover and mediation paths. *Sustain. Prod. Consum.* **2022**, *32*, 306–317. [CrossRef]

International Journal of
Environmental Research and Public Health

MDPI

Article

Digitalization, Electricity Consumption and Carbon Emissions—Evidence from Manufacturing Industries in China

Qian Zhang and Qizhen Wang *

Business School, Nanjing Xiaozhuang University, Nanjing 211171, China
* Correspondence: wqz@njxzc.edu.cn

Abstract: The development of China's manufacturing industry is constrained by factors such as energy and resources, and low-carbon development is arduous. Digitalization is an important method to transform and upgrade traditional industries. Based on the panel data of 13 manufacturing industries in China from 2007 to 2019, a regression model and a threshold model were used to empirically test the impact of digitalization and electricity consumption on carbon emissions. The research results were as follows: (1) The digitalization level of China's manufacturing industry was steadily increasing; (2) The proportion of electricity consumption in China's manufacturing industries in the total electricity consumption hardly changed from 2007 to 2019, basically maintaining at about 6.8%. The total power consumption increased by about 2.1 times. (3) From 2007 to 2019, the total carbon emissions of China's manufacturing industry increased, but the carbon emissions of some manufacturing industries decreased. (4) There was an inverted U-shaped relationship between digitalization and carbon emissions, the higher the level of digitalization input, the greater the carbon emissions of the manufacturing industry. However, when digitalization develops to a certain extent, it will also suppress carbon emissions to a certain extent. (5) There was a significant positive correlation between electricity consumption and carbon emissions in the manufacturing industry. (6) There were double energy thresholds for the impact of labor-intensive and technology-intensive manufacturing digitalization on carbon emissions, but only a single economic threshold and scale threshold. There was a single scale threshold for capital-intensive manufacturing, and the value was −0.5352. This research provides possible countermeasures and policy recommendations for digitalization to empower the low-carbon development of China's manufacturing industry.

Keywords: digitalization; electricity consumption; carbon emissions; manufacturing industries

Citation: Zhang, Q.; Wang, Q. Digitalization, Electricity Consumption and Carbon Emissions—Evidence from Manufacturing Industries in China. *Int. J. Environ. Res. Public Health* **2023**, *20*, 3938. https://doi.org/10.3390/ijerph20053938

Academic Editors: Taoyuan Wei and Qin Zhu

Received: 5 February 2023
Revised: 20 February 2023
Accepted: 20 February 2023
Published: 22 February 2023

1. Introduction

Under the goal of carbon neutrality, traditional high-carbon emission industries such as steel and cement have received greater attention, but the digital economy is becoming a new driving force for the high-quality development of China's economy and plays a very important role in carbon emission reduction. In 2020, the total scale of China's digital economy reached 39.2 trillion yuan, with a growth rate three times that of China's GDP and a contribution of 38.6% to GDP. As a new mode of production, digital technology will help China achieve the goals of carbon peaking and carbon neutrality and at the same time, provide fast funding channels for the development of low-carbon cities [1]. Although the development of China's digital economy is showing a steady upward trend, there is still a phenomenon of regional development imbalance [2].

China is experiencing an unprecedented process of digitalization and modernization, and its manufacturing industry is also speeding up adjustment, optimization, and upgrading. In 2021, the added value of China's manufacturing industry was 31.4 trillion yuan, accounting for nearly 30% of the world's total. The energy utilization rate industry continued to rise. The comprehensive energy consumption of steel and other units has dropped by more than 9% compared with 2012. Digitalization and manufacturing are integrated and developed, and the energy efficiency of infrastructure is also continuously optimized. From 2007 to 2019, the level of digitalization in China's manufacturing industry steadily increased (See Figure 1 for more details).

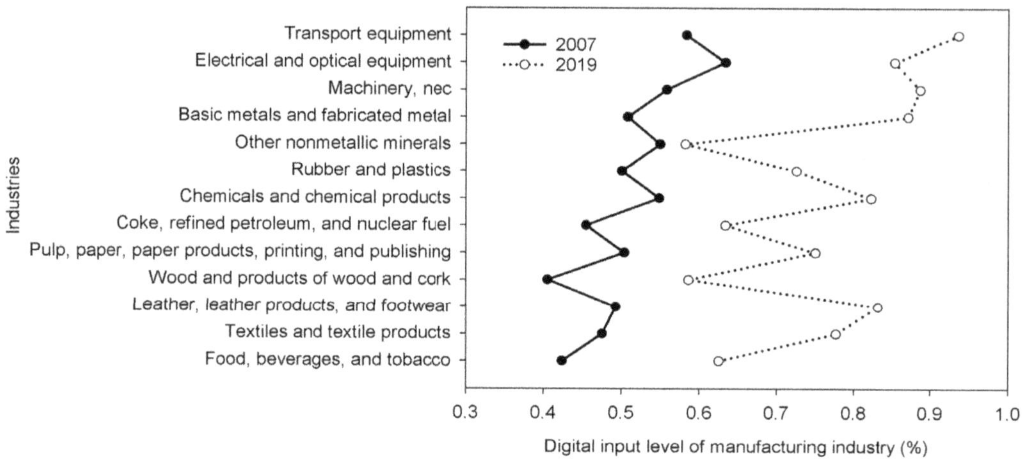

Figure 1. Digitalization level of China's 13 manufacturing industries in 2007 and 2019 (%).

An important path in achieving carbon emission reduction is transforming the energy industry. In 2020, carbon emissions from national energy consumption accounted for 85% of total carbon emissions, and those from the power sector accounted for 40% of the total. The digital economy mainly empowers the energy sector at three levels to help achieve carbon emission reduction goals. First, from the perspective of the energy supply side, the trend and fluctuation of power demand can be monitored and controlled in real-time through digital technology to achieve the optimal allocation of resources and improve energy utilization efficiency. Second, from the perspective of the energy demand side, digital technology can monitor the disclosure of information such as carbon emissions. It is possible to measure and source carbon emissions, helping companies achieve demand-side management of carbon emissions at a lower cost. This can further improve the carbon emissions trading market. Third, from the perspective of energy trading, digital technology can solve the time and space barriers in the transaction, which can optimize the matching of the supply and demand sides and then improve the energy transactions efficiency.

The rapid development of China's manufacturing industry is accompanied by a large amount of energy consumption and carbon emissions. Electricity is an important source of energy and is clean, but the process of producing electricity is not. From 2019 to 2021, China accounted for almost all of the growth in global carbon emissions from the power and heat sectors. The CO_2 emissions from the power and heating sector increased by 6.9% in 2021, due to a sharp increase in global electricity demand. From 2007 to 2019, the proportion of electricity consumption of China's manufacturing industries in the total electricity consumption hardly changed, maintaining at about 6.8%. However, the total electricity consumption of the manufacturing industry has increased by about 2.1 times, from 122.3 billion kWh to 260.4 billion kWh (See Figure 2 for details). As the world's largest carbon emitter, China is actively taking responsibility for reducing carbon emissions. "Made in China (2025)" clearly states that by 2025, the added value of carbon emissions per unit of China's manufacturing industry should be reduced by 40% based on 2015. In recent years, although the carbon emissions of some manufacturing industries have decreased, the total carbon emissions have continued to increase (See Figure 3 for details).

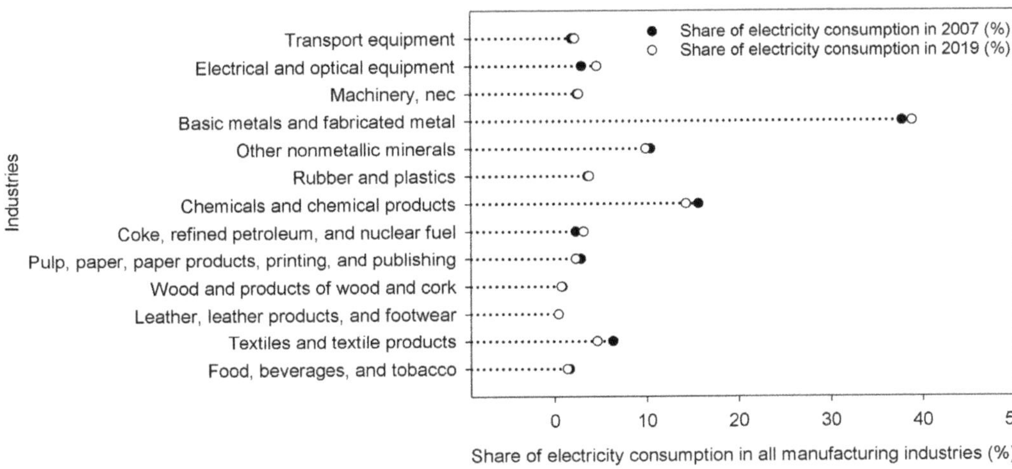

Figure 2. The proportion of electricity consumption in China's 13 manufacturing industries in 2007 and 2019.

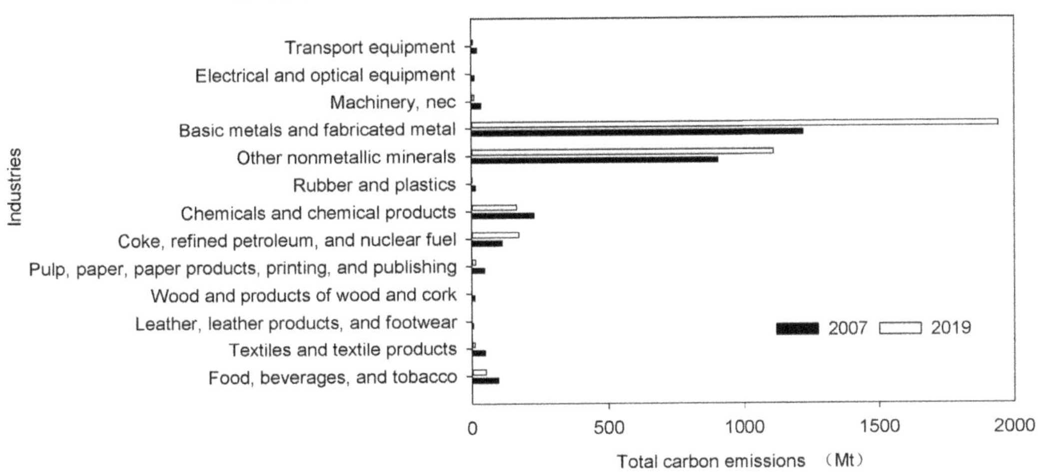

Figure 3. Total carbon emissions of China's 13 manufacturing industries in 2007 and 2019.

This paper aimed to analyze the impact of digitalization and electricity consumption on carbon emissions in China's manufacturing industry, accurately identify the influencing factors and threshold effects of carbon emissions in China's manufacturing industry, and then choose appropriate emission reduction paths and policies to help China's manufacturing industry successfully realize digitalization and low-carbon transformation goals. Possible marginal contributions are: First, the article analyzes the industry heterogeneity of China's manufacturing industry and takes into account the spatiotemporal factors of global value chain participation. Second, the article uses the threshold model to analyze the energy, economic, and scale effects of digitalization on China's manufacturing carbon emissions. Third, the article puts forward specific countermeasures for manufacturing carbon emission reduction from the aspects of demand, supply, and transaction sides.

In the process of digitalization and the rapid growth of energy consumption, China's manufacturing industry is facing greater pressure in reducing carbon emissions. This paper focuses on the relationship between digitalization, electricity consumption, and carbon emissions. It examines the impact of the digital economy on China's manufacturing carbon emissions and the threshold effect. Then we propose corresponding carbon emission reduction countermeasures. This paper provides theoretical and practical references for the government and enterprises. Furthermore, it has important practical significance for the development of a low-carbon society.

2. Literature Review

2.1. Understanding Digitalization

Digitalization involves digital technology and its integrated use in the production process [3]. The development of the digital economy includes two aspects. One is digital industrialization. Information technology has given birth to many new industries. The industry based on digital elements has promoted the industrial structure to be technology-intensive and environment-friendly. The second is industrial digitalization, which refers to the combination of traditional industries and digitalization, and the application of digitalization in production to promote the transformation and upgrading of traditional industries. The upgrading of industrial structures can promote the use of clean energy, replace traditional high-carbon emission energy with clean energy, and ultimately reduce carbon emissions [4]. The upgrading of industrial structures can stimulate the R&D and application of low-carbon technologies. Meanwhile, it can improve the energy structure to play a better substitution role and promote the green transformation of enterprises [5].

Digitalization has a greater impact on carbon emission efficiency. It can improve carbon productivity, and its impact on the central and western regions of China is significantly greater than that on the eastern regions. Furthermore, it mainly affects carbon productivity through technological innovation and industrial structure optimization and upgrading [6]. The promotion effect of digitalization on carbon emission reduction shows a trend of increasing with time and has a positive spatial spillover effect. Digitalization is becoming one of the new sources of energy to improve green development. With the technology accumulation, the coefficient of the impact of digitalization on total factor carbon productivity is getting higher and more significant [7].

However, the carbon-reducing effect of digitalization is controversial. Some scholars [8–10] believe that digitalization has a carbon emission reduction effect, while others [11,12] believe digitalization can promote carbon emissions. Therefore, we will discuss these questions in the next section. The topics are divided into the decoupling of digitalization and carbon emissions, the uncertainty of digitalization and carbon emissions, the relationship between digitalization and manufacturing carbon emissions, the specific path to realize digital carbon emission reduction, and finally puts forward the *hypotheses* that this paper wants to test.

2.2. Debate on the Relationship between Digitalization and Carbon Emissions

2.2.1. Digitalization and Carbon Emissions Are Gradually Decoupling

(1) Linear analysis

Based on the linear analysis between digitalization and carbon emissions, it is found that the two are slowly decoupling. Digitalization mainly promotes the low-carbon transformation of cities through innovation, and its impact on low-carbon development in cities will become stronger and stronger [13]. Digitalization uses the internet to reduce offline activities, travel, and carbon emissions. Meanwhile, it promotes the popularization of green and low-carbon behaviors, making low-carbon a daily behavior standard and also promoting the effective use of a green economy [4]. From a technical point of view, China's information and communications technology (ICT) industry helps reduce carbon emissions, and the ICT industry in the central region has a greater impact on CO_2 emissions than the eastern region [14]. The innovative development of ICT provides opportunities for the coordinated development of shared prosperity, energy conservation, and emission reduction. It effectively promotes carbon emission reduction by reducing energy consumption [15]. Meanwhile, improving energy structure and technological progress can effectively reduce carbon emission intensity [4]. Digitalization can significantly increase carbon productivity. Technological innovation, reduction of energy consumption intensity, and improvement of urban productivity are the main paths [16,17].

Digitalization has a significant driving effect on the coordinated governance of carbon dioxide and haze pollution, and there is a positive spatial spillover effect [18]. It mainly improves environmental pollution through technological innovation and optimal allocation of resources [19]. Furthermore, there is a long-term positive and significant relationship between internet use and carbon emissions, but no causal relationship exists. The rapid growth of the internet is not the main reason the environment is threatened. Therefore, promoting the development of the internet will not lead to environmental degradation [20].

(2) Nonlinear analysis

In order to more scientifically assess the relationship between the digital economy and carbon emissions, nonlinear analysis is becoming more popular. In the study of digitalization and carbon emissions, the development of regional digitalization has significantly reduced the intensity of carbon emissions. The relationship with carbon emissions presents an inverted U-shaped relationship that first rises and then declines. The specific transmission paths are mainly technological innovation, industrial structure, and energy structure. Digitalization has latecomer advantages in achieving carbon neutrality goals [21]. The empirical analysis finds that comprehensive infrastructure construction will increase energy intensity and thus hinder carbon emissions. However, information integration infrastructure is conducive to developing the tertiary industry, and the carbon emissions generated will be less than those generated by comprehensive infrastructure construction. This leads to an inverted U-shaped relationship between integrated infrastructure development and carbon emissions [22]. However, digitalization contributes to carbon emissions when green energy is less efficient and vice versa [23].

Besides, digitalization has spatial spillover effects on carbon emission reduction. Using the spatial Durbin model (SDM), it is found that digitization has a U-shaped spatial spillover emission reduction effect and presents an inverted U-shaped carbon emission reduction effect that is first promoted and then suppressed. Technological progress and economic growth are the main mechanisms [24]. Using the panel data of 277 cities in China from 2011 to 2019, an inverted U-shaped nonlinear relationship between digitization and carbon emissions was also found. The industrial structure upgrading makes the effect of digitalization on carbon emissions also follow the characteristics of the Environmental Kuznets Curve [25]. Digitalization has a significant negative direct effect on green total factor energy efficiency (GTFEE) through electrification, hollowing out of industrial scale, and industrial efficiency. However, with economic development, its impact on GTFEE gradually turns from negative to positive. Based on the SDM and threshold models, the inverted U-shaped relationship between digitalization and carbon emissions has been further verified [26].

As a new form of economy, digitalization is important for reducing carbon emissions in the transportation and logistics industries. It has a mitigating effect on carbon emissions in the transportation sector. It also accelerates carbon emissions in the transportation sector in the low-urbanization stage but reduces carbon emissions in the high-urbanization stage [27]. With the provincial panel data from 2005 to 2019, the nonlinear regression model and the quantile regression model were used to empirically test the U-shaped relationship between digitalization and carbon emissions in the logistics industry. In the first half of the U-shaped relationship, digitalization had both a restraining effect and a significant evolutionary effect on carbon emissions in the logistics industry. As the quantile increases, the marginal impact of digitalization on carbon emission reduction in the logistics industry gradually decreased [28].

2.2.2. Digitalization Brings Uncertainty about Carbon Emissions

Although the above studies find that digitalization and carbon emissions are decoupling, the nexus between the two remains uncertain. In analyzing the carbon deduction effect of digitalization, considering the impact of digital demand and supply, digitalization may bring about 6% of carbon emissions [11]. With the promotion of digital demand and scale, between 2002 and 2007, the carbon emissions brought by digitalization rose from 210 Mt to 418 Mt. From 2007 to 2017, with the improvement of carbon efficiency and digital application structure, the carbon emission caused by digitalization has been alleviated. However, with the intermediary model and the panel threshold model, it was found that improving energy efficiency can promote carbon emission reduction, although digitalization increases carbon emissions. Nevertheless, digital development is not conducive to the improvement of energy efficiency. Considering energy efficiency, digital development has a significant double-threshold effect on carbon emissions, showing an N-shaped trend. Population expansion, coal-based energy consumption structure, and industrial structure were the main reasons for the increased carbon emissions [29].

Although digitalization can effectively reduce urban carbon emissions and improve total factor productivity, the improvement of energy efficiency, technological innovation, and industrial structure upgrading are the main reasons for the existence of digital low-carbon governance effects. However, digitalization can only promote the low-carbon transformation of old industrial bases. The urban development of traditional resource industries is path-dependent, and the effect of low-carbon governance is not obvious [30]. Although artificial intelligence can produce carbon emission reduction effects through industrial structure, information infrastructure, and green technology innovation, these are only for big cities and cities with better infrastructure and advanced technology. There are differences in the development of the digital economy among different countries, especially in hyper-digitalized and under-connected countries. Although digitalization reduces total carbon emissions, it increases carbon emissions per capita [31].

2.3. *Digitalization, Electricity Consumption and Manufacturing Carbon Emissions*

2.3.1. Digitalization and Manufacturing Carbon Emissions

Digitalization is a very important industrialization process. Influenced by Industry 4.0, it directly affects all production processes and manufacturing sectors. Therefore, it is imperative to increase the productivity and sustainability of the manufacturing sector [32]. Manufacturing digitalization is an important enabling factor for improving competitive advantage [33]. Industry 4.0 has become a continuous and predicted outcome of past industrial ages. From a technological point of view, it can be considered an increase in digitalization and automation, as well as an increase in communication enabled by the creation of digital value chains [34]. However, digital traceability should be used to improve the traceability and added value of products, shorten the production cycle and promote the manufacturing industry to adapt to the requirements of Industry 4.0 [35].

The reduction of carbon emissions is the main reason driving the growth of TFP and technical efficiency in the context of the deep integration of the digital service industry and the manufacturing industry. Industrial integration and carbon emissions show a U-shaped relationship. The integration of capital-intensive, technology-intensive, and labor-intensive manufacturing industries and digital services promotes the growth of total factor productivity but first suppresses and then promotes carbon emissions [36]. Digitalization has a positive effect on TFP in time and space and can promote the development of manufacturing. Labor-intensive and capital-intensive industries have the same characteristics as the total sample [37].

The development of digitalization reduces the carbon emission intensity of enterprises and improves the efficiency of resource allocation. However, the market drive has improved the ability of digital carbon emission reduction, while government regulation has reduced the ability of digital carbon emission reduction [38]. Through the sample of heavily polluting enterprises listed on China's A-shares, it is found that digitalization can significantly improve the efficiency of energy conservation and emission reduction of enterprises, especially for mining and manufacturing industries. Digitalization has promoted cleaner production through technological innovation, easing financing constraints, and promoting market competition. This promotion effect is more significant in areas with more developed economies and less government financial pressure [12]. From a global perspective, investment in manufacturing digitization has reduced carbon emission intensity. The industry spillover effect becomes more significant over time. From the perspective of industry heterogeneity, the carbon emission reduction effect of digitalization in pollution-intensive manufacturing is more obvious [39].

Based on this, this paper proposes the following hypothesis:

Hypothesis 1. *There is an inverted U-shaped relationship between digitalization and China's manufacturing carbon emissions.*

2.3.2. Electricity Consumption and Carbon Emissions

Increased electricity use adversely impacts carbon emissions [40–45]. The continuing increase in electricity consumption is one of the main sources of carbon emissions [40]. Electricity-related carbon emissions release more than 40% of global and Chinese carbon emissions [44]. Electricity is usually an important energy source for a country, and the demand for electricity often increases energy consumption and pollution. Through the econometric analysis of the co-integration panel, it is proved that the adverse effect of electricity consumption on carbon emissions exists. However, electricity output from renewable sources can ease the pressure on carbon emissions. Overall, electricity consumption and generation are the main sources of carbon emissions [41].

China is also committed to research on carbon reduction of electricity consumption. Regression analysis of 25 years of relevant data from 123 countries found that using renewable energy had a carbon emission reduction effect [42]; however, there was no causal relationship between electricity consumption and carbon emissions in China [46]. In 2020, 66% of China's power generation came from coal, and coal consumption accounted for 61% of energy consumption. A reduction in coal consumption would result in a 51% reduction in overall carbon emissions [43]. Based on the carbon release effect of electricity consumption, how to reduce carbon emissions through technology has attracted the attention of the power sector. Previous studies have often neglected electricity-related carbon emissions induced by electricity consumption. Through the comprehensive application of IPCC's carbon emission accounting method, taking Shanghai as the research object, it was again verified that electricity consumption and population size have indeed promoted carbon emissions. However, the increase in electricity efficiency and the decline in carbon emission intensity offset the increase in carbon emissions [45]. As the digital transformation deepens, the promotion effect of energy consumption per capita on carbon emissions is weaker, while the effect of renewable energy on carbon emission reduction is stronger [47].

Hypothesis 2. *Electricity consumption contributes to China's manufacturing carbon emissions.*

2.4. The Specific Path of Digital Empowerment for Carbon Emissions Reduction

Digitalization empowers carbon emissions mainly through the following paths. First, it mainly relies on industrial progress and energy consumption optimization to curb carbon emissions and has significant spatial spillover effects on neighboring provinces [48]. In the short term, increased energy consumption and non-green technological progress are the main paths to increasing carbon emissions. However, technological progress and industrial structure upgrading are the main paths for long-term carbon emission reduction [49]. Second, it plays a role mainly through resource flow and energy flow. From the perspective of element resource misallocation (capital misallocation and labor misallocation), digitalization can improve carbon emission efficiency in both southern and northern China. Meanwhile, it has a long-term positive impact on the carbon emission rate by mitigating factor misallocation [50]. Digital financial inclusion is an important factor for digitalization to affect carbon emissions [51]. Third, from the perspective of digital transformation, digital infrastructure, digital trade competitiveness, digital technology, and energy consumption have significant threshold effects on carbon emissions. Although trade brings economic benefits, it also implies the environmental costs of carbon emissions [44]. Smart city construction will also significantly reduce corporate carbon emission intensity [52]. Finally, improved energy efficiency helps reduce carbon emissions in manufacturing and transportation [53].

The above analysis found that there are many related studies. However, the relationship between digitalization and carbon emissions is still inconclusive. Furthermore, although electricity consumption boosts carbon emissions, there are not many specific analyses on the manufacturing industry. Based on this, this paper builds the analysis framework shown in Figure 4 to capture the direct effects of digitalization on carbon emissions and related threshold effects (See Figure 4 for more details).

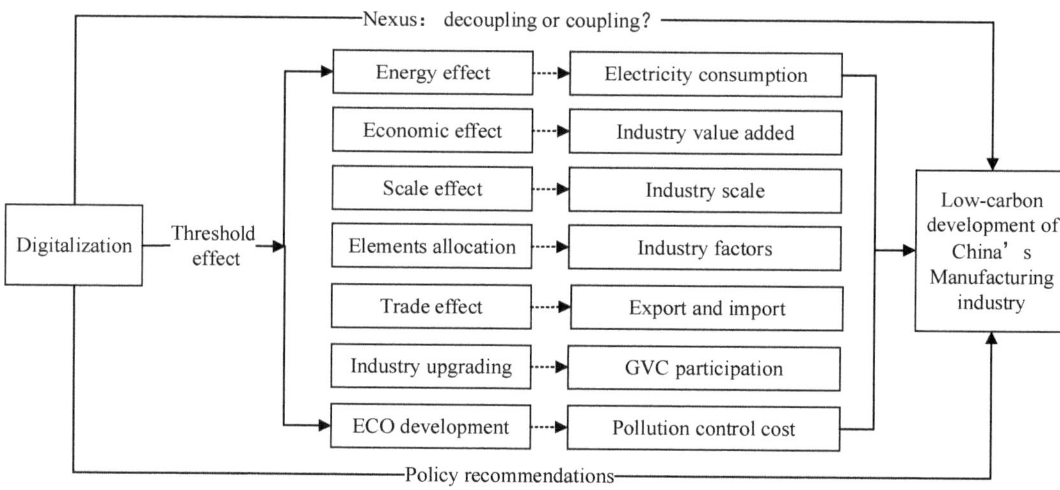

Figure 4. Research framework and theoretical mechanism.

3. Methodology and Data Source

3.1. Variables and Data Sources

According to the results of our literature review, we found that the carbon emissions of the manufacturing industry are mainly affected by factors such as energy consumption, scale effects, industrial structure transformation and upgrading, international trade and division of labor under an open economy. At the same time, factor input plays a very important role in exerting the competitive advantage of digital transformation in the manufacturing industry. Factor input can promote product upgrading, reduce undesired output, and reduce pollution. Based on the theoretical mechanism, this paper takes carbon emissions as the explained variable and digital input level and electricity consumption as the core explanatory variables. The latter is used to measure the energy effect. Taking into account other different effects and data availability, control variables, including industry factors, industry scale, export dependence, import dependence, the industry added value, participation in global value chains, the proportion of industrial waste gas treatment cost, are selected. The variable GVC-pat was used to measure the upgrading of industrial structure because participation in GVC will promote the change of industrial structure [54]. Besides, China is facing pressure to reduce carbon emissions, but at the same time it is facing the heavy responsibility of economic development. At the time of carbon emission reduction, we must consider the cost of pollution control. Table 1 presents all selected variables and data sources.

Table 1. Variables and data sources.

	Abbreviation	Variable	Measurement	Data Source
Dependent variable	Carbon	Total carbon emissions	Data aggregation	CEADs
Core explanatory variable	Dig	Digital level	Value added of C14 (electrical and optical equipment) and C27 (post and telecommunications)/Gross Value Added in the Manufacturing Sector	ADB MRIO2021
	Dig^2	Square of Dig	Dig×Dig	ADB MRIO2021
	Er	Proportion of electricity consumption in the industry	Electricity consumption of each manufacturing industry/total electricity consumption of manufacturing industries	IFIND
Control variable	Ind	Industry factors	Ratio of capital factor to labor factor (Manufacturing industry paid-in capital divided by manufacturing employees)	CISY, CSTSY
	Scale	Industry scale	The output value of each manufacturing industry divided by the total output value	ADB MRIO2021
	Exp	Export dependence	Exports divided by total industry output	ADB MRIO2021
	Imp	Import dependence	Imports divided by total industry input	ADB MRIO2021
	Vab	Industry added value	Value-added of each manufacturing industry	ADB MRIO2021
	GVC-pat	Participation in global value chains	The sum of GVC forward participation and backward participation	UIBE GVC
	Ratiowair	Proportion of industrial waste gas treatment cost	Industrial waste gas treatment cost divided by total industry treatment cost	CESY

Note: 1. Digitalization input level (Dig): Based on the practice [55,56], in this paper C14 (electrical and optical equipment) and C27 (post and telecommunications) industries in the ADB MRIO2021 input-output table are used as the basic sectors of the digitalization. The ratio of the added value of the two divided by the total added value of the manufacturing industry is used to measure the level of digital input in the manufacturing industry. 2. CEADS: Carbon Emission Accounts & Datasets; IFIND: Financial data terminal of Tonghuashun; CISY: China Industry Statistical Yearbook; CSTSY: China Science and Technology Statistical Yearbook; CESY: China Environmental Statistical Yearbook. 3. The carbon emission data in this paper is the total carbon emission of various energy-related manufacturing industries. The data of electricity consumption is directly derived from the total electricity consumption of various industries in the IFIND database.

3.2. China's Manufacturing Sector Segmentation

The manufacturing industries selected in this article were the 13 manufacturing sub-sectors covered by the ADB MRIO2021. The sub-sectors with missing data were deleted. To connect with the manufacturing industry in China Statistical Yearbook and China Industrial Statistical Yearbook, this paper processed relevant data, and the processing methods are shown in Table 2.

The processing method for the industry: 1. Merge the rubber and plastics industries into rubber and plastics; 2. Consolidate automobile manufacturing and railway, ship, aerospace, and other transportation equipment into transportation equipment manufacturing; 3. Consolidate the food processing industry, food manufacturing, beverage manufacturing, and tobacco processing industries into food, beverage, and tobacco; 4. Merge the paper and paper products industry and the printing industry, the reproduction of recording media into pulp, paper, paper products, printing, and publishing; 5. The ferrous metal smelting and rolling processing industry, non-ferrous metal smelting and rolling processing industry, and metal products industry were combined into basic metals and fabricated metal.

Table 2. Manufacturing Industry Segmentation.

Code in This Paper	ADB MRIO Code	Industry	Industry Category
n1	c3	Food, beverages, and tobacco	LI
n2	c4	Textiles and textile products	LI
n3	c5	Leather, leather products, and footwear	LI
n4	c6	Wood and products of wood and cork	LI
n5	c7	Pulp, paper, paper products, printing, and publishing	LI
n6	c8	Coke, refined petroleum, and nuclear fuel	CI
n7	c9	Chemicals and chemical products	TI
n8	c10	Rubber and plastics	LI
n9	c11	Other nonmetallic minerals	CI
n10	c12	Basic metals and fabricated metal	CI
n11	c13	Machinery, nec	CI
n12	c14	Electrical and optical equipment	TI
n13	c15	Transport equipment	TI

Note: LI = labor intensive, CI = capital intensive, TI = technology intensive.

3.3. Model Construction

3.3.1. Panel Regression Model

In order to verify the impact of digital investment and electricity consumption on carbon emissions, this paper selected the panel data of 13 industries in China's manufacturing industry from 2007 to 2019, and finally built the following panel data model:

$$Carbon_{it} = \alpha + \beta Dig_{it} + \beta Dig_{it}^2 + \rho Er_{it} + \gamma Cont_{it} + \mu_i + \delta_t + \varepsilon_{it} \quad (1)$$

where $Carbon_{it}$ represents the total carbon emission of manufacturing industry i in period t, and α represents the constant item. Dig represents the core explanatory variable digital input level. To verify the nonlinear impact of digitalization on carbon emissions, the variable Dig^2 was included to represent the square of the digital input level; Er represents the proportion of electricity consumption of each industry in the total manufacturing electricity consumption; *Cont* represents the control variable; μ_i and δ_t represent the individual and period effects respectively; and ε_{it} represents the stochastic disturbance term.

3.3.2. Threshold Effect Model

The effect of digitalization on carbon emission reduction mainly depends on a sound digital ecology and industrial layout. Bridging the existing digital division is an important problem to be solved in the process of digitalization. Digital technology empowers carbon emission reduction mainly through energy structure, economic base, and industrial scale. Therefore, this paper used the threshold effect model to further explore the energy, economic, and scale effects of digitalization on China's manufacturing carbon emissions. The specific threshold model was constructed as follows:

$$Carbon_{it} = \alpha + \beta Dig_{it} \cdot I(thr_{it} < \gamma) + \delta Dig_{it} \cdot I(\gamma \leq thr_{it}) + \theta Cont_{it} + \mu_i + \varepsilon_{it} \quad (2)$$

where $Carbon_{it}$ represents the explained variable. *Dig* represents the level of digitalization, and *thr* represents the threshold variables, which are the proportion of electricity consumption, the economic added value of the manufacturing industry, and the industry scale; $I(\cdot)$ represents the indicator function; γ represents the threshold value to be estimated; *Cont* represents each control variable, and the control variables are consistent with the benchmark model; μ_i represents the industry effect; and ε_{it} represents the stochastic disturbance term.

4. Results

4.1. Descriptive Statistics of Variables

To avoid the influence of different dimensions of variables, all the variables were standardized, and Table 3 displays their descriptive statistics. In this paper, variance inflation factor (VIF) and tolerance (1/VIF) were used to test whether there was multicollinearity among the variables. The higher the VIF value, the more serious the multicollinearity. The tolerance was generally between 0 and 1. The smaller the tolerance, the more serious the collinearity. The average value of the selected variable VIF in this paper was 8.99, and the tolerance was 0.22, indicating that there was no multicollinearity among the variables.

Table 3. Descriptive statistics of variables.

Variable	Obs	Mean	Std. Dev.	Min	Max	Unit
Carbon	169	265.4977	512.6949	0.7527	1942.2960	Mt
Dig	169	0.2275	0.5115	0.0071	2.3739	
Er	169	6.8330	9.9424	0.3462	40.2882	%
Ind	169	19.3438	14.1929	2.8405	78.5029	RMB 10,000/person
Scale	169	0.0500	0.0406	0.0024	0.1730	
Exp	169	0.1188	0.0978	0.0231	0.4605	
Imp	169	0.0632	0.0378	0.0187	0.1985	
Vab	169	1,306,574	867,336	132,607	3,704,156	Million-yuan
GVC-pat	169	0.3109	0.1046	0.1031	0.6068	
Ratiowair	169	4.2115	7.2584	0.0411	30.1847	%

4.2. Estimations Results

Column (1) of Table 4 shows the model estimation results. There is an inverted U-shaped relationship between digitalization and carbon emissions, the higher the level of digitalization, the greater the carbon emissions of the manufacturing industry. Nevertheless, when digitization develops to a certain extent, it will curb carbon emissions. However, electricity consumption and carbon emissions in the manufacturing industry show an increasing trend with a positive correlation.

The higher the electricity consumption, the more carbon is emitted. This is because electricity consumption is a demand for energy. High power usage leads to high power consumption and more energy consumption [57]. Although electricity is a secondhand energy source, it can be regarded as a clean energy source. However, the generation process of electricity is accompanied by primary energy use such as oil, coal, natural gas, etc., which will bring pollution; however, the process of generating electricity from water, nuclear energy, and wind energy is clean. China's economic development needs energy, and its electricity demand will grow by 10% in 2021. Coal is required to meet the 56% increase in electricity demand as this growth exceeds the growth of low-emissions supply. Although China is vigorously developing new energy sources, new energy sources are far from meeting today's market demand, and coal investment is still the main force for power generation. The demand for electricity consumption directly or indirectly promotes electricity production, thereby causing carbon emissions in various industries [58]. China's non-clean energy power generation accounts for about 65% in 2021. Because of the energy structure and energy efficiency, the power sector still has a certain distance from zero emissions [59]. The reasons may be: first, people's living standards are increasing, and the demand for electricity is increasing, resulting in more carbon emissions from electricity; second, the loss in the power transmission process is large, and energy waste is serious, which also brings about an increase in carbon emissions.

Manufacturing industry factors have a negative correlation with carbon emissions, which is significant at the 5% level. The higher the ratio of paid-in capital to labor factors in the manufacturing industry, the lower the total carbon emissions. The production strategy of a capital-constrained manufacturing producer would be less sensitive to carbon prices and would re-engineer older products at higher quality levels. However, manufacturers with high industry factors and abundant funds will be more sensitive to carbon prices and carbon emissions when formulating production strategies. Therefore, there will be a more obvious carbon emission reduction effect [60].

Table 4. Estimation results.

Variable	Core Explanatory Variable	
	(1)	(2)
	Dig-1	Dig-2
Dig	0.453 ***	0.652 **
	(3.35)	(2.87)
Dig^2	−0.331 **	−0.596 **
	(−2.81)	(−2.66)
Er	0.435 **	0.319 *
	(3.19)	(2.52)
Ind	−0.134 **	−0.196 ***
	(−3.23)	(−4.54)
Scale	−0.278 ***	−0.216 ***
	(−5.64)	(−4.67)
Exp	−0.139 ***	−0.122 **
	(−3.50)	(−3.15)
Imp	0.273 ***	0.328 ***
	(3.82)	(4.25)
Vab	0.0383	0.0840 *
	(0.74)	(2.19)
GVC−pat	−0.354 ***	−0.316 ***
	(−5.02)	(−4.81)
Ratiowair	0.725 ***	0.748 ***
	(6.08)	(6.40)
Obs	169	169
R^2	0.8954	0.8941

Note: * $p < 0.05$, ** $p < 0.01$, *** $p < 0.001$; t statistics in parentheses.

The manufacturing industry scale is negatively correlated with carbon emissions and is significant at the 1% level. It shows that China's manufacturing industry has the potential to reduce carbon emissions. While the scale of the industry is expanding, the industrial structure is also constantly being adjusted. Additionally, the proportion of high-emission enterprises will become lower and lower. The scale of the manufacturing industry and carbon emissions will eventually be decoupled. Large-scale production under economic agglomeration is conducive to improving the emission reduction effect of diversification [61].

The increase in export dependence can promote carbon emission reduction and is significant at the 1% level. The carbon emission factors brought about by import and export trade are becoming more significant. The reduction in exports is mainly due to consumption-based carbon emissions [62,63]. As China's exports have shifted from primary industrial products to high-tech products, the proportion is increasing. The technical effect of export products is becoming more and more obvious. Compared with the extensive production method, exports are conducive to technological innovation and the exertion of carbon emission reduction effects.

The increase in import dependence promotes carbon emissions and is significant at the 1% level. The diversification of imported and exported products will affect energy-related consumption. One of the main goals of manufacturing development in developing countries is to increase energy efficiency. Imports can introduce new and improved technologies, update production methods, and further promote green production. However, the diversification of imported products also makes it easier to obtain cheaper intermediate products. If these intermediate products include building materials, mechanical appliances, and transportation equipment, they will increase the total carbon emissions of developing countries [64]. Besides, to meet domestic demand by importing more energy-intensive products will bring carbon emissions into the importing country and increase the carbon emissions of the importing country. In the process of reducing carbon emissions, special attention should be paid to the hidden carbon emissions in trade and the resulting environmental game between governments [61].

The output value of the manufacturing industry is positively correlated with carbon emissions, but insignificant. It shows that the decoupling of carbon emissions and economic development has not been verified. China's economic development model has changed from a planned traditional economy to a socialist market economy. Under the tide of de-industrialization, China has gradually increased its efforts in the reform and opening-up of the economy. In the context of expanding domestic demand, infrastructure construction has increased, and large-scale manufacturing clusters have gradually formed in the coastal areas. China has gradually become the world's largest exporter. Under the goal of manufacturing transformation and upgrading, the relationship between industry-added value and carbon emissions has gradually improved. However, increasing infrastructure construction gradually increases carbon emissions [65].

The increase in the global value chain (GVC) participation can promote carbon emission reduction and is significant at the 1% level. With the continuous improvement of global integration, participation in the GVC has become the new normal of the international labor division. The low status of China's manufacturing participation will lead to the high carbonization of enterprises' production methods. The rise of GVC participation greatly impacts the manufacturing industry's carbon emissions. The increase in intermediate product exports can promote the rise of GVC status, which is more conducive to reducing carbon emissions [66,67].

The cost of industrial waste gas treatment is positively correlated with carbon emissions and is significant at the 1% level. China is faced with the dual tasks of reducing carbon emissions and developing the economy. On the one hand, to obtain environmental governance performance, there may be a development path of pollution first and then governance. On the other hand, pollution control equipment may have high energy consumption. In addition, considering the spatial spillover effects of environmental governance, pollution governance does increase local carbon emissions [68].

4.3. Robustness Test and Heterogeneity Analysis

4.3.1. Robustness Test: Core Explanatory Variables Replacement

This paper adopted the method of [69] to measure the digital input level. The digital input level was measured by the complete consumption coefficient; that is, the direct consumption coefficient plus the indirect consumption coefficient. This article started with the industries that the digital economy relies on. This includes the investment in software services and digital economic hardware facilities. Then, based on the input-output model, the complete consumption coefficient of each manufacturing industry was calculated; that is, the total input amount of the industries relying on the digital economy that needs to be consumed to produce the final products. This paper evaluated the digitalization level of the industry by calculating the complete consumption coefficient of the digital industry in various industries in the manufacturing industry.

The calculation method of the direct consumption coefficient was as follows. Variable a_{ij} $(i,j = 1,2,3,\ldots, n)$ indicates the value of the goods or services of the department i directly consumed by the unit total output of department j in the production process. It is usually expressed as matrix A. The complete consumption coefficient is the sum of direct consumption and indirect consumption, usually expressed by B. The larger the direct coefficient, the stronger the direct dependence of department j on department i.

$$a_{ij} = x_{ij}/X_j \quad (i,j = 1,2,3,\ldots,n) \tag{3}$$

$$b_{ij} = a_{ij} + \sum_{k=1}^{n} a_{ik}a_{kj} + \sum_{s=1}^{n}\sum_{k=1}^{n} a_{is}a_{sk}a_{kj} + \ldots \tag{4}$$

where in the first item, a_{ij} represents the direct consumption of the j product department to the i product department. The second item $\sum_{k=1}^{n} a_{ik}a_{kj}$ indicates the first round of indirect consumption of product department j to product department i; $\sum_{s=1}^{n}\sum_{k=1}^{n} a_{is}a_{sk}a_{kj}$ is the second round of indirect consumption, and so on. Round $n + 1$ is the indirect consumption of round n (Please see Formulas (3) and (4)). Therefore, the full consumption coefficient was used to measure the digital input level of each manufacturing industry, which was brought into the benchmark model. The model estimates are in Column (2) in Table 4, which proves the model estimation results were robust.

4.3.2. Heterogeneity Analysis

In this paper, the manufacturing industry was further divided into capital-intensive, labor-intensive, and technology-intensive. Afterward, two digital evaluation indicators were used for model estimation. Table 5 shows the estimated results. The empirical results show that although the relationship between capital-intensive digitization and carbon emissions shows a U-shaped nonlinear relationship, it was not significant. The higher the level of capital-intensive digitalization, the lower the carbon emissions will be. There was a U-shaped nonlinear relationship between labor-intensive, technology-intensive, and carbon emissions. However, for all manufacturing industries, the higher the electricity consumption, the greater the carbon emissions, and this result was significant at the 1% level.

Table 5. Heterogeneity analysis results.

Variable	Capital Intensive		Labor Intensive		Technology Intensive	
	Core Explanatory Variable: Dig					
	(1)	(2)	(3)	(4)	(5)	(6)
	Dig-1	Dig-2	Dig-1	Dig-2	Dig-1	Dig-2
Dig	−2.251 **	−0.049	−1.136	−0.080 ***	−0.081	−0.098
	(−3.23)	(−0.10)	(−1.20)	(−4.92)	(−1.36)	(−0.45)
Dig^2	3.261	0.117	3.299	0.077 ***	0.047	0.050
	(1.97)	(0.24)	(1.03)	(4.57)	(1.13)	(0.31)
Er	1.302 ***	1.543 ***	0.217 ***	0.220 ***	0.422 ***	0.412 ***
	(5.50)	(5.84)	(6.85)	(8.05)	(6.70)	(7.29)
Controls	Yes	Yes	Yes	Yes	Yes	Yes
Obs	52	52	78	78	39	39
R^2	0.9720	0.9613	0.9182	0.9357	0.9818	0.9823

Note: ** $p < 0.01$, *** $p < 0.001$; t statistics in parentheses.

4.4. Endogeneity

In this paper, the method of IV-2SLS was used to alleviate the possible endogeneity problems in variables. In this paper, the first-order lag item of the core explanatory variable (Dig) was used as an instrumental variable, and it was brought into the benchmark model for verification. The results prove that the model was robust. There was a positive linear relationship between electricity consumption and carbon emissions (See Table 6 for details).

4.5. Threshold Effect

Table 7 shows the estimated results of the threshold test and the threshold value. The results show that the digitalization of labor−intensive industries had double thresholds for energy effects on carbon emissions, and the thresholds were −0.5478 and 0.0042, respectively. Moreover, there was a single economic threshold and industry scale effect threshold for the impact of labor-intensive industry digitization on manufacturing carbon emissions, and the threshold values were −0.4233 and −0.8088, respectively. Capital-intensive industries had a single scale threshold, and the threshold value was −0.5352; Technology-intensive industries had double energy thresholds, and the threshold values were −0.2248 and 0.8068 respectively. In addition, there was a single economic threshold and industry scale threshold simultaneously, and the threshold values were −0.2786 and 0.1690, respectively.

Table 6. Estimation results of IV-2SLS model.

Variable	Dependent Variable: Carbon	
	Core Explanatory Variable	
	(1)	(2)
	Dig-1	Dig-2
Dig	0.526 ***	0.674 **
	(3.64)	(3.05)
Dig^2	−0.388 **	−0.617 **
	(−3.13)	(−2.84)
Er	0.453 ***	0.319 **
	(3.41)	(2.61)
Ind	−0.131 **	−0.197 ***
	(−3.26)	(−4.70)
Scale	−0.288 ***	−0.216 ***
	(−5.95)	(−4.82)
Exp	−0.146 ***	−0.123 **
	(−3.76)	(−3.27)
Imp	0.276 ***	0.329 ***
	(3.99)	(4.41)
Vab	0.0311	0.0837 *
	(0.62)	(2.25)
GVC-pat	−0.365 ***	−0.317 ***
	(−5.30)	(−4.99)
Ratiowair	0.719 ***	0.748 ***
	(6.22)	(6.61)
Obs	156	156
R^2	0.8952	0.8941

Note: * $p < 0.05$, ** $p < 0.01$, *** $p < 0.001$; *t* statistics in parentheses.

Table 7. Threshold effect test results.

Industry	Threshold	Energy Threshold		Economic Threshold		Scale Threshold	
		Threshold Value	*p* Value	Threshold Value	*p* Value	Threshold Value	*p* Value
Labor intensive	Single	−0.5478	0.0100	−0.4233	0.0533	−0.8088	0.0300
	Double	0.0042	0.0300	0.2929	0.2867	−0.9171	0.2000
Capital intensive	Single	0.3053	0.3833	−0.1759	0.6300	−0.5352	0.0000
	Double	−0.4059	0.3967	−0.7593	0.9467	2.2503	0.1900
Technology intensive	Single	−0.2248	0.0000	−0.2786	0.0000	0.1690	0.0000
	Double	0.8068	0.0000	−0.4057	0.6233	1.3146	0.1167

Note: The dependent variable is Carbon.

Table 8 shows the results of estimating the three groups with thresholds. For labor-intensive industries, when power consumption reaches the threshold, digitalization can promote China's manufacturing industry's carbon emissions. However, when the power consumption exceeds the threshold, although the coefficient of digital investment is still positive, the coefficient becomes smaller. This means that electricity consumption has less impact on contributing to carbon emissions as technology improves. The same effect exists for economic value added; when the economic value added exceeds the threshold, the impact of digital investment on carbon emissions will gradually become smaller. In terms of the industry scale, before and after reaching the threshold, the expansion of the industry scale has a restraining effect on the carbon emissions of labor-intensive industries, and the restraining effect is gradually improved. For capital-intensive industries, the expansion of the industry scale has a positive effect on the carbon emissions of capital-intensive industries, but as the industry scale exceeds a certain threshold, this promotion effect is gradually alleviated. For technology-intensive industries, before and after electricity consumption reaches a threshold, digital investment can curb China's manufacturing carbon emissions. In contrast, as electricity consumption exceeds a certain threshold, this inhibitory effect will gradually slow down. Economic value-added can promote the increase of carbon emissions, but as the economic value-added exceeds the threshold, the promotion effect of digitalization on carbon emissions is greatly alleviated. The scale of the industry also promotes the increase of carbon emissions, but as the scale of the industry exceeds a certain threshold, the impact on carbon emissions gradually weakens.

Table 8. Estimation results of the threshold model.

	Dependent Variable: Carbon						
Industry	**Labor Intensive**			**Capital Intensive**	**Technology Intensive**		
	Energy Threshold	**Economic Threshold**	**Scale Threshold**	**Scale Threshold**	**Energy Threshold**	**Economic Threshold**	**Scale Threshold**
Dig	0.5628	0.7665	−0.7336	4.8318 ***	−0.0696	0.5713 ***	0.2219 **
[<qx1]	(0.8022)	(0.7134)	(0.8482)	(1.6403)	(0.1150)	(0.1283)	(0.0892)
Dig	0.4316	0.5810	−0.6251	2.7802 *	−0.1030	0.0543	0.1301 ***
[>qx1]	(0.8009)	(0.7124)	(0.8383)	(1.5653)	(0.1443)	(0.0678)	(0.0463)
Controls	Yes	Yes	Yes	Yes	Yes	Yes	Yes
Obs	78	78	78	52	39	39	39
R^2	0.2206	0.2206	0.5048	0.5974	0.4361	0.9490	0.5852

Note: *** significant at 1% level, ** significant at 5% level, * significant at 10% level. qx1 represents the threshold value (standard errors in parentheses).

5. Discussion and Conclusions

5.1. Discussion

Existing studies have analyzed the correlation between digitalization and carbon emissions. For example, Yang et al. [21] found that digitalization can reduce carbon emission intensity, and this effect has obvious regional heterogeneity. Wang et al. [70] found that digital technology is an important path to achieving carbon neutrality. Moreover, digitalization mainly reduces carbon emissions by optimizing resource allocation and reducing energy consumption costs. Although the existing research is sufficient, the analysis of digital investment and electricity consumption in the manufacturing industry has been little, thereby buttressing the innovation of our research. Besides, this research takes into account the factors that the manufacturing industry participates in international competition, that is, the degree of participation in the global value chain. Only by participating in international competition can we force the transformation and upgrading of industrial structure, which is ultimately conducive to carbon emission reduction [54].

The estimation results of this research are consistent with results in [21,57]. The higher the level of digitization, the greater the carbon emissions of manufacturing. The demand for electricity contributes to carbon emissions. The main reasons are as follows. First, in the short term, the increase in energy use and non-green technologies use have caused an increase in carbon emissions. In the long run, the upgrading of the industrial structure and the spillover of technology have made digitization show a strong carbon emission reduction effect [71]. Second, the increase in demand for electricity consumption has brought about an increase in electricity supply and coal energy consumption, making electricity consumption a major contributor to carbon emissions [40,43].

5.2. Conclusions

This paper mainly used the panel regression model to analyze the impact of digital input and electricity consumption on China's manufacturing industry's carbon emissions and industry heterogeneity. It used the threshold model to analyze the energy, economic, and scale effects of the manufacturing industry. The conclusions were as follows:

(1) The level of digital investment in China's manufacturing industry is rising steadily. From 2007 to 2019, the proportion of electricity consumption of China's manufacturing industries in the total electricity consumption hardly changed, maintaining at about 6.8%. However, the total electricity consumption of the manufacturing industry increased by about 2.1 times. From 2007 to 2019, the total carbon emissions of China's manufacturing industry increased, but the carbon emissions of some manufacturing industries decreased.

(2) Hypothesis 1 was verified. There was an inverted U-shaped relationship between digitization and carbon emissions. The higher the level of digitization, the greater the carbon emissions of manufacturing; however, when digitization develops to a certain extent, it will curb carbon emissions.

(3) Hypothesis 2 was verified. There was a significant positive correlation between electricity consumption and carbon emissions in the manufacturing industry. The higher the electricity consumption, the more energy is consumed and the more carbon emissions are generated.

(4) The impact of digitalization of labor-intensive and technology-intensive manufacturing on carbon emissions da double energy threshold, a single economic threshold, and an industry scale threshold. There was a single scale threshold for capital-intensive manufacturing, and the threshold value was −0.5352.

According to these research results, this paper puts forward the following policy implications.

(1) From the perspective of the supply side, energy structure transformation and industrial optimization and upgrading should be promoted to improve energy utilization efficiency. Optimizing the energy structure is an important way for the low-carbon development of the manufacturing industry [65]. Carbon emissions have a great impact on developing a green economy. Carbon dioxide emissions do reduce environmental performance and green economic performance. However, digital development, technological innovation, and industrial structure upgrades can promote green economic performance [72]. China is the largest energy consumer and carbon emitter in the world. Optimizing industrial structure, reducing population size, and adjusting energy structure can indeed promote carbon emission reduction [73].

(2) From the perspective of the demand side, the total energy consumption should be reasonably controlled to improve the efficiency of energy-intensive utilization through digitization. Digital technology is an effective path to achieving carbon neutrality, and it mainly reduces carbon emissions by optimizing resource allocation and reducing energy consumption costs [70]. The development of the manufacturing industry has increased the importance of digital elements and processes in strategy and planning. The concepts of digitization and automation are distinct yet interrelated. Both can be used directly in the manufacturing field. Digitization can generate large amounts of data and form network integration. Automation can improve inefficient production steps and increase the consistency of the production process. The development of China's manufacturing industry should improve the ability of independent innovation and take the road of innovation and development [74]. In terms of dependence on foreign trade, the self-sufficiency rate of China's high-end chemical products is insufficient, and government policy support is needed to increase production capacity and reduce dependence on foreign trade.

(3) From the transaction point of view, the market mechanism is the most direct path to achieve the optimal allocation of resources and the most cost-effective way to achieve carbon emission reduction. Digitalization's suppression of carbon emissions has a phased feature. Under the blockchain technology of big data, the establishment of a carbon emission trading rights platform should be improved [75]. Carbon emission trading and carbon tax mechanism are the main ways to achieve carbon emission reduction. Excessive carbon

pricing will reduce the economic advantages of carbon transaction costs for high-emission manufacturing enterprises [76]. Digital technology has played an important role in reducing carbon emissions from regional trade and improving the energy efficiency of traded products. Trade has the potential to affect embodied carbon emission flows and embodied carbon emission intensity. The share of carbon emissions from trade is declining, and the manufacture of computers and electro-optical products is the main source of embodied carbon emissions [77]. In China, the ICT sector can generate a large volume of emissions through the demand for carbon-intensive intermediate inputs in the non-ICT sector. Moreover, the power and basic materials sectors are significant sources of carbon emissions; therefore, addressing ICT-related carbon emissions requires a targeted, integrated carbon management strategy that combines supply chain and economic drivers [78]. Consequently, the realization of low-carbon development in the manufacturing industry requires the joint participation of the government, market, enterprises, and technical service departments.

The limitations of this article are: First, due to the availability of data, this article only analyzes the situation of 13 manufacturing industries in China from 2007 to 2019. Second, there is no further in-depth analysis of the energy efficiency and energy consumption structure of the manufacturing industry. This is also our future research direction.

Author Contributions: Conceptualization, Q.W. and Q.Z.; methodology, Q.Z.; formal analysis, Q.Z.; investigation, Q.W.; resources, Q.W.; data curation, Q.Z.; writing—original draft preparation, Q.Z.; writing—review and editing, Q.Z.; visualization, Q.W.; supervision, Q.W. All authors have read and agreed to the published version of the manuscript.

Funding: This study was funded by the High-level Cultivation Project of Nanjing Xiaozhuang University (Grant No. 2022NXY15) and the National Social Science Foundation of China (Grant No. 21BJY085).

Institutional Review Board Statement: Not applicable.

Informed Consent Statement: Not applicable.

Data Availability Statement: The data involved in this study are all from public data.

Acknowledgments: Our deepest gratitude goes to the editors and anonymous reviewers for their careful work and thoughtful comments.

Conflicts of Interest: The authors declare no conflict of interest.

References

1. Yao, F.; Xue, L.; Liang, J. Research on coupling coordination and influencing factors between Urban low-carbon economy efficiency and digital finance—Evidence from 100 cities in China's Yangtze River economic belt. *PLoS ONE* **2022**, *17*, e0271455. [CrossRef]
2. Su, J.; Su, K.; Wang, S. Evaluation of digital economy development level based on multi-attribute decision theory. *PLoS ONE* **2022**, *17*, e0270859. [CrossRef]
3. Björkdahl, J. Strategies for Digitalization in Manufacturing Firms. *Calif. Manag. Rev.* **2020**, *62*, 17–36. [CrossRef]
4. Xie, Y. The Effect and Mechanism of Digital Economy on Regional Carbon Emission Intensity. *Contemp. Econ. Manag.* **2022**, *44*, 68–78. [CrossRef]
5. Xie, W.; Gao, K.; Yu, J. Digital Economy, Industrial Structure Upgrading and Carbon Emissions. *Stat. Decis.* **2022**, *38*, 114–118. [CrossRef]
6. Yu, S.; Fan, X.; Jiang, H. Research on the Impact of Digital Economy Development on Carbon Productivity Improvement. *J. Stat. Inf.* **2022**, *37*, 26–35.
7. Han, D.; Ding, Y.; Shi, Z.; He, Y. The impact of digital economy on total factor carbon productivity: The threshold effect of technology accumulation. *Environ. Sci. Pollut. Res.* **2022**, *29*, 55691–55706. [CrossRef]
8. Wang, J.; Dong, K.; Sha, Y.; Yan, C. Envisaging the carbon emissions efficiency of digitalization: The case of the internet economy for China. *Technol. Forecast. Soc. Chang.* **2022**, *184*, 121965. [CrossRef]
9. Salahuddin, M.; Alam, K.; Ozturk, I. The effects of Internet usage and economic growth on CO_2 emissions in OECD countries: A panel investigation. *Renew. Sustain. Energy Rev.* **2016**, *62*, 1226–1235. [CrossRef]
10. Balogun, A.L.; Marks, D.; Sharma, R.; Shekhar, H.; Balmes, C.; Maheng, D.; Arshad, A.; Salehi, P. Assessing the Potentials of Digitalization as a Tool for Climate Change Adaptation and Sustainable Development in Urban Centres. *Sustain. Cities Soc.* **2020**, *53*, 101888. [CrossRef]
11. Zhou, X.; Zhou, D.; Zhao, Z.; Wang, Q. A framework to analyze carbon impacts of digital economy: The case of China. *Sustain. Prod. Consum.* **2022**, *31*, 357–369. [CrossRef]

12. Zhang, R.; Fu, W.; Kuang, Y. Can Digital Economy Promote Energy Conservation and Emission Reduction in Heavily Polluting Enterprises? Empirical Evidence from China. *Int. J. Environ. Res. Public Health* **2022**, *19*, 9812. [CrossRef]
13. Wang, L.; Chen, L.; Li, Y. Digital economy and urban low-carbon sustainable development: The role of innovation factor mobility in China. *Environ. Sci. Pollut. Res.* **2022**, *29*, 48539–48557. [CrossRef]
14. Zhang, C.; Liu, C. The impact of ICT industry on CO_2 emissions: A regional analysis in China. *Renew. Sustain. Energy Rev.* **2015**, *44*, 12–19. [CrossRef]
15. Xu, Q.; Zhong, M. Shared prosperity, energy-saving, and emission-reduction: Can ICT capital achieve a "win-win-win" situation? *J. Environ. Manag.* **2022**, *319*, 115710. [CrossRef]
16. Zhao, X.; Dong, Y.; Gong, X. The Digital Economy and Carbon Productivity: Evidence at China's City Level. *Sustainability* **2022**, *14*, 10642. [CrossRef]
17. Guo, F.; Yang, S.; Ren, Y. The Digital Economy, Green Technology Innovation and Carbon Emissions: Empirical Evidence from Chinese City-level Data. *J. Shaanxi Norm. Univ. (Philos. Soc. Sci. Ed.)* **2022**, *51*, 45–60. [CrossRef]
18. Zhou, B.; Zhao, H.; Yu, J.; He, T.; Liu, J. Does the growth of the digital economy boost the efficiency of synergistic carbon-haze governance? evidence from China. *Front. Environ. Sci.* **2022**, *10*, 984591. [CrossRef]
19. Che, S.; Wang, J. Digital economy development and haze pollution: Evidence from China. *Environ. Sci. Pollut. Res.* **2022**, *29*, 73210–73226. [CrossRef]
20. Erdmann, L.; Hilty, L.M. Exploring the Macroeconomic Impacts of Information and Communication Technologies on Greenhouse Gas Emissions. *J Ind Econ.* **2010**, *14*, 826–843. [CrossRef]
21. Yang, Z.; Gao, W.; Han, Q.; Qi, L.; Cui, Y.; Chen, Y. Digitalization and carbon emissions: How does digital city construction affect China's carbon emission reduction? *Sustain. Cities Soc.* **2022**, *87*, 104201. [CrossRef]
22. Wang, N.; Zhu, Y. The Integration of Traditional Transportation Infrastructure and Informatization Development: How Does It Affect Carbon Emissions? *Energies* **2022**, *15*, 7535. [CrossRef]
23. Yu, Z.; Liu, S.; Zhu, Z. Has the Digital Economy Reduced Carbon Emissions?: Analysis Based on Panel Data of 278 Cities in China. *Int. J. Environ. Res. Public Health* **2022**, *19*, 11814. [CrossRef]
24. Li, Z.; Wang, J. How does the development of digital economy affect spatial carbon emissions under the background of economic agglomeration? *J. Xian Jiaotong Univ. (Soc. Sci.)* **2022**, *42*, 87–97. [CrossRef]
25. Ge, L.; Mo, L.; Huang, N. Development of Digital Economy, Upgrading of Industrial Structure and Urban Carbon Emission. *Mod. Financ. Econ.* **2022**, *42*, 20–37. [CrossRef]
26. Zhao, S.; Peng, D.; Wen, H.; Wu, Y. Nonlinear and spatial spillover effects of the digital economy on green total factor energy efficiency: Evidence from 281 cities in China. *Environ. Sci. Pollut. Res.* **2022**, 1–21. [CrossRef]
27. Lee, C.; Yuan, Y.; Wen, H. Can digital economy alleviate CO_2 emissions in the transport sector? Evidence from provincial panel data in China. *Nat. Resour. Forum* **2022**, *46*, 289–310. [CrossRef]
28. Li, J.; Wang, Q. Impact of the Digital Economy on the Carbon Emissions of China's Logistics Industry. *Sustainability* **2022**, *14*, 8641. [CrossRef]
29. Zhang, L.; Mu, R.; Zhan, Y.; Yu, J.; Liu, L.; Yu, Y.; Zhang, J. Digital economy, energy efficiency, and carbon emissions: Evidence from provincial panel data in China. *Sci. Total. Environ.* **2022**, *852*, 158403. [CrossRef]
30. Zhang, J.; Fu, K.; Liu, B. Can Digital Economy Promote Low-carbon Transformation of Cities from the Perspective of Dual Objective Constraint. *Mod. Financ. Econ.* **2022**, *42*, 3–23. [CrossRef]
31. Chen, P.; Gao, J.; Ji, Z.; Liang, H.; Peng, Y. Do Artificial Intelligence Applications Affect Carbon Emission Performance?—Evidence from Panel Data Analysis of Chinese Cities. *Energies* **2022**, *15*, 5730. [CrossRef]
32. Branca, T.A.; Fornai, B.; Colla, V.; Murri, M.M.; Streppa, E.; Schröder, A.J. The Challenge of Digitalization in the Steel Sector. *Metals* **2020**, *10*, 288. [CrossRef]
33. Buer, S.-V.; Strandhagen, J.W.; Semini, M. The digitalization of manufacturing: Investigating the impact of production environment and company size. *J. Manuf. Technol. Manag.* **2020**, *32*, 621–645. [CrossRef]
34. Papadopoulos, T.; Singh, S.P.; Spanaki, K.; Gunasekaran, A.; Dubey, R. Towards the next generation of manufacturing: Implications of big data and digitalization in the context of industry 4.0. *Prod. Plan. Control.* **2021**, *33*, 101–104. [CrossRef]
35. Beliatis, M.; Jensen, K.; Ellegaard, L.; Aagaard, A.; Presser, M. Next Generation Industrial IoT Digitalization for Traceability in Metal Manufacturing Industry: A Case Study of Industry 4.0. *Electronics* **2021**, *10*, 628. [CrossRef]
36. Cui, S.; Li, G.; Liu, J. Can economic growth and carbon emissions reduction be owned: Evidence from the convergence of digital services and manufacturing in China. *Environ. Sci. Pollut. Res.* **2022**, *30*, 20415–20430. [CrossRef]
37. Meng, F.; Zhao, Y. How does digital economy affect green total factor productivity at the industry level in China: From a perspective of global value chain. *Environ. Sci. Pollut. Res.* **2022**, *29*, 79497–79515. [CrossRef]
38. Chen, P. Relationship between the digital economy, resource allocation and corporate carbon emission intensity: New evidence from listed Chinese companies. *Environ. Res. Commun.* **2022**, *4*, 075005. [CrossRef]
39. Fang, H.; Jiang, C.; Hussain, T.; Zhang, X.; Huo, Q. Input Digitization of the Manufacturing Industry and Carbon Emission Intensity Based on Testing the World and Developing Countries. *Int. J. Environ. Res. Public Health* **2022**, *19*, 12855. [CrossRef]
40. Akpan, G.E. Electricity Consumption, Carbon Emissions and Economic Growth in Nigeria. *Int. J. Energy Econ. Policy* **2012**, *2*, 292–306.

41. Ehigiamusoe, K.U. A disaggregated approach to analyzing the effect of electricity on carbon emissions: Evidence from African countries. *Energy Rep.* **2020**, *6*, 1286–1296. [CrossRef]
42. Sovacool, B.K.; Schmid, P.; Stirling, A.; Walter, G.; MacKerron, G. Differences in carbon emissions reduction between countries pursuing renewable electricity versus nuclear power. *Nat. Energy* **2020**, *5*, 928–935. [CrossRef]
43. Wang, Z.-X.; Zhang, J.-J.; Pan, L.; Yang, F.; Shi, L.-N. Estimate of China's energy carbon emissions peak and analysis on electric power carbon emissions. *Adv. Clim. Chang. Res.* **2014**, *5*, 181–188. [CrossRef]
44. Wei, W.; Hao, S.; Yao, M.; Chen, W.; Wang, S.; Wang, Z.; Wang, Y.; Zhang, P. Unbalanced economic benefits and the electricity-related carbon emissions embodied in China's interprovincial trade. *J. Environ. Manag.* **2020**, *263*, 110390. [CrossRef]
45. Wei, W.; Zhang, P.; Yao, M.; Xue, M.; Miao, J.; Liu, B.; Wang, F. Multi-scope electricity-related carbon emissions accounting: A case study of Shanghai. *J. Clean. Prod.* **2019**, *252*, 119789. [CrossRef]
46. Cowan, W.N.; Chang, T.; Inglesi-Lotz, R.; Gupta, R. The nexus of electricity consumption, economic growth and CO_2 emissions in the BRICS countries. *Energy Policy* **2014**, *66*, 359–368. [CrossRef]
47. Zhang, H. Pathways to carbon neutrality in major exporting countries: The threshold effect of digital transition. *Environ. Sci. Pollut. Res.* **2022**, *30*, 7522–7542. [CrossRef]
48. Hao, X.; Wen, S.; Li, Y.; Xu, Y.; Xue, Y. Can the digital economy development curb carbon emissions? Evidence from China. *Front. Psychol.* **2022**, *13*, 938918. [CrossRef]
49. Li, Z.; Wang, J. The Dynamic Impact of Digital Economy on Carbon Emission Reduction: Evidence City-level Empirical Data in China. *J. Clean. Prod.* **2022**, *351*, 131570. [CrossRef]
50. Ge, W.; Xu, Y.; Liu, G.; Shen, B.; Su, X.; Liu, L.; Yang, X.; Ran, Q. Exploring the Impact of the Digital Economy on Carbon Emission Efficiency Under Factor Misallocation Constraints: New Insights From China. *Front. Environ. Sci.* **2022**, *10*, 953070. [CrossRef]
51. Xu, W.; Zhou, J.; Liu, C. The impact of digital economy on urban carbon emissions: Based on the analysis of spatial effects. *Geogr. Res.* **2022**, *41*, 111–129.
52. Liu, Y.; Li, Q.; Zhang, Z. Do Smart Cities Restrict the Carbon Emission Intensity of Enterprises? Evidence from a Quasi-Natural Experiment in China. *Energies* **2022**, *15*, 5527. [CrossRef]
53. She, Q.; Wu, L. The Carbon Emission Reduction Effect of Digital Economy Development. *Econ. Surv.* **2022**, *39*, 14–24. [CrossRef]
54. Wiryawan, B.A.; Aginta, H.; Fazaalloh, A.M. Does GVC participation help industrial upgrading in developing countries? New evidence from panel data analysis. *J. Int. Trade Econ. Dev.* **2022**, 1–18. [CrossRef]
55. Zhang, Y.; Ling, D.; Liu, H. Does the digital economy promote the upgrading of global value vain in China's manufacturing? *Stud. Sci. Sci.* **2022**, *40*, 57–68. [CrossRef]
56. Zhou, R.; Tang, D.; Da, D.; Chen, W.; Kong, L.; Boamah, V. Research on China's Manufacturing Industry Moving towards the Middle and High-End of the GVC Driven by Digital Economy. *Sustainability* **2022**, *14*, 7717. [CrossRef]
57. Ali, S.S.S.; Razman, M.R.; Awang, A. The Nexus of Population, GDP Growth, Electricity Generation, Electricity Consumption and Carbon Emissions Output in Malaysia. *Int. J. Energy Econ. Policy* **2020**, *10*, 84–89. [CrossRef]
58. Qu, S.; Liang, S.; Xu, M. CO_2 Emissions Embodied in Interprovincial Electricity Transmissions in China. *Environ. Sci. Technol.* **2017**, *51*, 10893–10902. [CrossRef]
59. Liu, Q.; Zheng, X.-Q.; Zhao, X.-C.; Chen, Y.; Lugovoy, O. Carbon emission scenarios of China's power sector: Impact of controlling measures and carbon pricing mechanism. *Adv. Clim. Chang. Res.* **2018**, *9*, 27–33. [CrossRef]
60. Wang, Y.; Chen, W.; Liu, B. Manufacturing/remanufacturing decisions for a capital-constrained manufacturer considering carbon emission cap and trade. *J. Clean. Prod.* **2017**, *140*, 1118–1128. [CrossRef]
61. Peng, H.; Wang, Y.; Hu, Y.; Shen, H. Agglomeration Production, Industry Association and Carbon Emission Performance: Based on Spatial Analysis. *Sustainability* **2020**, *12*, 7234. [CrossRef]
62. Khan, Z.; Ali, M.; Liu, J.; Shahbaz, M.; Yang, S. Consumption-based carbon emissions and trade nexus: Evidence from nine oil exporting countries. *Energy Econ.* **2020**, *89*, 104806. [CrossRef]
63. Liddle, B. Consumption-based accounting and the trade-carbon emissions nexus. *Energy Econ.* **2018**, *69*, 71–78. [CrossRef]
64. Hu, G.; Can, M.; Paramati, S.R.; Doğan, B.; Fang, J. The effect of import product diversification on carbon emissions: New evidence for sustainable economic policies. *Econ. Anal. Policy* **2020**, *65*, 198–210. [CrossRef]
65. Jin, B.; Han, Y. Influencing factors and decoupling analysis of carbon emissions in China's manufacturing industry. *Environ. Sci. Pollut. Res.* **2021**, *28*, 64719–64738. [CrossRef]
66. Liu, H.; Zong, Z.; Hynes, K.; De Bruyne, K. Can China reduce the carbon emissions of its manufacturing exports by moving up the global value chain? *Res. Int. Bus. Finance* **2019**, *51*, 101101. [CrossRef]
67. Liu, C.; Zhao, G. Can global value chain participation affect embodied carbon emission intensity? *J. Clean. Prod.* **2020**, *287*, 125069. [CrossRef]
68. Zhang, Z.; Zhang, J.; Feng, Y. Assessment of the Carbon Emission Reduction Effect of the Air Pollution Prevention and Control Action Plan in China. *Int. J. Environ. Res. Public Health* **2021**, *18*, 13307. [CrossRef]
69. Yang, X.; Hou, W.; Wang, Y. Digital Investment and Employment Structure Changes within China's Industry: "Upgrade" or "Polarization". *J. Shanxi Univ. Financ. Econ.* **2022**, *44*, 58–68. [CrossRef]
70. Wang, H.; Li, Y.; Lin, W.; Wei, W. How does digital technology promote carbon emission reduction? Empirical evidence based on e-commerce pilot city policy in China. *J. Environ. Manag.* **2023**, *325*, 116524. [CrossRef]

71. Pan, M.; Zhao, X.; Lv, K.; Rosak-Szyrocka, J.; Mentel, G.; Truskolaski, T. Internet Development and Carbon Emission-Reduction in the Era of Digitalization: Where Will Resource-Based Cities Go? *Resour. Policy.* **2023**, *81*, 103345. [CrossRef]
72. Jia, L.; Hu, X.; Zhao, Z.; He, B.; Liu, W. How Environmental Regulation, Digital Development and Technological Innovation Affect China's Green Economy Performance: Evidence from Dynamic Thresholds and System GMM Panel Data Approaches. *Energies* **2022**, *15*, 884. [CrossRef]
73. Kong, H.; Shi, L.; Da, D.; Li, Z.; Tang, D.; Xing, W. Simulation of China's Carbon Emission based on Influencing Factors. *Energies* **2022**, *15*, 3272. [CrossRef]
74. Zhou, J. Digitalization and intelligentization of manufacturing industry. *Adv. Manuf.* **2013**, *1*, 1–7. [CrossRef]
75. Zhang, J.; Li, J.; Ye, D.; Sun, C. The Impact of Digital Economy of Resource-Based City on Carbon Emissions Trading by Blockchain Technology. *Comput. Intell. Neurosci.* **2022**, *2022*, 6366061. [CrossRef]
76. Huisingh, D.; Zhang, Z.; Moore, J.C.; Qiao, Q.; Li, Q. Recent advances in carbon emissions reduction: Policies, technologies, monitoring, assessment and modeling. *J. Clean. Prod.* **2015**, *103*, 1–12. [CrossRef]
77. Dong, X.; Jiang, Q.; Wang, J. Assessing Embodied Carbon Emission and Its Intensities in the ICT Industry: The Global Case. *Front. Energy Res.* **2021**, *9*, 685021. [CrossRef]
78. Zhou, X.; Zhou, D.; Wang, Q.; Su, B. How information and communication technology drives carbon emissions: A sector-level analysis for China. *Energy Econ.* **2019**, *81*, 380–392. [CrossRef]

International Journal of
Environmental Research and Public Health

MDPI

Article

Exploration of Spatio-Temporal Characteristics of Carbon Emissions from Energy Consumption and Their Driving Factors: A Case Analysis of the Yangtze River Delta, China

Weiwu Wang [1,2,3,*], Huan Chen [1], Lizhong Wang [1,2], Xinyu Li [1], Danyi Mao [1] and Shan Wang [1]

1 College of Civil Engineering and Architecture, Zhejiang University, Hangzhou 310058, China; chenhuan01@zju.edu.cn (H.C.); lizhongwang@zju.edu.cn (L.W.); lixinyu1001@zju.edu.cn (X.L.); maodanyi0122@zju.edu.cn (D.M.); wangshan33@zju.edu.cn (S.W.)

2 China of Institute of Urbanization, Zhejiang University, Hangzhou 310058, China

3 Center for Balance Architecture, Zhejiang University, Hangzhou 310058, China

* Correspondence: weiwuwang@zju.edu.cn

Abstract: For the Yangtze River Delta (YRD) region of China, exploring the spatio-temporal characteristics of carbon emissions from energy consumption (CEECs) and their influencing factors is crucial to achieving carbon peaking and carbon neutrality as soon as possible. In this study, an improved LMDI decomposition model based on the Tapio model and Kaya's equation was proposed. Combined with the improved LMDI and k-means cluster analysis methods, the energy structure, energy intensity, unit industrial output value and population size were selected as the driving factors, and the contribution of each driving factor to the CEECs of prefecture-level cities was quantitatively analyzed. Our study found that: (1) By 2020, the total amount of CEECs in the 26 prefecture-level cities in the YRD will stabilize, while their intensity has shown a downward trend in recent years. (2) The decoupling relationship between CEECs and economic development generally showed a trend from negative decoupling to decoupling. The dominant factor in decoupling was generally the shift of *DEL* values towards urbanization rate and energy intensity and the open utilization of energy technologies. (3) From 2000 to 2010, the dominant factors affecting CEECs in 26 cities were energy intensity and energy structure, followed by industrial output value and urbanization rate. In general, the promotion effect of economic development on carbon emissions in the YRD region was greater than the inhibitory effect. After 2010, the restrictive effect of various factors on CEECs increased significantly, among which the role of gross industrial output was crucial. The research results can provide a scientific policy basis for the subsequent spatial management and control of carbon emission reduction and carbon neutrality in the YRD region at a finer scale.

Keywords: carbon emissions from energy consumption; decoupling elasticity; spatio-temporal characteristics; improved LMDI model; k-means clustering; map visualization

Citation: Wang, W.; Chen, H.; Wang, L.; Li, X.; Mao, D.; Wang, S. Exploration of Spatio-Temporal Characteristics of Carbon Emissions from Energy Consumption and Their Driving Factors: A Case Analysis of the Yangtze River Delta, China. *Int. J. Environ. Res. Public Health* **2022**, *19*, 9483. https://doi.org/10.3390/ijerph19159483

Academic Editors: Taoyuan Wei and Qin Zhu

Received: 17 June 2022
Accepted: 27 July 2022
Published: 2 August 2022

1. Introduction

In recent years, Chinese local governments, relevant departments and critical industries have worked together to reduce carbon emissions and have set emission reduction targets in their mid- and long-term economic development plans [1]. China has set a goal of peaking carbon emissions by 2030 and is striving to achieve carbon neutrality by 2060 [2]. However, China's current emission reduction plan targets are mostly set at the national and provincial levels. During the implementation of the actual emission target, due to the huge differences in regional economic development, industrial structure and energy structure, the different dynamic factors affecting carbon emissions are also complex and changeable. Therefore, analyzing the spatio-temporal characteristics, decoupling relationships and driving factors of carbon emissions in critical regions or cities in China is of great practical significance for effectively formulating carbon emission mitigation targets for regions or cities [3].

As the most active urban agglomeration in China, the YRD region is not only the region with the fastest economic growth, the largest economic aggregate and the greatest economic potential in China, but also a critical control area for achieving carbon peaking and carbon neutrality goals [4]. The YRD urban agglomeration is an essential industrial intersection between the "Belt and Road Initiative" and the Yangtze River economic belt. It is also a demonstration area for high-quality integrated development in China. In 2020, the total GDP will account for about 15% of the country's total [5], the urban population will account for about 17% of the country's total urban population and the total carbon emissions will account for about 16% of the country's total carbon emissions [6]. It is a crucial control area for achieving carbon peaking and carbon neutrality [7], and it is also an essential leader in China's achieving the dual-carbon goal.

The decomposition analysis of the CEECs' driving factors is the basis for achieving the regional carbon emission target. In recent years, many studies have explored the path of effective emission reduction and low-carbon green economic development by analyzing the decoupling relationship between economic growth, energy consumption and CO_2 emissions. Currently, decomposition analysis research mainly focuses on the CEECs from various industries such as the transportation industry, textile industry, manufacturing industry, non-metallic minerals and residential CEECs. The research on the decomposition analysis of decoupling relationships mainly focuses on the CEECs in various industries such as transportation [8], textile [9], manufacturing [10], non-metallic minerals [11] and real estate [12]. Ma et al. [13] used the LMDI decomposition analysis method to study seven energy consumption sectors in China and found that eliminating excess capacity and promoting structural transformation has become the only way for China to reduce emissions. Xu et al. [14] analyzed the decomposition of CEECs based on the two dimensions of China's various periods and industries and studied the main factors that promote and suppress CEECs in China at different stages. Yang et al. [15] emphasized that economic energy consumption is the biggest driver of carbon emission growth in China and pointed out that introducing electricity import measures can buffer the impact of the carbon emission intensity of annual energy consumption. Liu et al. [16] classified the emission reduction of each city from the perspective of periods and groups based on the decomposition analysis of CEECs at the provincial level in China. As the world's second-largest economy and a developing country with a large population base, high degree of industrialization and high dependence on coal, China has a long way to go to reduce emissions. Therefore, understanding the various characteristics of CEECs and the driving factors of CEECs has always been the direction for China's cities to follow when exploring low-carbon economic development and ecological construction.

According to the environmental Kuznets curve (EKC) hypothesis, there is an inverted U-shaped relationship between environmental pressure and economic growth. That is, in the early stage of economic development, environmental quality deteriorates with economic growth. Environmental quality gradually improves when economic development reaches a certain level [17]. The "decoupling" theory is a basic theory proposed by the Organization for Economic Cooperation and Development to describe the connection between economic growth and resource consumption or environmental pollution (Paris: OECD, 2002). Decoupling of carbon emissions, which, in essence, is to measure whether economic growth is at the cost of resource consumption and environmental damage, can be used to describe the relationship between changes in CO_2 emissions and economic growth. When economic growth is achieved, if the growth rate of CO_2 emissions is negative or lower than the economic growth rate, it can be regarded as decoupling. That is, based on economic growth, energy consumption is gradually reduced [18]. Zhang [19] introduced the decoupling index in energy and environment research in 2000, and Freitas et al. [20] used this method to explore the decoupling between economic activity and CEECs in Brazil from 2004 to 2009. In addition, in 2005, Tapio [21] proposed a theoretical framework for decoupling, which defined the difference between "decoupling", "connection" and "negative decoupling" and then divided them into "weak", "strong", "extended" and "decline". Some existing studies used decomposition analysis models and clustering methods to study the decoupling relationship between carbon emissions and China's economic development. Zhang et al. [22] used the LMDI method to decompose the decoupling index between economic growth and energy consumption in China from 1991 to 2012. The results showed that economic activity negatively impacted the decoupling. Chang et al. [23] used a new regional classification framework combining factor analysis and Ward clustering to divide 30 provinces in China into four regions and investigated the differences in the impacts of population size, per capita GDP, energy structure and energy intensity on CO_2 emissions. In summary, the existing literature has achieved important results regarding the decoupling relationship between economic development and CEECs. As an essential demonstration area for economic development and CO_2 emission control in China, the YRD region is committed to promoting China's deep decarbonization and achieving high-quality economic development.

2. Literature Review

2.1. Empirical Study on Energy Carbon Emissions in the YRD Region

In recent years, some studies have gradually been carried out on low-carbon emission reduction in the YRD region. By formulating a green economy model, the World Resources Institute (WRI) predicted that the YRD region will reach a carbon peak in 2024 under the background of a green economy, with a peak value of about 1.793 billion tons of carbon dioxide equivalent and the peak forecast will be reduced by 90 million tons of carbon dioxide equivalent, which is two years earlier than the current background and policy deduction to achieve the carbon peak, and this benefit can lead the national carbon reduction action (WRI, China, 2021). At the same time, many researchers are also actively committed to research on emission reduction in the YRD region. Song et al. [24] calculated the annual variation of energy consumption in the YRD region from 1995 to 2010 and proposed ways to control carbon emissions in the YRD region against the background of sustained economic growth. Gong et al. [25] used the STIRPAT model to quantitatively analyze the relationship between CO_2 emissions in the YRD and influencing factors such as population, per capita GDP, foreign direct investment and technological progress. Pei et al. [26] investigated the carbon footprint from fossil energy consumption and the decoupling relationship between carbon footprint pressure and economic growth in the YRD combined with land resource constraints. They combined gap analysis and social network analysis. Shen et al. [27] found that the YRD highlights the strong collaborative development ability and driving ability of developed cities, thus generating the greatest potential to reduce CO_2 emissions in the short and medium terms. In summary, it can be seen that the research on carbon emissions in the YRD region has become a typical research model relating to low-carbon emissions in China's urban agglomeration. However, the influencing factors of the decoupling relationship between all carbon emissions and economic development in the YRD region and their temporal and spatial distribution still need further research. Therefore, an in-depth study of energy consumption and carbon emissions in the YRD region to provide policymakers with information to achieve carbon emission reduction targets is crucial for helping China's economically developed regions achieve future emission reduction targets.

2.2. Methods for Identifying Drivers of Carbon Emissions

At present, the application of the LMDI method in the field of energy and environment can be mainly divided into three directions: (1) The LMDI method is directly used to explore the influencing factors of carbon emissions. For example, Quan et al. used the LMDI method to decompose the carbon emission factors of China's logistics industry from 2000 to 2016 into five dimensions, carbon emission factors, energy intensity, energy structure, economic development level and population size, and the carbon emission contribution rates were analyzed separately [28]. (2) As an evaluation model, LMDI is combined with other models to decompose and analyze the data obtained by other models and conduct in-depth evaluation and analysis. The current application in the field of carbon emission research is mainly combined with the decoupling model to further study the low-carbon development of cities in a quantitative manner. For example, Wang et al. [29] combined the Tapio decoupling index and the LMDI model to decompose the factors affecting energy consumption and carbon emissions and put forward the focus of promoting green and low-carbon development and the transformation of Qinghai Province. (3) As a previous decomposition model, the relationship of influencing factors obtained by decomposition is mainly used for subsequent analysis, such as peak prediction, situation simulation, etc. For example, Zhang et al. [30] combined the scenario analysis method with the Monte Carlo prediction method, using LMDI to decompose the driving factors of China's total water consumption to predict the trend of China's water consumption change before 2030 and then judge the peak time point and occurrence of water consumption. Gu et al. [31] combined LMDI with a system dynamics model (SD method) to quantify and estimate emission reduction potential in Shanghai, China. In addition, the application of LMDI is not limited to the traditional energy field but also includes economic cost estimation and related patent research. For example, Zhang et al. [32] investigated the relevant biogas user data in 19 villages in China in 2015, quantified the gap between the theoretical cost and actual cost of CO_2 emission reduction per unit and analyzed the main factors affecting the cost with the help of the LMDI model.

Our main research purpose is to go deep into 26 prefecture-level cities in the YRD region to reveal the spatio-temporal changes in the decoupling relationship between carbon emissions and economic development. At the same time, LMDI is introduced as an evaluation model, the data obtained by the decoupling model are decomposed and analyzed and the decoupling elasticity of urbanization rate, energy intensity, unit industrial output value and energy structure is deeply evaluated. Based on previous research, the k-means clustering analysis method is further used to cluster the results, and the leading factors of carbon emissions in cities at various levels in the YRD region are summarized. Specifically, this study aims to reveal the spatio-temporal characteristics and influencing factors of CEECs in prefecture-level cities in the YRD region, the spatio-temporal characteristics and decoupling factors of CEECs from 2000 to 2020 and the contribution rate of each driving factor to CEECs. On the one hand, this will help the YRD region to decompose the national emission reduction targets into local-level cities and, on the other hand, formulate effective emission reduction measures according to local conditions to achieve the goal of pilot testing in the YRD region.

The remainder of this paper is organized as follows. In the next section, we present an overview of the study area. In Section 4, we offer a flowchart of the method used in this study and the data sources. Section 5 reports the main results. Section 6 presents the conclusions and policy implications.

3. Study Area

This paper selected the YRD region as the research object. The research area covers 26 cities, including Shanghai, Nanjing, Wuxi, Changzhou, Suzhou, Nantong, Yancheng, Yangzhou, Zhenjiang, Taizhou, Hangzhou, Ningbo, Jiaxing, Huzhou, Shaoxing, Jinhua, Zhoushan, Taizhou, Hefei, Wuhu, Ma'anshan, Tongling, Anqing, Chuzhou, Chizhou and Xuancheng (Figure 1).

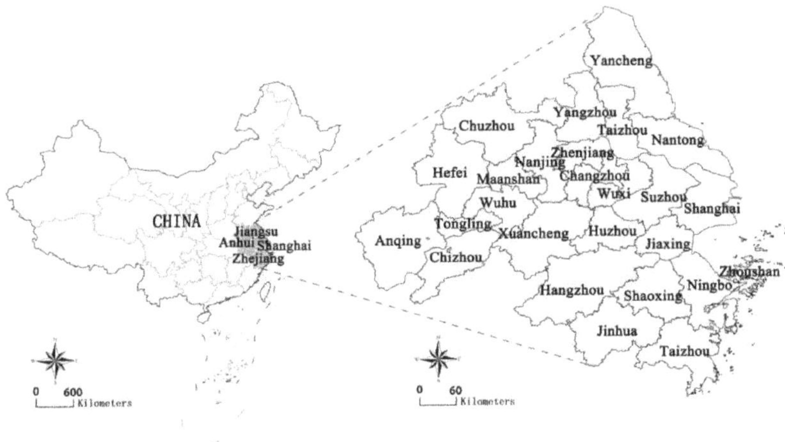

Figure 1. 26 prefecture-level cities in the YRD.

The YRD region is located in East China, covering 211,700 square kilometers, accounting for about 2.2% of China's land area. In 2019, the GDP of the YRD region was about CNY 20 trillion, accounting for about 20% of China's GDP. At the end of the year, the resident population was about 160 million, representing more than one-tenth of the country's resident population. As an important functional area leading China's regional economic development, the YRD region has a vast economic hinterland with a modern transportation network and advantageous industrial clusters centered on electronics, automobiles, modern finance and other industries, as well as rich scientific and educational resources.

4. Methods

4.1. Data Sources

This paper includes three parts of research data. The first is urban statistical data from the 2006–2019 "China Urban Statistical Yearbook", "China Urban Construction Statistical Yearbook", "China Energy Statistical Yearbook", "Jiangsu Statistical Yearbook", "Zhejiang Statistical Yearbook" and "Anhui Statistical Yearbook", as well as statistical yearbooks and social development statistical reports of 26 cities, including the main energy consumption statistics of the above-scale industries in the local-level cities from 2000 to 2020, industrial output value, urbanization rate, comprehensive energy consumption and other indicators. Except for a few city data from corporate websites, statistical data were provided by CNKI China Social and Economic Big Data Research Platform, and the link is https://data.cnki.net/HomeNew/index (accessed on 31 January 2022).

4.2. Methodology

The specific analysis process is shown in Figure 2. It mainly included using IPCC urban carbon emission calculation, the improved LMDI model based on the Kaya equation and Tapio to analyze each factor's decoupling elasticity and the k-means cluster analysis and map classification to visualize carbon emission characteristics and carbon emission impact factors for regional management and control analysis.

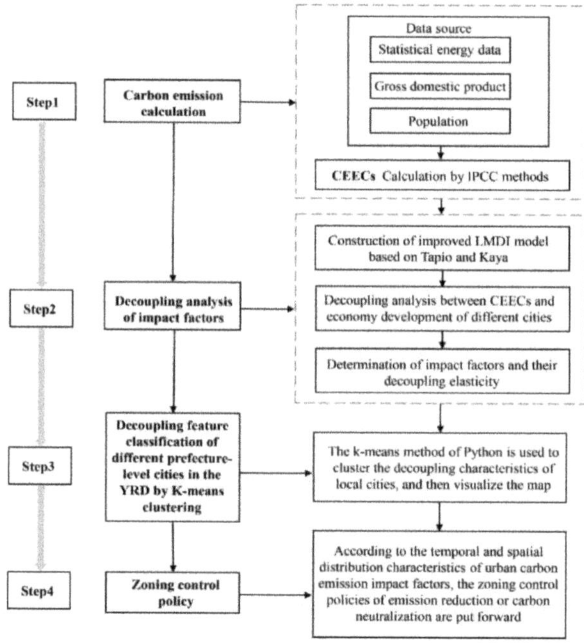

Figure 2. Flowchart of the methodology used in this study.

4.2.1. CEEC Model

Referring to the method proposed by Ang [33], the CEEC calculation formula is as follows:

$$C^{mn} = \sum_i E_i^{mn} \times ei \times pi \times 44/12 \quad (1)$$

where m represents the city m of the 26 cities in the YRD, n represents the node year n from 2000 to 2020, C^{mn} represents the CEEC of the mth city in the node year n, i represents the ith type of fossil energy, E_i^{mn} represents the consumption of the ith fossil energy in the mth city in the node year n, ei represents the conversion coefficient of standard coal to the ith energy, which is taken from the General Principles of Comprehensive Energy Consumption Calculation (GB/T2589- 2020), and pi represents the carbon emission factor, which is taken from the IPCC reference value. In the equation, 44/12 represents the molecular weight ratio of carbon dioxide to carbon. The standard coal reduction coefficients and carbon emission coefficients for 21 fossil energy sources are shown in Table 1.

Table 1. Calculation parameters of carbon emissions.

Energy Types	Standard Coal Coefficient/(kgce/kg)	Carbon Emission Coefficient/(kg/kgce)
Raw coal (tons)	0.7143	0.7559
Washed coal (tons)	0.9	0.7559
Other washed coal (tons)	0.2857	0.7559
Coal products (tons)	0.2857	0.7559
Coke (tons)	0.9714	0.855
Other coking products (tons)	0.9714	0.855
Coke oven gas (10,000 cubic meters)	0.6	0.3548
Blast furnace gas (10,000 cubic meters)	0.1286	0.3548
Converter gas (10,000 cubic meters)	0.2571	0.3548
Producer gas (10,000 cubic meters)	0.1786	0.3548

Table 1. *Cont.*

Energy Types	Standard Coal Coefficient/(kgce/kg)	Carbon Emission Coefficient/(kg/kgce)
Natural gas (10,000 cubic meters)	1.2	0.4483
LNG (tons)	1.7572	0.4483
Crude oil (tons)	1.429	0.5857
Gasoline (tons)	1.4714	0.5538
Kerosene (tons)	1.4714	0.5714
Diesel (tons)	1.4571	0.5921
Fuel oil (tons)	1.4286	0.6185
LPG (tons)	1.7143	0.5042
Refinery dry gas (10,000 cubic meters)	1.5714	0.4602
Other petroleum products (tons)	1.7	0.5857
Other fuels (tons)	1	0.7561

4.2.2. The Improved LMDI Model Based on Kaya Equation and Tapio

In recent years, many scholars have used the concept of decoupling and its indicators to reflect the relationship between economic growth and CO_2 emissions and have used decoupling elasticity as the main tool to measure the low-carbon status of various regions [14]. In our study, the Tapio model was selected to calculate the decoupling elasticity value, and 26 prefecture-level cities in the YRD were classified and analyzed.

The formula for calculating the elasticity coefficient of the Tapio model is as follows:

$$D^m = \frac{\Delta C^m_{t2-t1} / C^m_{t1}}{\Delta GIO^m_{t2-t1} / GIO^m_{t1}} \quad (2)$$

where m represents the city m of the 26 cities in YRD, D^m represents the decoupling elasticity coefficient of the city m between the two node years $t1$ and $t2$ and ΔC^m and ΔGIO^m represent the changes in carbon emissions and gross industrial output value of the city m in node year $t2$ relative to node year $t1$, respectively. Parameters $t1$ and $t2$ represent the base node year and the end node year in the study from 2000 to 2020, respectively.

According to the magnitude of the decoupling elasticity and the positive and negative conditions of ΔC and ΔGIO, Tapio divides the decoupling state into eight decoupling states, as shown in Table 2 [21].

Table 2. Tapio decoupling system.

Condition		$\Delta CO_2/CO_2$	$\Delta GIO/GIO$	Elasticity (D)	Significance
Decoupling entry	Enhance	<0	>0	$D < 0$	In the most ideal state, the carbon emission growth index shows an inverse relationship with economic growth.
	Weaken	>0	>0	$0 < D < 0.8$	The growth rate of carbon emissions is lower than that of economic growth.
	Decline	<0	<0	$D > 1.2$	Carbon emissions decay faster than economic recession.
Negative decoupling	Enhance	>0	<0	$D < 0$	In the most unsatisfactory state, economic growth is negative, and carbon emissions still tend to rise.
	Weaken	<0	<0	$0 < D < 0.8$	Carbon emissions decay faster than economic recession.
	Increase	>0	>0	$D > 1.2$	Carbon emissions are growing faster than economic growth.

Table 2. *Cont.*

Condition		$\Delta CO_2/CO_2$	$\Delta GIO/GIO$	Elasticity (D)	Significance
Connect	Increase	>0	>0	$0.8 < D < 1.2$	Carbon emissions grow at the same time as the economy, at the same rate and in a linear relationship.
	Decline	<0	<0	$0.8 < D < 1.2$	Carbon emissions and the economy decline at the same time, at the same speed and in a linear relationship.

The Kaya identity proposed by Kaya [34] can be used to investigate the influencing factors of changes in greenhouse gas emissions at the national or regional level. The role of this identity is to describe the relationship between social, economic, energy, carbon emissions and other factors with simple mathematical relationships from an overall, macroscopic perspective.

We selected energy structure, energy intensity, the ratio of industrial output value to urbanization rate and urbanization rate as the analysis objects to construct the energy consumption carbon emissions of typical cities in the YRD. The Kaya identity can be expressed as follows:

$$\begin{aligned} C &= \tfrac{C}{TOE} \times \tfrac{TOE}{GIO} \times \tfrac{GIO}{POP} \times POP \\ &= ES \times EI \times EL \times P \end{aligned} \tag{3}$$

where C represents the carbon emissions from energy consumption, TOE represents the total energy consumption, GIO represents the gross industrial production and POP represents the urbanization rate. ES, EI, EL and P on the right-hand side of the equation represent energy structure, energy intensity, the industrial output value of "unit urbanization rate" and urbanization level, respectively.

Referring to the previous study, the formula for the LMDI method is as follows:

$$\begin{aligned} \Delta C &= C_{t2} - C_{t1} \\ &= \Delta C_{ES} + \Delta C_{EI} + \Delta C_{EL} + \Delta C_P \end{aligned} \tag{4}$$

where ΔC_{ES} represents the energy structure effect, ΔC_{EI} represents the energy intensity effect and ΔC_{EL} represents the unit industrial output value effect; ΔC_P represents the population scale effect.

Combined with the above decoupling model, the logarithmic decomposition model of the LMDI method can be expressed as follows:

$$\begin{aligned} D &= \tfrac{GIO^m_{t1}}{\Delta GIO^m \times C^m_{t1}} \times \Delta C^m_{t2-t1} \\ &= \tfrac{GIO^m_{t1}}{\Delta GIO^m \times C^m_{t1}} \times \left(\Delta C^m_{ES} + \Delta C^m_{EI} + \Delta C^m_{EL} + \Delta C^m_P\right) \end{aligned} \tag{5}$$

$$DES^m = \frac{GIO^m_{t1}}{\Delta GIO^m \times C^m_{t1}} \times \sum\nolimits_{im} \frac{\Delta C^m_{t2-t1}}{lnC^m_{t2} - lnC^m_{t1}} \times ln\frac{ES^m_{t2}}{ES^m_{t1}} \tag{6}$$

$$DEI^m = \frac{GIO^m_{t1}}{\Delta GIO^m \times C^m_{t1}} \times \sum\nolimits_{im} \frac{\Delta C^m_{t2-t1}}{lnC^m_{t2} - lnC^m_{t1}} \times ln\frac{EI^m_{t2}}{EI^m_{t1}} \tag{7}$$

$$DEL^m = \frac{GIO^m_{t1}}{\Delta GIO^m \times C^m_{t1}} \times \sum\nolimits_{im} \frac{\Delta C^m_{t2-t1}}{lnC^m_{t2} - lnC^m_{t1}} \times ln\frac{EL^m_{t2}}{EL^m_{t1}} \tag{8}$$

$$DP^m = \frac{GIO^m_{t1}}{\Delta GIO^m \times C^m_{t1}} \times \sum\nolimits_{im} \frac{\Delta C^m_{t2-t1}}{lnC^m_{t2} - lnC^m_{t1}} \times ln\frac{P^m_{t2}}{P^m_{t1}} \tag{9}$$

where m represents the city m of the 26 cities in the YRD, DES^m represents the decoupling elasticity of energy structure of city m, DEI^m represents the decoupling elasticity of energy intensity of city m, DEL^m represents the decoupling elasticity of unit industrial output value of city m, DP^m represents the decoupling elasticity of population size of city m and parameters $t1$ and $t2$ represent the base node year and the end node year in the study from 2000 to 2020, respectively.

4.2.3. K-Means Method

After decoupling the CEECs of each prefecture-level city, the next important task was to fully reveal the spatial and temporal differences in the driving factors of CEECs in each prefecture-level city. The main idea of k-means clustering method is to use k centroids to cluster multiple discrete data points. The essence is to group the points with higher similarity into a group and separate the points with lower similarity. The method converges to the optimal solution by continuously updating the position of the group centroid. In our study, Python 3.7 was combined with the elbow method described above to determine the optimal number of sets. Then, the k-means method for clustering was used to set the optimal number of groups found by the elbow method as the number of clusters to obtain the final specific number of groups.

5. Results and Analysis

5.1. Variation Analysis of Total CEECs and CEEC Intensity in the YRD

As shown in Figure 3, there were differences in the total amount of CEECs and the intensity of CEECs in the YRD region from 2000 to 2020. The total amount of CEECs increased from 569.75 million tons to 1221.28 million tons, with an average annual growth rate of about 3.89%, and CEEC was basically stable with small fluctuations after 2011. Between 2000 and 2020, the CEEC intensity in the YRD region decreased from 3.44 tons/CNY 10,000 to 1.37 tons/CNY 10,000, an average annual decrease of 4.50%. It shows that the energy utilization rate in the YRD region is increasing year by year, and the dependence of economic development on energy consumption is constantly weakening. However, in 2019–2020, the carbon emission intensity increased significantly from 0.87 tons/CNY 10,000 to 1.37 tons/CNY 10,000, and the increase in carbon emissions in 2019–2020 was small. It can be speculated that the reason for the sharp rise in carbon emission intensity is that the YRD region was affected by the epidemic. Overall, CEECs in the YRD region increased rapidly in the early stage, slowed down and stabilized in the later stage and the carbon emission intensity gradually decreased. The structural improvement of energy was an important factor. That is to say, the proportion of clean energy, such as natural gas, gradually increased, and the proportion of traditional energy, such as gasoline and diesel, gradually decreased.

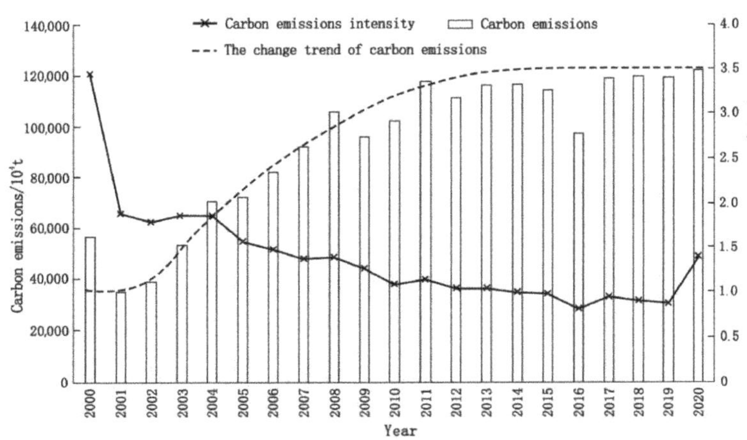

Figure 3. Status of energy carbon emissions in the YRD from 2000 to 2020.

5.2. Spatial and Temporal Distribution Characteristics of Total CEECs

As shown in Figure 4, there were significant spatio-temporal differences in CEECs among the cities in the YRD. Overall, the spatial distribution of carbon emissions was characterized by high levels in the northeast and low levels in the southwest. In 2000, Nanjing and Ningbo had the highest carbon emissions, and the type of CEEC was super heavy (>40 million tons). The CEECs of some cities in the southwestern and southern parts of the YRD region (Chizhou, Xuancheng, Jinhua, etc.) were relatively low, which is related to their low level of economic development and limited industrial technology. In 2005, the carbon emissions of all cities increased to varying degrees, and the number of cities whose carbon emissions were classified as heavy or above increased from two to seven. In particular, the CEECs of some cities in the middle of the YRD region (Nanjing, Wuxi, Suzhou, Taizhou, etc.) increased by more than 5000×10^4 t. The increase in total CEECs in the YRD during this period was related to the fact that cities vigorously developed traditional industries and pursued GDP growth too much, leading to extensive economic development. In 2010, except for in Huzhou, Ningbo, Tongling and Taizhou, carbon emissions decreased; the carbon emissions of other cities still increased to varying degrees, and the number of cities with super-heavy carbon emissions increased from six to seven. In 2015, the CEECs of Hangzhou, Huzhou, Shaoxing, Jinhua, Taizhou, Wuxi, Yangzhou, etc., dropped significantly. The CEECs of other cities began to decline as a whole, and the increase in CEECs narrowed significantly. Nanjing became the only place with an increase in CEECs of more than 5000×10^4 t at city level. This is because, from 2010 to 2015, measures such as low-carbon city construction, low-carbon policies and industrial upgrading in the YRD region were gradually implemented. In 2020, the carbon emissions of Hangzhou, Shaoxing, Jinhua, Ma'anshan, Tongling, Wuxi, Nantong, Taizhou and Shanghai will decrease. However, the overall CEEC increase in the YRD region was about 2053 $\times 10^4$ t, and showing a rebounding trend. So, from 2015 to 2020, the YRD region needs to pay more attention to reducing emissions of super-heavy cities with carbon emissions (especially in Anqing, Ma'anshan, Changzhou, Wuxi, Zhenjiang, Nanjing and Ningbo).

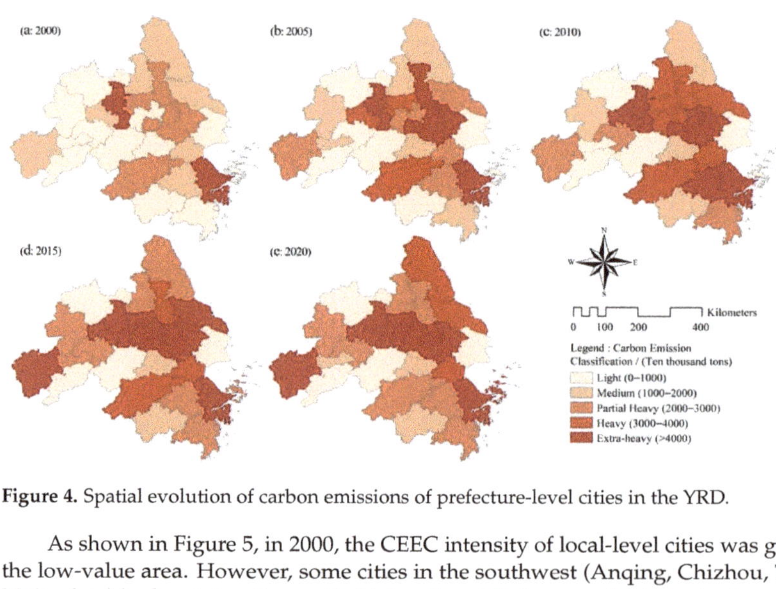

Figure 4. Spatial evolution of carbon emissions of prefecture-level cities in the YRD.

As shown in Figure 5, in 2000, the CEEC intensity of local-level cities was greatest in the low-value area. However, some cities in the southwest (Anqing, Chizhou, Tongling, Ma'anshan) had an intensity mostly higher than 6.5t, because the industry types of these cities mostly consisted of traditional industries, and their economic development depended on high-energy-consuming industries such as petrochemicals, building materials and steel. With time, the high-value areas of carbon emission intensity in the YRD region gradually decreased, and the CEEC intensity of Hefei, Chuzhou, Ningbo and other places dropped significantly, which indicates that the YRD region has taken measures such as industrial supply-side reform and production process improvement in recent years. This promotes the economy of the YRD to move in the direction of low carbonization and high quality. Although carbon emission intensity declined to a certain extent, the trend of a substantial increase in carbon increment and total carbon is not optimistic. It is still necessary to continue to reduce CEEC intensity (especially in Anqing and Ma'anshan) to further reduce the increment of CEECs and control the total amount of CEECs.

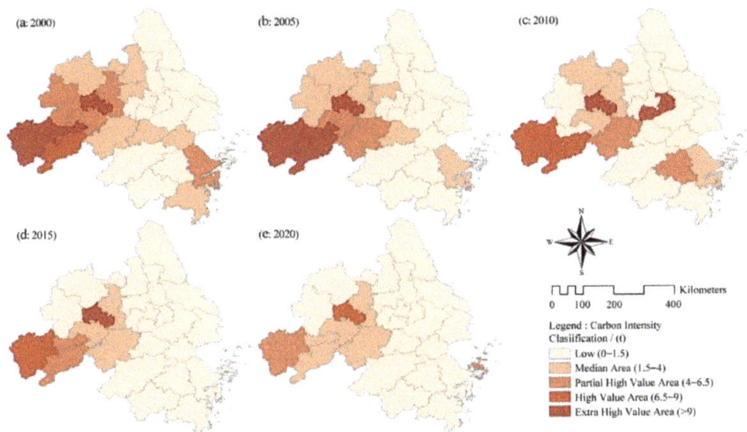

Figure 5. Spatial evolution of carbon intensity of prefecture-level cities in the YRD.

5.3. The Decoupling between CEECs and Economic Growth

According to the changing characteristics of CEECs in the cities in the YRD region, the changes in CEECs and the gross industrial output value of the cities in the YRD region from 2000 to 2020 were calculated. Moreover, through the Tapio model, the decoupling relationship of the four periods was obtained (Figure 6).

From 2000 to 2005, the economic growth and CEECs of most cities in the YRD region were in a state of negative decoupling growth, and the overall decoupling elasticity coefficient was high (Figures 6 and 7). For example, the decoupling elasticity of Yangzhou reached 4.425, which means that the decoupling state was poor. This shows that during this period of economic growth, CEECs were also increasing, but the growth rate of CEECs was much greater than that of economic growth.

From 2005 to 2010, the number of cities in a state of negative decoupling growth continued to increase among local-level cities in the YRD region, reaching 16. However, some cities began to show weakening decoupling and growth in connection, and the decoupling elasticity coefficient decreased significantly. For example, Huzhou changed from the negative decoupling growth state to the decoupling enhanced state, and the decoupling elasticity reached −16.169, which is the most ideal decoupling state. This is because, after the "Eleventh Five-Year Plan" (2001–2005), the YRD region responded to the national call to gradually improve the extensive economic growth model and continuously improve the efficiency of energy utilization. Among the cities, those in the decoupling stage were mainly located in Shanghai, central Zhejiang and central Jiangsu, and those in the negative decoupling stage were mainly located in Anhui, Jiangsu and northern Zhejiang.

From 2010 to 2015, which was China's "Twelfth Five-Year Plan", China introduced a large number of emission reduction measures and eliminated outdated production capacity. The total carbon emissions in the YRD region showed a stagnant trend (Figure 2), and many cities showed a state of increasing decoupling. For example, Wuxi actively developed zero-carbon technology, and its decoupling state changed from the negative decoupling growth state in 2000–2005 to the decoupling enhancement state. The decoupling elasticity reached −36.98, which is the most ideal decoupling state. Therefore, the economic growth of the YRD region and CEECs continued to show an increase in decoupling in the long run or a decoupling state alternating between increased decoupling and negative decoupling growth. Among the cities, those in the decoupling stage were mainly located in Shanghai, Shaoxing, Hangzhou, Taizhou, etc. (southern Zhejiang Province, central Zhejiang Province and northern Anhui Province), and those in the negative decoupling stage were mainly located in Wuhu, Anqing, Hefei, Nanjing, etc. (Anhui Province and eastern Jiangsu Province).

From 2015 to 2020, the decoupling index showed large fluctuations. The decoupling relationship between CEECs and economic development in many cities (e.g., Shanghai, Hangzhou, Wuxi, etc.) was further improved, and the decoupling elasticity index continued to decline. However, there were still cities with a high proportion of secondary industries (e.g., Zhenjiang, Anqing, Chizhou) where the decoupling relationship between CEECs and economic development deteriorated to varying degrees. At the same time, the types of decoupling state also increased from three to six, indicating that differences in the level of economic development and economic structure lead to differences in carbon emissions and their decoupling relationship with economic development in different cities in the province.

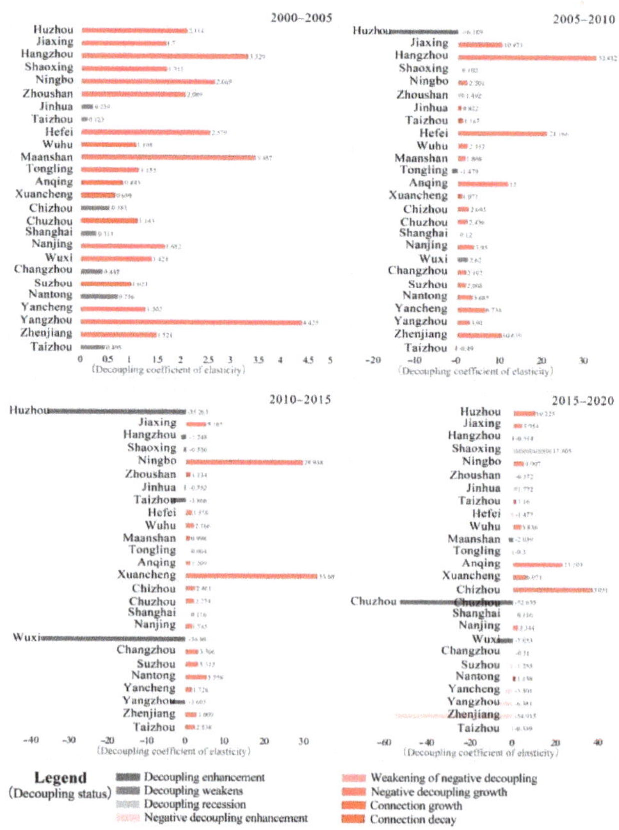

Figure 6. Decoupling of CEEC and industrial output value of prefecture-level cities in the YRD region from 2000 to 2020.

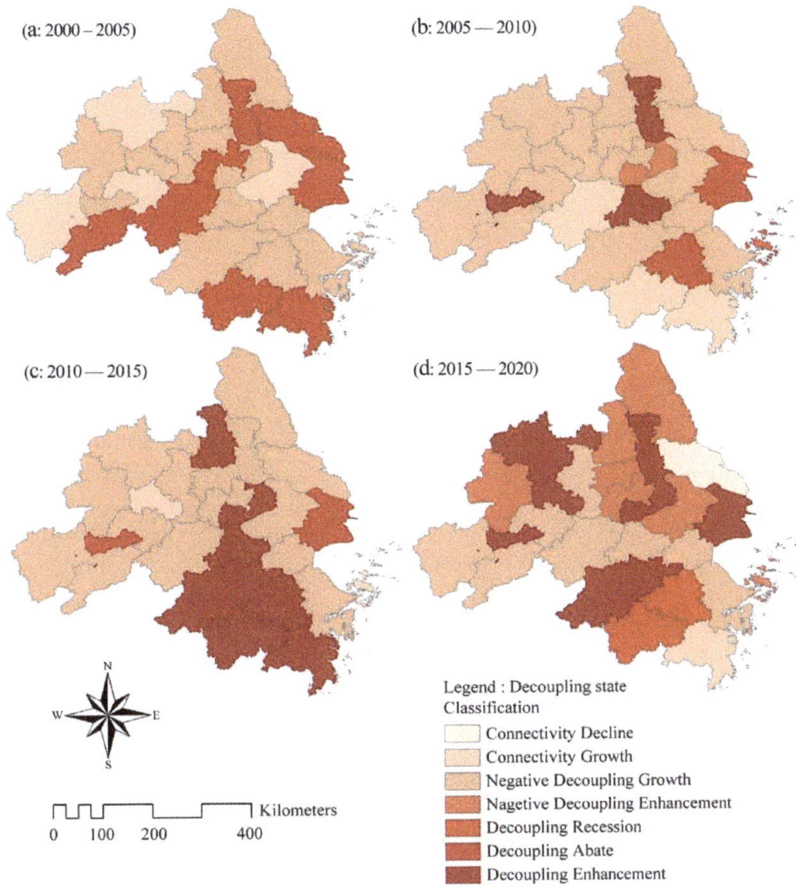

Figure 7. Spatial distribution characteristics of decoupling between CEEC and gross industrial out-put value of prefecture-level cities in the YRD.

To sum up, the YRD region has a relatively large proportion of the prefecture-level cities in the state of negative decoupling growth, which shows that an increase in carbon emissions accompanies the economic growth of the YRD region, and the increase in carbon emissions is greater than that of the economy. The rate may be related to the rapid accumulation of cities and towns and the rapid expansion of industries in the YRD. From 2015 to 2020, the number of prefecture-level cities in a state of negative decoupling growth will decrease thanks to the proposal of the "dual-carbon" target in the YRD region and the formulation of related carbon reduction policies. For example, Shanghai proposed to achieve carbon peaking in 2025, Jiangsu Province stated that it would be the first in the county to achieve carbon peaking and the YRD urban agglomeration proposed to achieve carbon peaking in the midterm of the "15th Five-Year Plan". In the future, the YRD region should further respond to the national call, actively develop the "national low-carbon city pilot", shut down high-consumption and high-polluting enterprises and carry out industrial upgrading. Through coordinated development, efforts will be made to reverse the relationship between economic growth and CEECs in more cities in the YRD region to decouple growth.

5.4. Decomposition Analysis of Influencing Factors of CEEC Decoupling

Equations (6)–(9) were used to decompose the decoupling elasticity of CEECs in each city in the YRD region to obtain the carbon emissions of four factors: energy structure (*DES*), energy intensity (*DEI*), unit industrial output (*DEL*) and urbanization rate (*DP*). The results are shown in Figure 8.

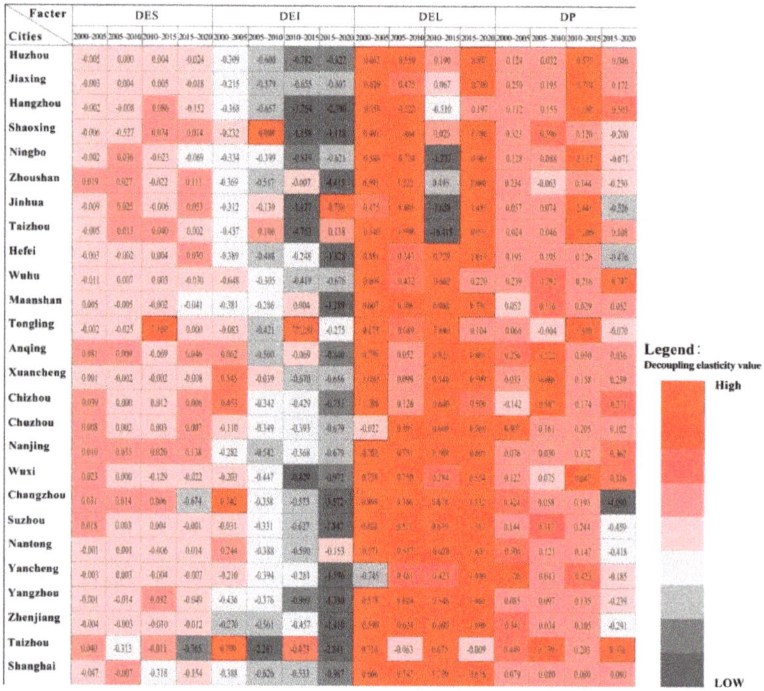

Figure 8. Decomposition of influencing factors of decoupling carbon emissions in prefecture-level cities in the YRD from 2000 to 2020.

Overall, from 2000 to 2020, the decoupling elasticity of CEECs for the four factors in the 26 cities was significantly similar (Figure 8). Among them, the decoupling elasticity of *DEL* value was the largest, which was positively correlated with energy carbon emissions as a whole, and its elasticity coefficient was 0.531. The second was the *DP* and *DES*; the impact of *DEI* was the smallest, and its elasticity coefficient was −0.621, decreasing yearly. It can be seen that the "dual-carbon strategy" in the YRD region needs to focus on coordinating the relationship between the quality of industrial production and operation and environmental benefits. In future energy consumption planning, it is necessary to gradually pay attention to the impact of *DP* and *DES* transformation.

5.4.1. The Impact of *DES* on Decoupling of CEECs

From 2000 to 2005, the *DES* of Zhoushan, Ma'anshan, Anqing, Xuancheng, Chizhou, Chuzhou, Nanjing, Wuxi, Changzhou and Suzhou had an impact on economic development. The effect of the carbon emission decoupling relationship was inhibition, and the *DES* of other prefecture-level cities promoted carbon emission decoupling. This shows that from 2000 to 2005, the *DES* of nearly half of the cities in the YRD region did not improve, and the traditional *DES* with high energy consumption was not conducive to local green development.

From 2005 to 2010, for Jiaxing, Ningbo, Jinhua, Taizhou, Wuhu, Anqing, Chuzhou, Nanjing, Changzhou, Suzhou, Nantong and Yancheng, the impact of the *DES* on the decoupling relationship between economic development and carbon emissions was inhibited, and the rest of the prefecture-level cities were promoted. Compared with 2000–2005, the number of cities in which the *DES* inhibited the decoupling of carbon emissions increased by half during this period. This shows that the urban *DES* in the YRD region was not significantly improved at this stage. The traditional *DES* with high energy consumption still inhibited the city's green development.

From 2010 to 2015, the *DES* of Huzhou, Jiaxing, Hangzhou, Shaoxing, Taizhou, Hefei, Wuhu, Tongling, Chizhou, Chuzhou, Nanjing, Changzhou, Suzhou and Yangzhou had an impact on economic development and carbon emissions. The effect of the decoupling relationship was inhibited, and the remaining prefecture-level cities were promoted. Among them, the absolute value of *DES* in Tongling was the largest, reaching 7.2. During this stage, the number of prefecture-level cities whose *DES* had an inhibitory effect on carbon emission decoupling increased significantly.

From 2015 to 2020, for Taizhou, Tongling, Anqing, Chizhou, Chuzhou, Nanjing, Changzhou, Suzhou, Yancheng, Yangzhou and Zhenjiang, the impact from the DES on the decoupling relationship was inhibition, and the rest of the prefecture-level cities were promoted.

In terms of the number of prefecture-level cities, nearly half of the prefecture-level cities in the YRD region (most of which were located in Anhui and Jiangsu provinces) had a *DES* that inhibited their decoupling of carbon emissions. Still, the absolute value of their *DES* was small. That is, the degree of influence of inhibition was small. From 2000 to 2020, the absolute value of *DES* in prefecture-level cities in the YRD was relatively small, reflecting that the *DES* had an excellent potential for accelerating low-carbon green development in the YRD region. The development of new energy and vigorous promotion of clean energy may become a major starting point for the green development of the region.

5.4.2. The Impact of *DEI* on Decoupling of CEECs

DEI refers to the ratio of energy consumption to economic output. From 2000 to 2005, except for Anqing, Xuancheng, Chizhou, Changzhou, Nantong and Taizhou, the impact of *DEI* on the decoupling relationship between economic development and CEECs was inhibited, and the remaining prefecture-level cities were all up for promotion. This shows that from 2000 to 2005, the influencing factor for the low-carbon development of most prefecture-level cities in the YRD region was not *DEI*. From 2005 to 2010, except for Shaoxing, Taizhou, Zhoushan and Wuxi, the impact of *DEI* on the decoupling relationship between economic development and CEECs was inhibited, and other prefecture-level cities were promoted. From 2010 to 2015, only Ma'anshan and Tongling had an inhibitory effect on the decoupling relationship between economic development and CEECs, while other prefecture-level cities promoted it. Among them, the absolute value of *DEI* in Tongling City was the largest, as high as 27.21. It can be seen that between 2005 and 2015, the industry in the YRD region was undergoing continuous transformation; energy consumption changed from low-end extensive to green and intensive and energy efficiency was relatively high. From 2015 to 2020, the *DEI* of Shaoxing, Zhoushan, Taizhou, Hefei, Suzhou, Nantong, Yancheng, Yangzhou and Zhenjiang were decoupled from economic development and CEECs. The influence of the relationship was inhibited, and the rest of the prefecture-level cities were promoted. Although the number of prefecture-level cities increased, the degree of inhibition of the *DES* decreased, and energy utilization began to develop into intensive and efficient development.

5.4.3. The Impact of *DEL* Value on CEEC Decoupling

From 2000 to 2005, only Chuzhou and Yancheng contributed to the decoupling relationship between economic development and CEEC, and the rest of the prefecture-level cities were inhibited. Among them, the absolute value of *DEL* in Chizhou was the largest, reaching 1.3. From 2005 to 2010, only Zhoushan, Wuxi and Taizhou's *DEL* value contributed to the decoupling relationship between economic development and carbon emissions. The prefecture-level cities were all suppressed, and the absolute value of *DEL* in Shaoxing was the largest, as high as 1.5. This shows that from 2000 to 2010, the industry in the YRD region was in an inefficient development period before the transformation; the unit output value was not high, and the industrial development model was not unbranded and clustered.

From 2015 to 2020, the impact of *DEL* value on the decoupling relationship was manifested in the promotion of prefecture-level cities, including Shaoxing, Zhoushan, Jinhua, Hefei, Changzhou, Suzhou, Nantong, Yancheng, Yangzhou, Zhenjiang and Taizhou. Among them, Changzhou had the largest absolute value of *DEL*, reaching 5.1. This shows that industry in the YRD was developing and transforming in the direction of intensification. Among the cities, the northeastern and southern cities of the YRD (mainly located in Jiangsu Province and Zhejiang Province) had a relatively rapid industrial transformation and development process, which had a relatively significant role in promoting low-carbon economic development.

5.4.4. The Impact of *DP* on CEEC Decoupling

The *DP* of most prefecture-level cities had an inhibitory effect on CEECs. From 2000 to 2020, only the *DP* of Chizhou, Tongling and Ningbo contributed to the decoupling relationship between economic development and CEECs, while the rest of the prefecture-level cities were inhibited. It shows that the rapid urbanization process had a certain inhibitory effect on the low-carbon sustainable development of the YRD urban agglomeration. Behind the increasing *DP* was the extensive urbanization development in the form of extension and spread. Transforming the surging urban population into the driving force for the low-carbon development of the city depended on the industrial upgrading and transformation of the city. As one of the regions with the most active economic development in China, the YRD urban agglomeration will be one of the main areas of low-carbon and green development in the future. The focus of future development will be the elimination, transformation, extension and transfer of the original development methods of some extensive cities in order to promote the transformation and upgrading of their own industries.

5.4.5. Drivers of CEECs by Stage

Based on the calculation results of the decoupling elastic coefficients of the four driving factors, the decomposition results of carbon emissions were used as the basis for k-means clustering. The 26 prefecture-level cities were divided into six groups to explore the clustering characteristics of the decoupling elastic coefficients of each city group. The clustering results are shown in Figure 9 and Table 3.

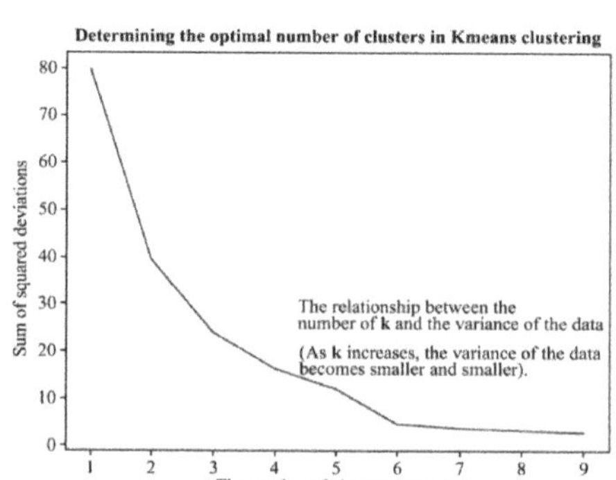

Figure 9. Optimal value of elbow method.

Table 3. Analysis of the main driving factors of each group of urban areas.

Groups	Period 2000–2005 Main Factor	Average Uncoupling Elasticity	2005–2010 Main Factor	Average Uncoupling Elasticity	2010–2015 Main Factor	Average Uncoupling Elasticity	2015–2020 Main Factor	Average Uncoupling Elasticity
Group A	*DEL*	0.698	*DEL*	0.614	*DEI*	−0.601	*DEL*	2.289
Group B	*DEL*	0.576	*DEL*	0.718	*DP*	6.532	*DEL*	0.697
Group C	*DEL*	0.647	*DEL*	0.741	*DEI*	−0.521	*DEI*	−1.398
Group D	*DEL*	0.667	*DEI*	−0.442	*DEL*	0.56	*DEI*	−0.719
Group E	*DP*	2.243	*DEI*	4.115	*DEI*	−1.687	*DEI*	−4.542
Group F	*DEL*	3.933	*DEI*	−3.824	*DEI*	23.485	*DEI*	−7.164

The dominant driving factors and driving directions of the carbon emissions of the six groups A–F can be discussed for different periods. The results for the first five years (2000–2005) are shown in Figure 10. During this period, for each group, *DEL* had the largest impact on carbon emissions, followed by *DP* and *DEI*. Among them, *DEL* value and *DP* promoted the growth of carbon emissions, while *DEI* inhibited the growth of carbon emissions. The effect of *DES* was not obvious. It is worth noting that in Group C, the impact of *DEI* on carbon emissions was not negligible and almost corresponded to the impact of *DEL*. In general, because the YRD region paid attention to economic development during this period, secondary industry accounted for a large proportion of the industrial structure, and the energy needed was still dominated by coal. However, the energy utilization efficiency was low, so *DEI* was increasing. The impact of carbon emissions was obvious.

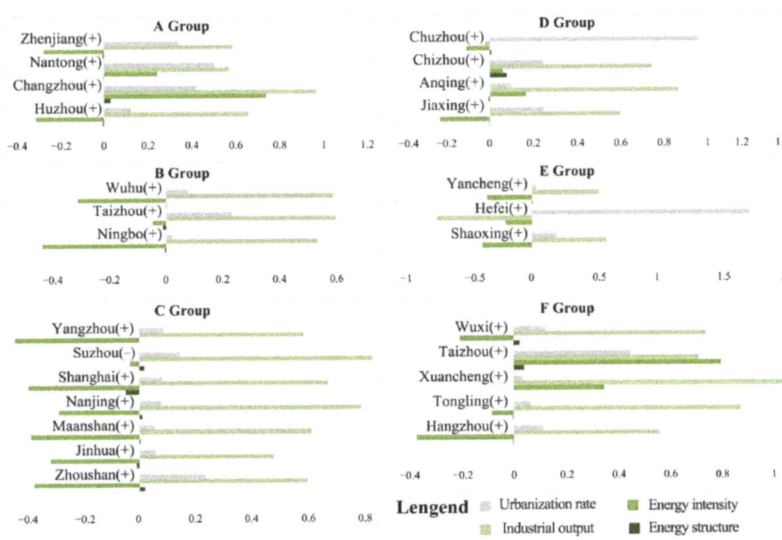

Figure 10. Cluster analysis results in the first period (2000–2005). Note: The symbol (+) after the city indicates an increase in carbon emissions and the symbol (−) indicates a decrease in carbon emissions, the same as in Figures 11–13.

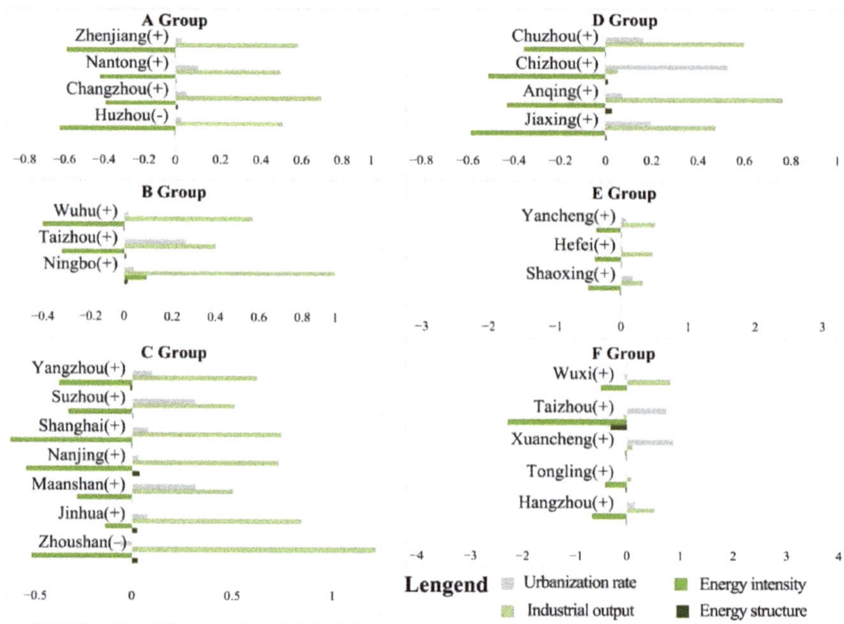

Figure 11. Cluster analysis results in the second five years (2005–2010).

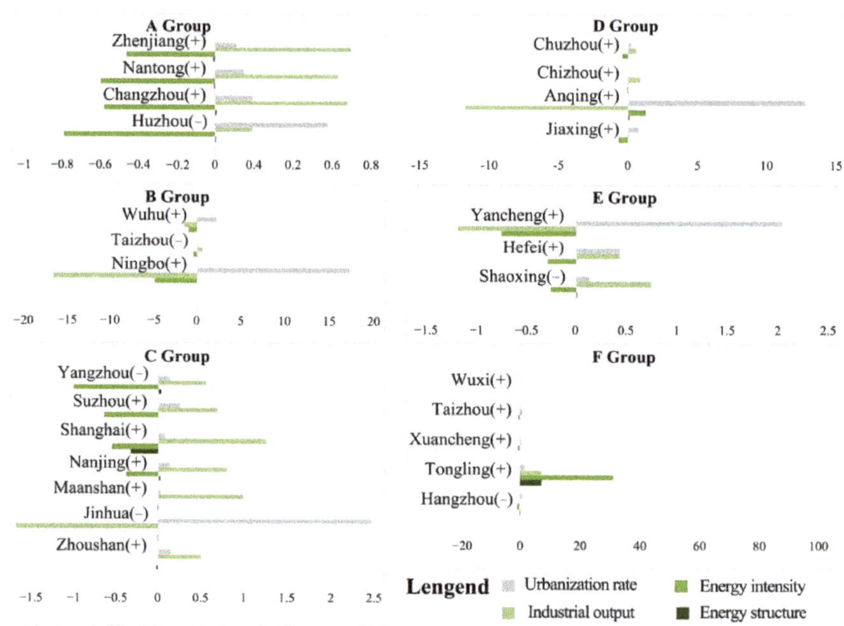

Figure 12. Cluster analysis results in the third five years (2010–2015).

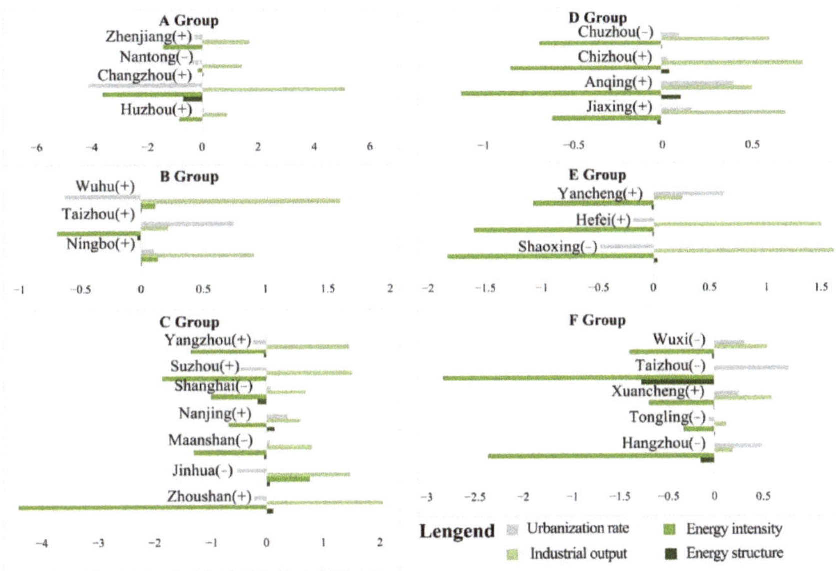

Figure 13. Cluster analysis results in the fourth period (2015–2020).

The decomposition results of the second five years (2005–2010) are shown in Figure 11. The main influencing factors in this period were *DEI* and *DEL*. The role of *DEL* was still to promote the growth of carbon emissions, and the role of *DEI* was to suppress the growth of carbon emissions. The obvious difference between this period and the previous period is that the impact of *DEI* greatly increased. In contrast, the impact of *DP* gradually decreased in some groups, especially in groups A, C and D. This characteristic was particularly obvious, which indicates an increased emphasis on energy consumption and environmental protection in the region, and the *DES* of some cities was adjusted during this period. In general, the role of restraining the development of carbon emissions in this period was significantly stronger than that in the first five years, the industrial structure advanced rapidly and the economy developed rapidly. However, the impact of *DEI* on CEECs still needs attention.

The decomposition results of the third five years are shown in Figure 12. The obvious difference between this period and the previous period is that the effect of suppressing carbon emissions became gradually stronger than that of promotion. *DEI* turned into a major factor contributing to the growth of carbon emissions, especially in groups A, C and E. In addition, the influence of *DP* gradually increased; in groups A and C, *DP* and *DEL* played an essential role in promoting carbon emissions. The biggest feature of Group F was that *DEI* and *DES* had a prominent impact on CEECs, while other influencing factors were not obvious.

The decomposition results of the fourth five years are shown in Figure 13. The role of *DEI* in carbon emissions was more critical during this period. The inhibitory effect of this period was similar to that of the previous period and was obviously stronger than the promotion effect. However, compared with the previous groups C, D and E, the intensity of the impact of *DEL* value on promoting carbon emissions increased. In contrast, the impact of *DP* decreased significantly, indicating that the industrial structure of the region still needs to be given attention. It is worth noting that the influence of *DES* increased during this period, especially in group F, which produced a significant inhibitory effect. That shows that the current economic consumption level and population structure changes in some cities had a higher multi-factor impact on carbon emissions than the single impact of population size changes. In general, the total CEECs in some regions declined during this period, but the per capita GDP increased yearly, and the economy developed healthily. However, it is still necessary to pay attention to the efficiency of energy utilization and to strengthen the upgrading of industrial structure and industrial layout to reduce CEECs.

Taken together, the carbon emission patterns and their underlying drivers vary by city group (Table 3). From 2000 to 2015, the correlation between carbon emissions and driving factors was in descending order of *DEL* value, *DP* and *DEI*, and the impact of *DES* could be ignored. The promotion effect of the overall impact factor was greater than the overall inhibitory effect, leading to an overall upward trend in energy carbon emissions in the YRD region in the early stage. However, from 2015 to 2020, the ranking of results was *DEI*, *DEL* value, *DP* and *DES*. The impact of the *DP* dropped significantly, while the impact of the *DES* gradually emerged. The overall inhibitory effect gradually increased and was significantly stronger than the promotion effect, attributed to the optimization of the *DES* and the positive impact of implementing low-carbon policies in recent years.

5.4.6. A Staged Analysis of the Composite Effects of CEEC Drivers

The composite effect analysis of the driving factors of CEECs in each period was conducive to revealing the internal mechanism of the temporal and spatial changes of CEECs in cities in the YRD from 2000 to 2020. We numerically numbered the CEECs' features across the four epochs and reclassified and visualized representations in Arcmap (Table 4 and Figure 14).

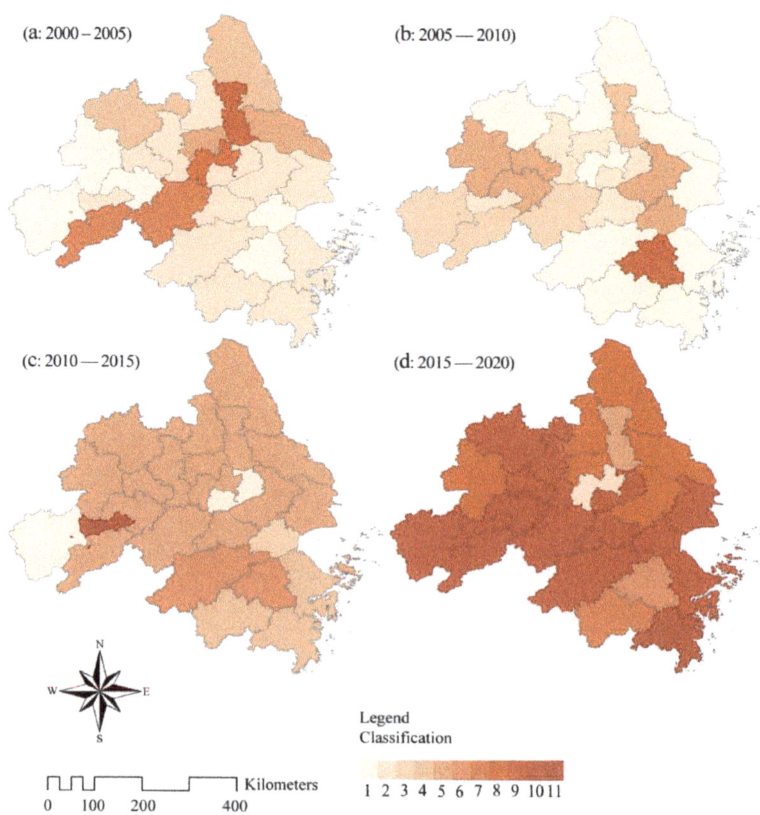

Figure 14. The characteristics of carbon emission composite impact factors in different periods.

Table 4. Classification of carbon emission characteristics.

Number	Carbon Emission Characteristics
1	*DEL* is the main influencing factor, and the *DP* is outstanding
2	*DEL* is the main influencing factor, and other factors have little influence
3	*DP* is the main influencing factor, and the *DEL* is outstanding
4	*DP* is the main influencing factors, and other factors have little influence
5	*DEL* and *DP* are the main influencing factors, and other factors have little influence
6	*DP* and *DES* are the main influencing factors, and the *DEL* is outstanding
7	*DEL* and *DEI* are the main influencing factors, and other factors have little influence
8	*DEL* and *DES* are the main influencing factors, and the *DP* is outstanding
9	*DEL*, *DP* and *DEI* are the main influencing factors
10	*DEL*, *DP* and *DES* are the main influencing factors
11	*DEL*, *DP*, *DEI* and *DES* are the main influencing factors

In the early stage, cities over-pursued economic development, and the drastic development model brought huge energy consumption. The per *DEL* value became the main and single driving factor in some economically developed cities (mainly concentrated in the more economically developed Zhejiang Province and southern Jiangsu Province and Shanghai: Yangzhou, Nanjing, Wuxi, Suzhou, Shanghai, Ningbo, Huzhou, Hangzhou, Taizhou, and Jinhua). Economically underdeveloped areas (Yancheng, Chuzhou, Tongling, Anqing) were mainly affected by the single factor of *DP*, and other factors were not obvious. At this stage, due to the singleness of the economic development model, about 54% of the cities were affected by the single factor of unit gross industrial output value or *DP*, and only Taizhou and Changzhou were affected by the three factors of unit gross industrial output value, *DP* and *DEI*. In the second and third stages, the influence of the *DP* gradually emerged, and the number of cities affected by the combined influence of the *DEL* value and the *DP* continued to increase. In particular, in the third stage, 17 cities (Hefei, Wuhu, Zhoushan, Yancheng, Chuzhou, Nantong, Zhenjiang, Chizhou, Xuancheng, Taizhou, Changzhou, Yangzhou, Ma'anshan, Nanjing, Suzhou, Shanghai and Huzhou) were affected by the unit industrial total. The dual factors influenced output value and *DP*. At the same time, some cities with prominent tertiary industries (Shaoxing, Hangzhou) were also gradually affected by the impact of the *DES*. At this stage, the number of cities whose carbon emissions were affected by two or three factors increased yearly. In the fourth stage, 15 cities (Anqing, Wuhu, Jiaxing, Zhoushan, Chuzhou, Chizhou, Xuancheng, Ma'anshan, Nanjing, Wuxi, Shanghai, Tongling, Ningbo, Huzhou and Hangzhou) were affected by the unit gross industrial output value, *DP* and the compound influence of the three factors of *DES*, and the four factors of Taizhou were outstanding. At this stage, carbon emissions needed to be controlled by focusing on *DEI* and *DES*.

In general, the carbon emissions of the more developed regions in the early stage were greatly affected by the unit gross industrial output value, and it was necessary to focus on controlling carbon emissions by monitoring the economic development trend. Some economically underdeveloped areas were mainly affected by the *DP*. It should be noted that the *DP* and the huge population base determine that the total carbon emissions of such cities will continue to rise for a long time. In the medium term, affected by the adjustment of economic structure, the cities gradually changed from the single-factor influence in the early stage to the influence of dual influence factors. At this stage, optimizing the industrial structure to control carbon emissions was necessary. In the later stage, the upgrading and transfer of related industries, the transformation of energy consumption forms and the upgrading of industrial layout structure brought diversified development to the city. The amount of carbon emissions tended to be stable, and the impact of a single factor on carbon emissions gradually turned into a composite impact of multiple factors. It can be seen that diversified development of the economy can effectively alleviate the pressure of carbon emissions in many aspects. Therefore, in the process of future economic development, we should pay more attention to the layout of various energy sources and industries, promote the utilization of new energy and renewable energy and promote the development of stagnant industries and the impact of multiple factors, especially *DEI* and *DES*, on carbon emissions so as to explore a reasonable and environmentally friendly carbon emission layout.

6. Conclusions and Policy Implications

By 2011, after experiencing a sharp increase in CEECs, the YRD region basically reached a CEECs carbon peak and entered a plateau. Facing the demand for medium and high economic growth in the YRD region in the near future, it will be difficult for the energy consumption and carbon emissions in the YRD region to decrease further. CEECs are expected to decline. Understanding the drivers of CEEC intensity in the 26 prefecture-level cities in the YRD is important for policy making, and decomposition analysis is a useful method for addressing quantitative changes in predetermined benefit factors. This study applied LMDI technology and used an improved Kaya identity to explore the driving factors of CEEC intensity in 26 prefecture-level cities in the YRD from 2000 to 2020. The main conclusions reached are as follows:

MC1: CEECs in the 26 prefecture-level cities in the YRD showed rapid growth in the early stage in the past 20 years, slowed down and stabilized in the later stage and fluctuated during 2009–2011 and 2012–2013. The overall CEEC intensity in the YRD region generally declined. However, there is significant room for reducing CEEC in the YRD. The CEECs in prefecture-level cities showed a more concentrated trend in the central region of the YRD and a trend that increased year by year. The CEEC intensity of the 26 prefecture-level cities was greatest in the low-value area, and the prefecture-level cities with higher intensity were mainly concentrated in the western part of the YRD. The overall carbon emission intensity gradually developed towards the low-value area.

MC2: Among the four driving factors selected and analyzed, the unit industrial output value had the greatest impact on the decoupling relationship between the economic development of prefecture-level cities in the YRD and CEECs, followed by energy intensity. The urbanization rate was the most widely influential factor in the decoupling relationship between economic development and CEECs in the 26 prefecture-level cities. At the same time, the decoupling relationship between CEECs and economic growth in various cities in the YRD region has significant spatial and temporal differences in geographic locations and development stages. In general, the growth rate of CEECs in the YRD region was greater than that of economic growth, which may be related to the rapid expansion of industries in the YRD region. The significant positive interaction between advanced energy structure and economic growth is gradually becoming more prominent.

MC3: The driving factors of the decoupling in the YRD region generally showed a trend changing from unit of gross industrial output value to urbanization rate and energy intensity. This shows that the development and utilization of energy technology and the transformation and upgrading of low-end industries are imminent. The cluster analysis results of the elastic value of driving factors showed that, from 2000 to 2010, the dominant factors affecting the CEECs of the 26 prefecture-level cities were mainly industrial output value and urbanization rate. Overall, the facilitation effect was greater than the inhibitory effect. After 2010, the role of various factors in restraining carbon emissions in the YRD region increased significantly, and the role of energy intensity is still crucial. From 2015 to 2020, the role of energy intensity in CEEC became the dominant factor, and the inhibitory effect gradually became stronger than the promotion effect. In addition, some cities had a relatively good level of urbanization, and the influence of energy structure increased during this period. However, there is still a lot of room for improvement in new energy utilization and low-carbon technology research and development.

MC4: The results of the phase analysis of the composite driving factors of carbon emissions showed that the urban carbon emissions in the YRD region from 2000 to 2010 were mainly affected by a single factor: the economic development area was affected by the industrial output value; the economically underdeveloped area was affected by the urbanization rate impact; carbon emissions showed an upward trend year by year. After

2010, carbon emissions gradually turned into the compound influence of multiple factors, and carbon emissions tended to stabilize. Therefore, in the process of future economic development, encouraging economic diversification, energy structure transformation and model upgrading can effectively alleviate the pressure of carbon emissions in many ways.

According to the discussion above, the YRD region should determine the path to achieve CEEC reduction goals based on the conditions and advantages of the 26 prefecture-level cities and the leading factors driving CEECs. Below are three main policy recommendations based on the above analysis.

PR1: Prefecture-level cities play an important role in China's administrative system. From the above analysis and the development of the literature, the regulation of provincial governance units urgently needs to be on the prefecture-level-city level. When formulating planning policies, relevant managers should base them on the prefecture-level-city level or even smaller-scale administrative units, thoroughly consider the differences in governance objects and formulate policies for differentiated carbon emissions or to achieve carbon neutrality goals.

PR2: Local-level cities should formulate specific low-carbon development policies based on the contribution rate of major impact factors to $CEEC_S$. In the future, the economy of the YRD region will continue to maintain a rapid growth rate. It is not realistic to reduce carbon emissions by reducing the speed of economic development, and prefecture levels should be able to adopt policies tailored to local conditions to reduce CEECs. For pollution-intensive industrial cities, it is necessary to take the lead in promoting the upgrade of the industrial structure. For prefecture-level cities with advanced technology, it is necessary to increase ecological innovation further. Effective carbon emissions trading markets and energy trading markets need to be established. Compared with the terminal processing of carbon emissions trading, energy trading based on source control is more in line with national requirements. For industries characterized by high energy consumption, high pollution, and high carbon emissions, enterprises need to formulate strict development plans and green models suitable for the three main industries. Enterprises should be encouraged to improve energy utilization efficiency and develop a circular economy with low extraction, high use and low emissions. At the same time, it is necessary to increase capital participation in green investment and promote the use of clean energy, such as hydro, wind, solar and many other new energy, renewable electricity and hybrid energy systems. Cities should implement differentiated carbon emission reduction measures. For high-carbon cities, it is necessary to focus on monitoring high-carbon industries and high-carbon enterprises, guide and encourage enterprises to conduct low-carbon operations and reduce carbon emission costs by participating in urban carbon trading. A low-carbon economy is not about no economy but about developing from a bigger economy to a better economy.

PR3: Local-level cities should use the development of new technologies to establish a low-carbon integrated interconnection and mutual assistance strategy network and platform. The YRD is an energy-deficient area, and, as a whole, it is necessary to speed up the research and development of clean energy. At the same time, in the process of industrial structure optimization and new urbanization, attention should be paid to promoting low-carbon industries and regenerative agricultural economic development models. To deal with the dual-carbon goal, we must establish the concept of urban and rural integration, equally treat and guide citizens to establish a zero-carbon life concept, accelerate the promotion of low-carbon lifestyles through modern media network platforms such as WeChat, Douyin and Kuaishou and create a low-carbon consumption atmosphere. Through the participation of the whole population, we will actively build a zero-waste and low-carbon city. We must vigorously develop the digital economy and platform economy, transfer high-energy-consuming industries in terms of industrial structure transformation, optimize the scientific research resources and foreign trade advantages of the YRD, rely on the scientific research resources and foreign trade advantages of the YRD, focus on developing the digital economy and platform economy and reduce dependence on energy consumption.

Our research deeply revealed the dominant factors, decoupling relationships and spatio-temporal variation law of carbon emissions in 26 prefecture-level cities in the YRD from 2000 to 2020. However, the research in this paper still had some limitations. As mentioned above, the influencing factors we chose mainly focused on the industrial development and urbanization of prefecture-level cities, which may not have been comprehensive, and the impact of technological innovation and policies was not quantitatively analyzed. In addition, due to the limitations of data acquisition, it is currently impossible to analyze the spatio-temporal differences within each prefecture-level city in more detail from the perspective of smaller, county-level cities.

Author Contributions: Conceptualization, W.W., H.C. and L.W.; methodology, W.W., H.C., X.L. and L.W.; formal analysis, W.W., H.C., X.L., D.M. and S.W.; writing—original draft preparation, W.W., H.C., X.L., D.M. and S.W.; writing—review and editing, W.W., H.C., X.L. and D.M.; software, S.W. and H.C.; supervision, W.W. and L.W. All authors have read and agreed to the published version of the manuscript.

Funding: This research was funded by The National Natural Science Foundation of China, grant number 51578482.

Institutional Review Board Statement: Not applicable.

Informed Consent Statement: Not applicable.

Data Availability Statement: The data are not publicly available due to privacy. The data presented in this paper are available on request from the corresponding author.

Conflicts of Interest: The authors declare no conflict of interest.

References

1. Li, L.C.; Chen, W.; Guo, K.M.; Yue, F.; Tang, Y. Policy analysis of carbon neutrality strategic actions in developed countries from the perspective of policy tools and its enlightenment to my country. *J. Intell.* **2021**, *40*, 63–71.
2. Yu, A.; Lin, X.; Zhang, Y.; Peng, L.H. Analysis of driving factors and allocation of carbon emission allowance in China. *Sci. Total Environ.* **2019**, *673*, 74–82. [CrossRef] [PubMed]
3. Yang, S.D.; Yang, X.; Wu, X.L.; Wu, Y.F.; Zhou, J. The impact of environmental regulation on the spatial and temporal differences of regional carbon emissions: An empirical analysis based on 32 prefecture-level cities in the three northeastern provinces. *J. Environ. Sci.* **2021**, *41*, 2029–2038.
4. Liu, Z.B.; Chen, L. Demonstration Value and Motivation Mechanism of Regional Integration Development in the Yangtze River Delta Region. *Reform* **2018**, *12*, 65–71.
5. Zhang, W.S.; Xu, Y.; Wang, C.; Streets, G. Assessment of the driving factors of CO_2 mitigation costs of household biogas systems in China: A LMDI decomposition with cost analysis model. *Renew. Energy* **2022**, *181*, 978–989. [CrossRef]
6. Zhu, X.H.; Zou, J.W.; Chao, F. Analysis of industrial energy-related CO_2 emissions and the reduction potential of cities in the Yangtze River Delta region. *J. Clean Prod.* **2017**, *168*, 791–802. [CrossRef]
7. Meng, H.; Huang, X.J.; Yang, H.; Chen, Z.G.; Yang, J.; Zhou, Y.; Li, J.B. The influence of local officials' promotion incentives on carbon emission in Yangtze River Delta, China. *J. Clean Prod.* **2019**, *213*, 1337–1345. [CrossRef]
8. Fan, F.Y.; Lei, Y.L. Responsive relationship between energy-related carbon dioxide emissions from the transportation sector and economic growth in Beijing Based on decoupling theory. *Int. J. Sustain. Transp.* **2017**, *11*, 764–775. [CrossRef]
9. Wang, L.L.; Li, Y.; He, W.W. The energy footprint of China's textile industry: Perspectives from de-coupling and decomposition analysis. *Energies* **2017**, *10*, 1461. [CrossRef]
10. Park, S.H.; Dissmann, B.; Nam, K.Y. A cross-country decomposition analysis of manufacturing energy-consumption. *Energy* **1993**, *18*, 843–858. [CrossRef]
11. Wang, J.; Hu, M.; Rodrigues, J.F.D. The evolution and driving forces of industrial aggregate energy intensity in China: An extended decomposition analysis. *Appl. Energy* **2018**, *228*, 2195–2206. [CrossRef]
12. Huo, T.F.; Ma, Y.L.; Yu, T.; Cai, W.G.; Liu, B.S.; Ren, H. Decoupling and decomposition analysis of residential building carbon emissions from residential income: Evidence from the provincial level in China. *Environ. Impact Assess. Rev.* **2021**, *86*, 106487. [CrossRef]
13. Ma, X.; Wang, C.; Dong, B.; Gu, G.; Chen, R.; Li, Y.; Zou, H.; Zhang, W.; Li, Q. Carbon emissions from energy consumption in China: Its measurement and driving factors. *Sci. Total Environ.* **2018**, *648*, 1411–1420. [CrossRef]
14. Xu, S.C.; He, Z.X.; Long, R.Y.; Chen, H. Factors that influence carbon emissions due to energy consumption based on different stages and sectors in China. *J. Clean. Prod.* **2016**, *115*, 139–148. [CrossRef]

15. Yang, J.; Cai, W.; Ma, M.; Li, L.; Liu, C.; Ma, X.; Li, L.; Chen, X. Driving forces of China's CO_2 emissions from energy consumption based on Kaya-LMDI methods. *Sci. Total Environ.* **2019**, *711*, 134569. [CrossRef]
16. Liu, D.; Cheng, R.; Li, X.; Zhao, M.M. On the driving factors of China's provincial carbon emission from the view of periods and groups. *Environ. Sci. Pollut. Res.* **2021**, *28*, 51971–51988. [CrossRef]
17. Grossman, G.M.; Krueger, A.B. *Environmental Impacts of a North American Free Trade Agreement (No. w3914)*; National Bureau of Economic Research: Cambridge, MA, USA, 1991.
18. OECD. *Indicators to Measure Decoupling of Environmental Pressures from Economic Growth*; OECD: Paris, France, 2002.
19. Zhang, Z.X. Decoupling China's carbon emissions increase from economic growth: An economic analysis and policy implications. *World Dev.* **2000**, *28*, 739–752. [CrossRef]
20. De Freitas, L.C.; Kaneko, S. Decomposing the decoupling of CO_2 emissions and economic growth in Brazil. *Ecol. Econ.* **2011**, *70*, 1459–1469. [CrossRef]
21. Tapio, P. Towards a theory of decoupling: Degrees of decoupling in the EU and the case of road traffic in Finland between 1970 and 2001. *Transp. Policy* **2005**, *12*, 137–151. [CrossRef]
22. Zhang, M.; Song, Y.; Su, B.; Sun, X.M. Decomposing the decoupling indicator between the economic growth and energy consumption in China. *Energy Effic.* **2015**, *8*, 1231–1239. [CrossRef]
23. Chang, K.L.; Du, Z.F.; Chen, G.J.; Zhang, Y.X.; Sui, L.L. Panel estimation for the impact factors on carbon dioxide emissions: A new regional classification perspective in China. *J. Clean Prod.* **2021**, *279*, 12363. [CrossRef]
24. Song, M.L.; Guo, X.; Wu, K.Y.; Wang, G.X. Driving effect analysis of energy-consumption carbon emissions in the Yangtze River Delta region. *J. Clean. Prod.* **2015**, *103*, 620–628. [CrossRef]
25. Gong, L.; Tu, H.Z.; Gong, C. Research on the influencing factors of energy consumption carbon emissions based on STIRPAT model: Taking the Yangtze River Delta as an example. *Ind. Technol. Econ.* **2018**, *37*, 95–102.
26. Ma, N.; Gao, X.; Tian, M.; Wang, C.; Zhang, Y.; Gao, P.P. Decoupling the Relationships between Carbon Footprint and Economic Growth within an Urban Agglomeration-A Case Study of the Yangtze River Delta in China. *Land* **2021**, *10*, 923. [CrossRef]
27. Shen, W.; Liang, H.; Dong, L.; Ren, J.; Wang, G. Synergistic CO_2 reduction effects in Chinese urban agglomerations: Perspectives from social network analysis. *Sci. Total Environ.* **2021**, *798*, 149352. [CrossRef]
28. Quan, C.G.; Cheng, X.J.; Yu, S.S.; Ye, X. Analysis on the influencing factors of carbon emission in China's logistics industry based on LMDI method. *Sci. Total Environ.* **2020**, *734*, 138473. [CrossRef]
29. Wang, M.; Feng, X.Z.; An, Q.; Zhuo, Y.; Zhao, M.X.; Du, X.L.; Wang, P. Research on Green and Low-Carbon Development Strategy of Qinghai Province Based on Decoupling Index and LMDI. *Prog. Clim. Change Res.* **2021**, *17*, 598–607.
30. Zhang, C.J.; Zhao, Y.; Shi, C.F.; Chiu, Y.H. Can China achieve its water use peaking in 2030? A scenario analysis based on LMDI and Monte Carlo method. *J. Clean. Prod.* **2021**, *278*, 123214. [CrossRef]
31. Gu, S.; Fu, B.T.; Thriveni, T.; Fujita, T.; Ahn, J. Coupled LMDI and system dynamics model for estimating urban CO_2 emission mitigation potential in Shanghai, China. *J. Clean. Prod.* **2019**, *240*, 118034. [CrossRef]
32. Guo, Y.; Cao, X.Z.; Wei, W.D.; Zeng, G. The impact of regional integration in the Yangtze River Delta on urban carbon emissions. *Geogr. Res.* **2022**, *41*, 181–192.
33. Ang, B.W. The LMDI Approach to Decomposition Analysis: A Practice Guide. *Energy Policy* **2005**, *33*, 867–871. [CrossRef]
34. Kaya, Y. Impact of Carbon Dioxide Emission on GNP Growth: Interpretation of Proposed Scenarios. Paris: Presentation to the Energy and Industry Subgroup, Response Strategies Working Group, IPCC. 1990. X'Pert Stress PW3208, Software for Residual Stress Analysis, PANalytical, The Netherlands. Available online: www.panalytical.com (accessed on 31 March 2022).

International Journal of
Environmental Research and Public Health

MDPI

Article

Forecasting Carbon Dioxide Price Using a Time-Varying High-Order Moment Hybrid Model of NAGARCHSK and Gated Recurrent Unit Network

Po Yun [1,*], Chen Zhang [2], Yaqi Wu [3] and Yu Yang [4]

1 School of Economics and Management, Hefei University, Hefei 230601, China
2 School of Management, Hefei University of Technology, Hefei 230601, China; sm.zhangchen@hfut.edu.cn
3 School of Economics, North Minzu University, Yinchuan 750021, China; yq111099@mail.hfut.edu.cn
4 School of Economics and Management, Anhui Jianzhu University, Hefei 230601, China; yangyu@ahjzu.edu.cn
* Correspondence: yunpo2010@mail.hfut.edu.cn; Tel.: +86-18155183561

Abstract: The carbon market is recognized as the most effective means for reducing global carbon dioxide emissions. Effective carbon price forecasting can help the carbon market to solve environmental problems at a lower economic cost. However, the existing studies focus on the carbon premium explanation from the perspective of return and volatility spillover under the framework of the mean-variance low-order moment. Specifically, the time-varying, high-order moment shock of market asymmetry and extreme policies on carbon price have been ignored. The innovation of this paper is constructing a new hybrid model, NAGARCHSK-GRU, that is consistent with the special characteristics of the carbon market. In the proposed model, the NAGARCHSK model is designed to extract the time-varying, high-order moment parameter characteristics of carbon price, and the multilayer GRU model is used to train the obtained time-varying parameter and improve the forecasting accuracy. The results conclude that the NAGARCHSK-GRU model has better accuracy and robustness for forecasting carbon price. Moreover, the long-term forecasting performance has been proved. This conclusion proves the rationality of incorporating the time-varying impact of asymmetric information and extreme factors into the forecasting model, and contributes to a powerful reference for investors to formulate investment strategies and assist a reduction in carbon emissions.

Keywords: carbon price forecasting; time-varying; high-order moment; NAGARCHSK; gate recurrent unit network

Citation: Yun, P.; Zhang, C.; Wu, Y.; Yang, Y. Forecasting Carbon Dioxide Price Using a Time-Varying High-Order Moment Hybrid Model of NAGARCHSK and Gated Recurrent Unit Network. *Int. J. Environ. Res. Public Health* **2022**, *19*, 899. https://doi.org/10.3390/ijerph19020899

Academic Editors: Taoyuan Wei and Qin Zhu

Received: 14 December 2021
Accepted: 12 January 2022
Published: 14 January 2022

1. Introduction

Dramatic increase in greenhouse gas emissions directly leads to the aggravation of negative environmental externality. The emission of pollutants such as carbon dioxide is a serious threat to human health, and it is unacceptable that the pollution of this harmful gas will continue in the foreseeable future. According to the report of the Word Bank, the global carbon emissions in 2018 were 2.1 times the 1960 level, the per capita emissions increased by nearly 67%, and the per capita energy consumption increased by nearly 60% (data from author's calculation based on the Wind database). Economic growth, CO_2 emissions and energy consumption are complementary [1,2]. Based on the scientific report released by the National Oceanic and Atmospheric Administration (NOAA) of the United States in 2020, the global CO_2 concentration was 280 ppm at the beginning of the first industrial revolution, that is, the CO_2 quality accounted for 2.8% of the global atmospheric quality. In 2017, this data increased to 400 ppm, and in 2019, it increased to 415 ppm. The continuous growth of carbon dioxide emissions not only leads to the rise of global temperature, triggers sea-level rise, aggravates glacier melting and other severe environmental problems, but also threatens human health, and ultimately affects the sustainability of global economy and human civilization. Therefore, it has become an urgent task for human society to effectively curb the global climate problems and reduce greenhouse gas emissions.

The establishment of carbon market is a market-oriented means for the international community to solve climate problems and reduce pollution emissions. Based on the "Kyoto Protocol" signed in 2005 and the "Paris Agreement" passed in 2015, the carbon allowance assets traded in the carbon market have commodity and financial attributes, and there exists three exclusive characteristics that cannot to be ignored compared with other capital markets. The first is the asymmetric distribution of market returns, the tail distribution has the characteristics of left deviation, and the skewness is negative [3–5]. The second is the high sensitivity to policy events or external events [6]. For example, the policy implementation of banning interterm storage of carbon quotas led to a serious decline in European carbon price at the end of 2007; the fall in carbon price caused by the expiration of the second phase of emission reduction in Europe at the end of 2012; the outbreak of COVID-19 virus led to global economic downturn and triggered a sharp drop in carbon price. The third is the time-varying characteristics of carbon price volatility [7–9]. Therefore, the research into carbon price prediction and pricing models need to reflect the above three indispensable characteristics. The price mechanism is the core of the carbon market to promote emission reduction of the whole society. Consequently, studying the pricing mechanism of the carbon market in this paper can better serve the emission reduction practice of entity enterprises and create a healthier social environment.

The structure of this paper is as follows: the second part is the literature review; the third section analyzes the econometric model; the fourth section is the empirical analysis and discussion; the last part summarizes the conclusion and the prospects.

2. Literature Review

Existing research methods on carbon price forecasting mainly focus on two aspects: one is the volatility modeling technology and the other is artificial intelligence-integrated technology.

2.1. Volatility Modeling Technology

As for volatility modeling technology, Byun and Cho [10] pointed out that the GARCH family model could better fit the carbon future returns than other volatility models. Conducting the asymmetric threshold GARCH model, Chevallier [11] concluded that the stock and bond market variables could effectively explain the asymmetric volatility of carbon future returns. Based on the autoregressive, comprehensive, moving average model, Dhamija et al. [12] found that the asymmetric ARIMA-GARCH model can fit the conditional return and variance of European carbon price. Using the multi-GARCH model, Oberndorfer [13] stated that the EUA (European Union Allowance, EUA) price was positively correlated with the electricity stock return, and the stock market return did not cause EUA market volatility. Based on the ARCH regression model, the crude oil, natural gas and coal returns have a significant effect on carbon price [14,15]. Testing the EGARCH model, Chevallier [16] maintained that the abnormal events, policy factors, compliance events and uncertainty after the Kyoto Protocol are evidence of the instability of carbon price. The time-varying GARCH model with generalized nonlinear parameters can effectively fit the carbon price for prediction [16]. Designing bilaterally modified variables, Ren et al. [17] point out that the AR-GARCH model can reveal the impacts of regulatory update events on the Chinese carbon market. Employing the dynamic nonlinear (DMA) model, Koop et al. [18] found that the pricing precision of the DMA model is superior to the time-varying parameter regression model (TVP). The European carbon price is characterized by heterogeneous volatility, the prediction performance of the GARCH model based on Markov regime switching is better than other GARCH models [19].

2.2. Artificial Intelligence-Integrated Technology

The volatility modeling technology represented by the GARCH family model usually requires the carbon price being subject to strict parameter assumptions and tail distribution, which means the application of the model has great limitations [18]. Artificial intelligence-integrated technology with the advantages of mapping nonlinear relations and without considering the tail distribution has been widely used in carbon price forecasting research. The BP neural network model with high-frequency data has a more accurate prediction performance on the CER (Certified Emission Reduction, CER) price than the GARCH family model [20]. Tiwari et al. [21] found that the time-varying Markov switching copula model can provide evidence of a time-varying tail-dependence structure, and AI (artificial intelligence) is an effective means to capture carbon price. The finite distributed lag (FDL) model based on a genetic algorithm (GA) has better performance on predicting carbon price than other GARCH models [22]. Based on the idea of ensemble learning, the EMD model (Empirical Mode Decomposition, EMD) is used to extract the intrinsic mode function (IMF) that represents the different coexisting oscillation modes of carbon series [23–25], and then a hybrid carbon price forecasting model integrating the variational mode decomposition (VMD) and optimal combination forecasting model (CFM) is constructed, the results suggesting the superiority of the proposed hybrid model for carbon price forecasting [26,27]. Conducting the EMD method, Wang et al. [28] proposed a new random forest-based nonlinear ensemble paradigm for carbon price prediction and proved the model's superiority in European carbon price forecasting. Furthermore, the hybrid carbon price forecasting model, for example the multiobjective grasshopper optimization algorithm model proposed by Hao et al. [29] and the wavelet least-square support vector machine (WLSSVM) model carried by Sun et al. [30] have been proven to have strong superiority and accuracy in carbon price prediction. Different from the prediction of EMD-type models, the LSTM (Long and Short-Term Memory network) model does not need to decompose the data frequency, and shows advantages in capturing the long-term lag return characteristics based on the special gate structure of forget gate, input gate and output gate [31,32]. The application of the LSTM model in predicting stock market index has stronger accuracy and robustness than other exponential smooth models and the ARIMA model [33,34]. Employing the models of ARIMA, CNN, GARCH and LSTM to extract the linear characteristics, hierarchical data structure, long memory characteristics and volatility characteristics of carbon return, respectively, the conclusion suggests that the hybrid model of ARIMA–CNN–LSTM and GARCH-LSTM contribute a lower prediction error [35,36]. Based on similar modeling ideas, the integrated models of EMD–LSTM and that composed of total average EMD with LSTM (MEEMD–LSTM) have also proven to have significant superiority in carbon price prediction [37,38].

The above literature provides valuable references for this paper. However, the most obvious defect is that the existing literature ignores the time-varying impact of market asymmetric information and extreme shock factors on carbon premium from the perspective of high-order moment (skewness and kurtosis). More importantly, the time-varying high-order moment characteristics have been ignored. In fact, studies have proven that the financial assets of low-order moment information cannot fully reflect the actual financial return distribution [39]. The innovation and contribution of this study is to construct a new hybrid carbon pricing model, NAGARCHSK-GRU, that reveals the time-varying high-moment volatility characteristics of carbon price. The proposed NAGARCHSK-GRU price-forecasting model combines the advantages of the NAGARCHSK model in parameter estimation of the time-varying, high-order moment characteristics and the superiority of GRU (Gated Recurrent Unit, GRU) network in nonlinear fitting and forecasting. The purpose for integrating the models of NAGARCHSK and GRU network is to improve the robustness and generalization ability of the proposed pricing model, and then to provide certain technical support for market participants to capture price information and predict carbon price.

3. Econometric Modeling

Based on the classical GARCH models, this paper first constructs constant and time-varying, high-order moment carbon price volatility methods to estimate the parameters of the proposed pricing model. Secondly, the multilayer GRU network model is designed to realize the nonlinear prediction based on the time-varying, high-order moment parameters estimated by the NAGARCHSK model.

3.1. High-Order Moment Volatility Model

3.1.1. Constant High-Order Moment Model

The constant high-order moment model assumes that the third-order moment skewness and the fourth-order moment kurtosis have no impact on the first-order moment return of carbon price, but assumes that they are constant. The common constant high-moment model is the GARCHSK ($q1$,$p1$;0,0;0,0) model with the constant high-order moment term. During modeling of the carbon price, we use the AR (R) model to describe the autocorrelation process of carbon price series, assuming the return series follows a first-order lag AR ®process:

$$R_t = \rho R_{t-1} + h_t^{1/2}\xi_t \tag{1}$$

where, $h_t^{1/2}$ is the conditional variance of carbon return; ξ_t means the conditional return item; ρ indicates the autocorrelation coefficient, and $\xi_t \sim N(0,1)$.

For modeling the conditional variance h_t process that with the characteristics of volatility clustering and asymmetric distribution, this paper uses the GARCH model and its derivative models to estimate the parameters of the proposed carbon pricing model. As the first-order GARCH model can simulate the financial return volatility, we introduce the following common forms of conditional variance based on the GARCH (1,1) model.

Conditional variance of GARCH (1,1) is:

$$h_t = \beta_0 + \beta_1 \varepsilon_{t-1}^2 + \beta_2 h_{t-1} \tag{2}$$

Conditional variance of TGARCH (1,1) is:

$$h_t^{1/2} = \beta_0 + \beta_1 |\varepsilon_{t-1}| + \beta_2 h_{t-1} + \beta_3 v_{t-1} |\varepsilon_{t-1}| \tag{3}$$

Conditional variance of NAGARCH (1,1) is:

$$h_t = \beta_0 + \beta_1 \left(\varepsilon_{t-1} + \beta_3 h_{t-1}^{1/2}\right)^2 + \beta_2 h_{t-1} \tag{4}$$

where, β_0 represents the constant term of the variance equation, skewness equation and kurtosis equation; β_1 and β_2 denote the ARCH and GARCH term coefficients of the high-order moment equation, respectively; ε represents the residual term; β_3 means the leverage coefficient, reflecting the impact of asymmetric information on the carbon returns; v_{t-1} is a dummy variable that controls the impact direction of asymmetric information, when $\varepsilon_{t-1} < 0$, $v_{t-1} = 1$;$\varepsilon_{t-1} > 0$, $v_{t-1} = 0$.

3.1.2. Time-Varying High-Order Moment Model

The constant high-order moment model regards the third-order moment skewness and fourth-order moment kurtosis as fixed constants and ignores the financial asset distribution characterization of leptokurtosis and fat-tail caused by the market asymmetric information and extreme factors, which make it difficult to meet the real asset volatility-modeling requirements. Therefore, this paper considers the third-order skewness and fourth-order moment kurtosis attributes with the exclusive features of time-varying volatility, so as to describe the shock of market asymmetric information and policy factors on carbon price. The specific form of the GARCHSK ($q1$,$p1$;$q2$,$p2$;$q3$,$p3$) model, considering the volatility of time-varying conditional variance, conditional skewness and conditional kurtosis, is as follows:

$$\begin{cases} R_t = \rho E_{t-1}(R_t) + \varepsilon_t = \mu_t + h_t^{1/2}\xi_t;\ \xi_t|I_{t-1} \sim F_n(0,1,s_t,k_t) \\ h_t = \beta_0 + \sum\limits_{i=1}^{q1} \beta_{1,i}\varepsilon_{t-i}^2 + \sum\limits_{j=1}^{p1} \beta_{2,j}h_{t-j} \\ s_t = \gamma_0 + \sum\limits_{i=1}^{q2} \gamma_{1,i}\xi_{t-i}^3 + \sum\limits_{j=1}^{p2} \gamma_{2,j}s_{t-j} \\ k_t = \delta_0 + + \sum\limits_{i=1}^{q3} \delta_{1,i}\xi_{t-i}^4 + \sum\limits_{j=1}^{p3} \delta_{2,j}k_{t-j} \end{cases} \quad (5)$$

The specific form of the NAGARCHSK ($q1$,$p1$;$q2$,$p2$;$q3$,$p3$) model with leverage effect that considers the volatility of the time-varying, high-order moment, is as follows:

$$\begin{cases} R_t = \rho E_{t-1}(R_t) + \varepsilon_t = \mu_t + h_t^{1/2}\xi_t;\ \xi_t|I_{t-1} \sim F_n(0,1,s_t,k_t) \\ h_t = \beta_0 + \sum\limits_{i=1}^{q1} \beta_{1,i}(\varepsilon_{t-1} + \beta_{3,i}h_{t-i}^{1/2})^2 + \sum\limits_{j=1}^{p1} \beta_{2,j}h_{t-j} \\ s_t = \gamma_0 + \sum\limits_{i=1}^{q2} \gamma_{1,i}\xi_{t-i}^3 + \sum\limits_{j=1}^{p2} \gamma_{2,j}s_{t-j} \\ k_t = \delta_0 + \sum\limits_{i=1}^{q3} \delta_{1,i}\xi_{t-i}^4 + \sum\limits_{j=1}^{p3} \delta_{2,j}k_{t-j} \end{cases} \quad (6)$$

where, I_{t-1} represents the information set when the carbon return volatility reaches the time of $t-1$; $E_{t-1}(R_t)$ is the corresponding conditional expected return that can be obtained steadily without risk impact under the certain I_{t-1} information set. The form AR(1) is used to depict the autoregressive carbon return process. $F_t(0,1,s_t,k_t)$ represents the fourth-order moment distribution type of the carbon return series based on the classical GARCH(1,1) model, and we can obtain $E_{t-1}(\xi_t) = 0$, $E_{t-1}(\xi_t^2)= 1$, $E_{t-1}(\xi_t^3) = s_t$, $E_{t-1}(\xi_t^4) = k_t$; s_t and k_t represents the skewness and kurtosis corresponding to standardized residual $\xi_t = h_t^{-1/2}\varepsilon_t$. $\beta_0, \beta_1, \beta_2, \beta_3$ denotes the coefficient of the conditional variance equation; $\gamma_0, \gamma_1, \gamma_2$ represents the coefficient of the conditional skewness equation; $\delta_0, \delta_1, \delta_2$ means the coefficient of the conditional kurtosis equation. ($q1$,$p1$);($q2$,$p2$);($q3$,$p3$) represents the lag order of the conditional variance, conditional skewness and conditional kurtosis equations for capturing the relationship between carbon return and its time-varying, conditional, high-order moment term.

For estimating the parameters of the time-varying, high-order moment model (NAGARCHSK), the Gram–Charlier expansion of normal density function is used and truncates it in the fourth moment. Then, the conditional probability density of the standard error can be obtained under the information set I_{t-1}:

$$\begin{aligned} f(\xi_t|I_{t-1}) &= g(\xi_t)\lambda(\xi_t)/\Gamma_t \\ &\frac{1}{\sqrt{2\pi}}e^{-\xi_t^2/2}(1 + \tfrac{s_t^*}{3!}(\xi_t^3 - 3\xi_t) + \tfrac{k_t^*-3}{4!}(\xi_t^4 - 6\xi_t^2 + 3))/(1 + \tfrac{s_t^*}{3!} + \tfrac{k_t^*-3}{4!}) \end{aligned} \quad (7)$$

where $\Gamma_t = 1 + \frac{s_t^*}{3!} + \frac{k_t^*-3}{4!}$.

Furthermore, the conditional distribution of ε_t is expressed as $h_t^{-1/2}f(\xi_t|I_{t-1})$, and the log likelihood function is expressed as:

$$LF(\varepsilon_t|I_{t-1},\theta) = -\frac{1}{2}\ln(2\pi) - \frac{1}{2}\ln h_t - \frac{1}{2}\xi_t^2 + \ln(\lambda^2(\xi_t)) - \ln(\Gamma_t) \quad (8)$$

By maximizing the likelihood function above, the consistency estimation of the parameter vector can be obtained, and the parameter estimation results of the conditional mean equation, conditional variance, conditional skewness and conditional kurtosis equations can also be obtained simultaneously. Where $\theta= [\beta,\gamma,\delta]' = [\beta_0,\beta_1,\beta_2,\beta_3;\gamma_0,\gamma_1,\gamma_2;\delta_0,\delta_1,\delta_2]'$

is the parameter vector, representing the parameter to be estimated in the time-varying, high-order moment, carbon price volatility model.

3.2. GRU Model

For mapping the nonlinear, time-varying, high-order moment shock of market asymmetric information and extreme events on carbon price, this paper constructs a multilayer GRU (Gated Recurrent Unit, GRU) model to predict and fit the carbon price with the characteristic of the time-varying, high-order moment. Different from the special input gate, forget gate and output gate structure of the LSTM (Long and Short-Term Memory network, LSTM), another feedforward network structure similar to the GRU network, the GRU model is constructed based on the gate structure of the LSTM and composed of update gate and reset gate [40]. The GRU training structure can be showed in Figure 1.

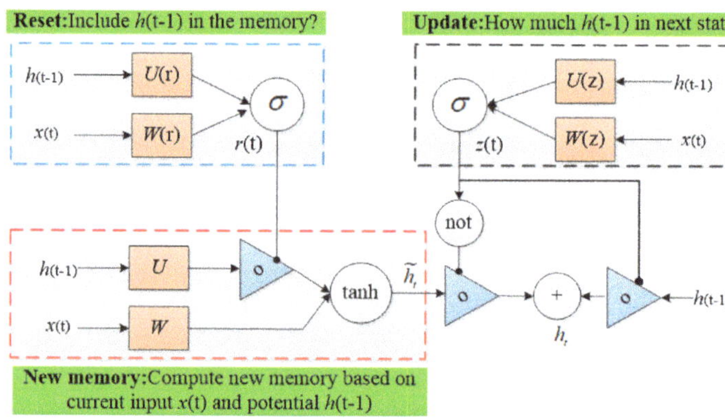

Figure 1. The model training structure of the GRU network.

Specifically, the update gate of GRU is combined of the input gate and forget gate of the LSTM network, and this function is used to determine the information to be discarded and the new information to be added. The reset gate determines the forgotten information in the past time series, which can help to capture the short-term dependency of the finance series. Unlike the LSTM model, which relies on the cell units to obtain the long-term information, the GRU network gets rid of the cell state instead of the hidden state, in order to transmit the previous information and obtain the long-term dependency. Although the debate about the model superiority of GRU and LSTM network continues, it is generally accepted that as an effective variant of LSTM network, the structure of the GRU network is simpler and requires fewer parameters and training samples. Therefore, some studies suggest that GRU is more effective than the LSTM model in solving the long dependency problem of RNN networks [41]. According to the above GRU model diagram, the forward propagation process of the GRU network is as follows:

Firstly, the state h_{t-1} transmitted from the previous network is combined with the input x_t of the current node to obtain the gate structure of the GRU network, that is, the reset gate r and update gate z. Where σ means the activation function, which converts the input data to a value in the range of 0–1 to act as a gating signal.

$$r_t = \sigma(W_r x_t + U_r h_{t-1}) \tag{9}$$

$$z_t = \sigma(W_z x_t + U_z h_{t-1}) \tag{10}$$

Secondly, after obtaining the gating signal, the reset gate is used to obtain the data after "reset". If the element value r_t in the reset gate is close to 0, it means the hidden state element related to the reset gate should be set to 0, that is, the hidden state information of the last time should be discarded. Further, the result of element multiplication is linked to the input of current time step, and the candidate hidden state $\widetilde{h_t}$ is calculated by the activation function tanh, and the element value ranges from −1 to 1. The calculation for candidate hidden state is:

$$\widetilde{h_t} = tanh(Wx_t + r_t U h_{t-1}) \tag{11}$$

Finally, the most critical process of training the GRU model is the update memory stage. The update gate z_t controls the forgotten information of the hidden layer h_{t-1} at the previous moment, and the new hidden layer information $\widetilde{h_t}$ needs to be added at the current moment. The update gate z_t is expressed as:

$$h_t = (1 - z_t)h_{t-1} + z_t\widetilde{h_t} \tag{12}$$

It is worth noting that the value of updated gating z_t is in a range from 0 to 1. The closer the gating value is to 1, the more data there is to be remembered, while the closer it is to 0, the more information is forgotten. The GRU model can realize data forgetting and memory at the same time by using update gating z_t, unlike the LSTM model that requires multiple gating.

3.3. NAGARCHSK-GRU Model

The proposed hybrid carbon price forecasting model combines the advantages of NAGARCHSK and GRU neural networks. Firstly, the NAGARCHSK model is better than the constant high-order moment models and other time-varying, high-order moment models in fitting the carbon price series with time-varying, high-order moment volatility characteristics. Therefore, we select the NAGARCHSK model to estimate the time-varying parameters of carbon price.

Secondly, we use these estimated parameters as network inputs, and use the GRU neural network to train the time-varying, high-order moment volatility characteristics of carbon price for improving the prediction accuracy. The basic idea of constructing the proposed NAGARCHSK-GRU carbon-forecasting model shown in Figure 2.

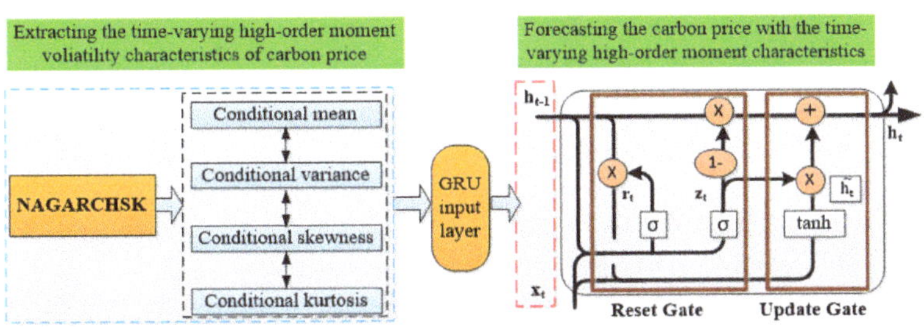

Figure 2. The structuring idea of the carbon price-forecasting hybrid model of NAGARCHSK–GRU.

3.4. Evaluation Criteria and the Benchmark Model

For evaluating the prediction performance of the proposed time-varying, high-order moment carbon-pricing model, this paper adopts the following criteria to measure the model performance.

$$RMSE = \sqrt{\frac{\sum_{i=1}^{T}(y_i - \hat{y}_i)^2}{T}} \tag{13}$$

$$MAE = \frac{1}{T}\sum_{i=1}^{T}|y_i - \hat{y}_i| \tag{14}$$

$$MAPE = \frac{1}{T}\sum_{i=1}^{T}\left|\frac{y_i - \hat{y}_i}{y_i}\right| \tag{15}$$

$$DA = \frac{1}{T}\sum_{i=1}^{T-1} a_i \text{ where } a_i = \begin{cases} 1, if(y_{i+1} - y_i) \times (\hat{y}_{i+1} - y_i) > 0 \\ 0, otherwise \end{cases} \quad (16)$$

where $Y = \{y_1, y_2, \cdots, y_T\}$ represents the carbon return series; $\hat{Y} = \{\hat{y}_1, \hat{y}_2, \cdots, \hat{y}_T\}$ represents the prediction return. T is a time series variable.

The values of root-mean-square error (RMSE), mean absolute error (MAE) and mean absolute percentage error (MAPE) range from 0 to 1, and a larger value means the deviation between predicted return and real return is greater, and the model performance is worse. The correct investment direction prediction can help investors make more valuable decisions. This paper uses DA (direction accuracy) index to measure the consistency probability of market trend and investors' prediction direction. The larger DA value means the predicted value of carbon return is closer to investors' psychological expectation.

For assessing the performance of the proposed pricing model, this paper also selects the following aggressors as the comparison benchmarks. The first one is the BP (back propagation network) model with the advantage of nonlinear mapping. The second is GBR (gradient boosting regression) model, which is a kind of integrated learning method. The third is MLP (multilayer perceptron); the parameter optimization of the MLP can improve the nonlinear mapping and carbon price-prediction accuracy. The fourth is the RNN (recurrent neural network) model, which is an artificial neural network with a tree structure, that has significant advantages in forecasting carbon price. The fifth is the LSTM (long and short-term memory network) model, which is another improved structure of the RNN model that shows superiority for solving the problems of gradient explosion.

4. Empirical Analysis and Discussion

4.1. The Data

This article selects the continuous futures contract of EUAf (European Union Allowance future, EUAf) from the European Energy Exchange as the representative variable of carbon assets. The data range from 22 June 2012 to 7 May 2021, with a total of 2274 data samples. The sample selection rule refers to the experience of Wen et al. [42], that is, continuous futures contracts with different maturity dates are connected according to the time sequence. Based on this, the sample of this article integrates the daily settlement price of the four futures contracts, DEC12, DEC16, DEC18 and DEC20. The reason for choosing EUAf is that the EUAf is the largest emission reduction quota in the world. The carbon futures trading of the European Energy Exchange accounts for about 70% of the global futures trading, and the EUAf trading volume is larger than EUAs (European Union Allowance spot, EUAs), the price discovery function is also relatively mature. It uses R_t to represent the carbon assets return:

$$R_t = 100 \times (lnP_t - lnP_{t-1}) \quad (17)$$

where P_t represents the carbon asset price, that is, the daily settlement price of EUAf continuous futures contracts.

4.2. Time-Varying High-Order Moment Characteristics Estimate

The study findings shown in Table 1 concluded that the ARCH and GARCH terms of all constant and time-varying high-order moment models are significant, indicating that the carbon return has obvious volatility clustering, which is not only caused by the variance shock, but also conditional skewness and conditional kurtosis, representing the impacts of asymmetric information and extreme factors on carbon return. All the volatility leverage coefficients β_3 are negative and significant, denoting that variance volatility has obvious asymmetry shock on carbon return, and the degree of negative impact is greater than the positive impact. This conclusion is completely consistent with the pioneering research results of Engle and Manganelli [43] that the negative VAR impact of the stock market is more significant. This finding indirectly proves that carbon assets have general financial attributes and common volatility characteristics.

The variance impact coefficient β_2 is smaller than the coefficient of the constant model, that is, with the addition of the conditional skewness and conditional kurtosis equations, the volatility clustering effect from the shock of variance term gradually decreases, for example, the β_2 coefficient of the GARCH, TGARCH and AGARCH models are 0.8824, 0.8875, 0.8904, respectively. When the time-varying conditional skewness and conditional kurtosis are added, the volatility clustering coefficients of AGARCH-K, NAGARCH-K and NAGARCHSK models are reduced to 0.8358, 0.8685, 0.7873, respectively. This conclusion is completely consistent with Harvey's [39] research.

This phenomenon shows that when the time-varying, high-order moment models no longer assume the skewness and kurtosis are constants, they can effectively identify the volatility clustering effect caused by asymmetric information and extreme shocks through the time-varying skewness and kurtosis equation. We can say that the impact of carbon return from the time-varying skewness and kurtosis is becoming more obvious, resulting in the variance impact coefficient decreasing as the skewness and kurtosis coefficient increases. This similar reason can be used to explain why the extreme impact coefficient δ_2 of the NAGARCHSK model is smaller than that of the AGARCH-K and the NAGARCH-K model.

Table 1. Parameter estimation of the high-order moment volatility model for carbon return.

Coefficient	Constant High-Order Moment Volatility Model			Time Varying High-Order Moment Volatility Model		
	GARCH	**TGARCH**	**NGARCH**	**AGARCH-K**	**NAGARCH-K**	**NAGARCHSK**
ρ	0.0794 (3.744)	0.0981 (4.562)	0.0916 (4.317)	0.0007 (0.011)	0.0007 (1.708)	0.0247 (0.005)
β_0	0.0012 (2.985)	0.0054 (4.4745)	0.0046 (3.951)	0.0715 (4.245)	0.0000 (2.177)	0.0483 (0.001)
β_1	0.1182 (7.794)	0.0899 (9.737)	0.1173 (17.87)	0.1463 (2.456)	0.1313 (6.428)	0.0587 (4.302)
β_2	0.8824 (70.919)	0.8875 (191.69)	0.8904 (185.7)	0.8358 (41.65)	0.8685 (50.659)	0.7873 (2.204)
β_3		−0.0632 (4.911)	−0.0536 (2.442)	−0.0029 (3.031)	−0.003 (3.848)	−0.0600 (2.964)
γ_0						0.7990 (0.807)
γ_1						0.0214 (2.831)
γ_2						0.0198 (5.325)
δ_0				0.6978 (0.012)	0.4452 (0.281)	0.0821 (0.064)
δ_1				0.3063 (3.086)	0.4265 (2.821)	0.6562 (1.987)
δ_2				0.5363 (3.161)	0.5698 (3.101)	0.201 (3.256)
Likelihood	5159.771	5070.972	5067.591	6469	6473	**4682**

Note: The bold indicates the model with the minimum maximum likelihood value and the best parameter estimation performance; the data in brackets indicate the t-statistic of parameter estimation of each model.

Compared with the constant model and other time-varying, high-order moment models, it should be noted that the maximum likelihood value of the NAGARCHSK model is the lowest, as shown in Table 1, therefore, this paper chooses the NAGARCHSK model to estimate the model parameters of time-varying conditional variance, conditional skewness and conditional kurtosis equations of the carbon return. Figure 3 shows that the conditional high-moment series of carbon assets have obvious volatility persistence effects, the risk of variance, skewness and kurtosis are large, and the high-moment volatility series also shows time-varying characteristics.

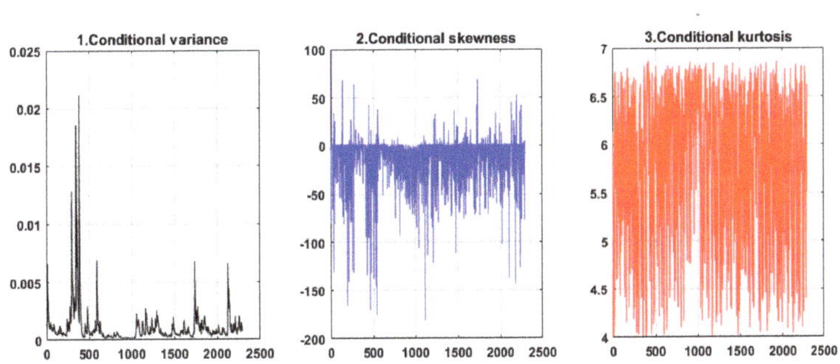

Figure 3. Time-varying, high-moment fluctuation of the carbon price identified by the NAGARCHSK model.

4.3. Predicting Results Analysis

We use the NAGARCHSK model to estimate the time-varying, high-order moment parameter characteristics that represent the shock from the asymmetric information and extreme external impact. Then, a multilayer GRU model is constructed to map and predict the carbon returns based on the obtained high-order moment series. The first 70% of samples of carbon return series are selected for model training, and the last 30% of samples for testing the prediction performance.

4.3.1. GRU Structure Construction

Input unit, output unit, number of hidden layers and hidden layer neurons are the basic structure of a deep-learning network. The input of the carbon price forecasting NAGARCHSK-GRU model is the time-varying conditional lagging mean, conditional variance, conditional skewness and conditional kurtosis of carbon returns estimated by the NAGARCHSK model, and the output is the carbon return series we need to predict. The hidden layer is a network structure for parameter optimization and feature learning. Fewer hidden layers may limit the learning ability of the forecasting model, which makes it difficult to reach the optimal solution. Research has found that a neural network with two hidden layers can already solve most problems [44]. Similarly, the designing of hidden layer neurons is to capture and map the input data. Although more neurons can improve the learning and generalization ability of the network, it may also consume more training time and lead to overfitting.

For determining the appropriate the GRU network structure, based on the experimental method, this paper measures the forecasting performance when the hidden layers number is 1, 2, 3, 4, 5, 6 and the hidden layers neuron nodes are 4, 8, 16, 32, 64, 128, respectively (as showed in Table 2). It is found that when there are two hidden layers in the NAGARCHSK-GRU model, and the neuron nodes in both layers are 16-16, the model's error criteria MSE, RMSE, and MAE values are 0.0006284, 0.0250681, and 0.1399925, respectively, which are the lowest of the whole experimental sample. Therefore, the network structure of the proposed NAGARCHSK-GRU forecasting model is designed as 4-16-16-1 for training the time-varying, high-order moment carbon series.

Table 2. Performance of the proposed NAGARCHSK-GRU: hidden layers and hidden nodes.

Hidden Layer	Node	NAGARCHSK-GRU			Hidden Layer	Nodes	NAGARCHSK-GRU		
		MSE	RMSE	MAE			MSE	RMSE	MAE
1	4	0.0010592	0.0325455	0.2575152	4	4	0.0007587	0.0275444	0.1993575
1	8	0.0007156	0.0267501	0.1769118	4	8	0.0006820	0.0261152	0.1653054
1	16	0.0007580	0.0275324	0.1745386	4	16	0.0007306	0.0270302	0.1982328
1	32	0.0011345	0.0336824	0.3124025	4	32	0.0005733	0.0239427	0.1742260
1	64	0.0013580	0.0368505	0.4301757	4	64	0.0007967	0.0282252	0.2044582
1	128	0.0026156	0.0511425	0.6136194	4	128	0.0011475	0.0338750	0.2910599
	Avg	0.0012735	0.0347506	0.3275272		Avg	0.0007815	0.0277888	0.2054400
2	4	0.0007035	0.0265229	0.1684342	5	4	0.0009507	0.0308342	0.2467976
2	8	0.0007153	0.0267458	0.1627419	5	8	0.0006662	0.0258103	0.1402034
2	16	**0.0006284**	**0.0250681**	**0.1399925**	5	16	0.0006943	0.0263491	0.1738529
2	32	0.0006848	0.0261686	0.1953299	5	32	0.0008011	0.0283045	0.2853657
2	64	0.0009183	0.0303032	0.2546803	5	64	0.0012432	0.0352584	0.2500270
2	128	0.0011984	0.0346173	0.4536587	5	128	0.0011692	0.0341942	0.3119049
	Avg	0.0008081	0.0282376	0.2291396		Avg	0.0009208	0.0301251	0.2346919
3	4	0.0008391	0.0289676	0.2116799	6	4	0.0007273	0.0269685	0.1978246
3	8	0.0007236	0.0268994	0.1764938	6	8	0.0007203	0.0268387	0.1932607
3	16	0.0008080	0.0284251	0.1900228	6	16	0.0006415	0.0253283	0.1762878
3	32	0.0008078	0.0284224	0.1932762	6	32	0.0006575	0.0256409	0.2317956
3	64	0.0011074	0.0332772	0.2451673	6	64	0.0009107	0.0301776	0.2643418
3	128	0.0016887	0.0410936	0.2753646	6	128	0.0013865	0.0372361	0.3278892
	Avg	0.0009958	0.0311809	0.2153341		Avg	0.0008406	0.0286984	0.2319000

Note: Bold numbers are the minimum MSE, RMSE and MAE, respectively.

4.3.2. Performance of the NAGARCHSK-GRU Model

For testing the prediction performance of the proposed NAGARCHSK-GRU model, this paper compares the prediction results of the proposed model and benchmark evaluation models. The results are shown in Table 3.

For high-order moment pricing models that consider time-varying conditional variance, conditional skewness and conditional kurtosis in Panel A, the NAGARCHSK-GRU model has significant advantages over other benchmark models in all the error evaluation criteria and market expected criteria. That is, the NAGARCHSK-GRU model has better prediction ability than other benchmark models (as shown in Figure 4).

Table 3. Performance of the proposed and benchmark model for forecasting the carbon price.

	Proposed Model	Benchmark Model				
Panel A: Pricing model considering the features of conditional variance, conditional skewness and conditional kurtosis						
	NAGARCHSK-GRU	NAGARCHSK-LSTM	NAGARCHSK-RNN	NAGARCHSK-MLP	NAGARCHSK-GRB	NAGARCHSK-BP
RMSE	**0.509902**	0.546867	3.348075	2.031849	2.033564	3.000422
MAE	**0.172333**	0.205202	2.413839	1.470269	1.471746	2.183393
MAPE	**0.594527**	1.219059	7.415169	0.705615	0.881136	3.450703
DA	**0.984211**	0.978947	0.742475	0.875	0.877551	0.729114
Panel B: Pricing model considering the features of conditional variance and conditional kurtosis						
	NAGARCHK-GRU	NAGARCHK-LSTM	NAGARCHK-RNN	NAGARCHK-MLP	NAGARCHK-GRB	NAGARCHK-BP
RMSE	2.034801	4.616283	3.347897	2.867784	2.036477	2.972087
MAE	1.473751	2.380641	2.413814	2.118404	1.475259	2.149908
MAPE	0.835218	3.237167	3.198746	3.293093	0.879472	2.526902
DA	0.731621	0.505615	0.583471	0.843725	0.715412	0.762763
Panel C: Pricing model without considering the feature of time-varying, high-order moment						
	GRU	LSTM	RNN	MLP	GBR	BP
RMSE	3.103279	3.623880	3.348127	3.107803	3.187022	3.822661
MAE	2.129058	2.434801	2.413955	2.261274	2.261036	2.194043
MAPE	7.316744	11.140187	7.395187	8.082685	8.035241	2.827125
DA	0.74368	0.536842	0.74	0.872774	0.875318	0.746231

Note: Bold numbers are the minimum MSE, RMSE, MAE, respectively, and the maximum of DA.

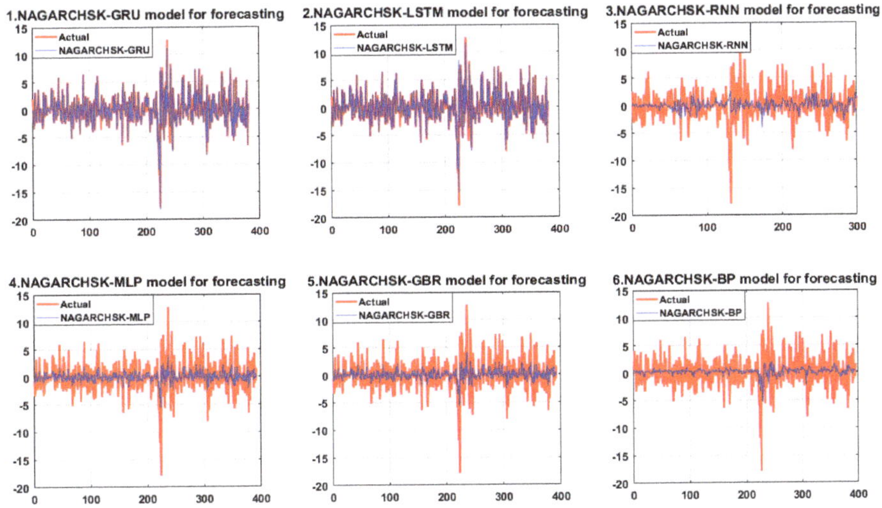

Figure 4. The forecasting performance of the proposed and benchmark model considering the features of time-varying conditional variance, conditional skewness and conditional kurtosis.

Specifically, as for the error evaluation criteria, the RMSE, MAE, and MAPE values of the NAGARCHSK-GRU model are 0.509902, 0.172333, and 0.594527, respectively, which are lower than those of benchmark models such as NAGARCHSK-LSTM, NAGARCHSK-RNN, NAGARCHSK-MLP, NAGARCHSK-GBR, and NAGARCHSK-BP. This result concludes that the NAGARCHSK-GRU model has better robustness and stability for fitting carbon price series with time-varying, high-order moment characteristics. For the market-expected criteria, the DA of the NAGARCHSK-GRU model was 0.984211, which is higher than that of the benchmark models NAGARCHSK-LSTM (0.978947), NAGARCHSK-GRB (0.877551), NAGARCHSK-MLP (0.875), NAGARCHSK-RNN (0.742475) and NAGARCHSK-BP (0.729114). This indicates that the NAGARCHSK-GRU model is in line with investors' psychological expectations for predicting carbon return, and the predicted returns are strongly consistent with the real return. As a result, the pricing model can provide technical support for investors to judge market conditions and formulate investment strategies.

In contrast, the error-evaluation criteria and market-expected criteria of the NAGARCHSK-RNN model shows the worst prediction effect of all models, that is, the RMSE, MAE and MAPE values are, respectively, 3.348075, 2.413839 and 7.415169, and the DA value is 0.742475. We can conclude that using the NAGARCHSK-RNN model it is difficult to map the carbon price series with the time-varying, high-order moment feature, and its predictive ability cannot meet investors' expectations.

For the high-order moment forecasting models considering time-varying conditional variance and conditional kurtosis, as shown in Panel B, the NAGARCHSK-GRU model still has obvious forecasting advantages in error evaluation criteria and is relatively better in market-expected criteria compared with other benchmark models. Specifically, the error indexes RMSE, MAE and MAPE of the NAGARCHSK-GRU model are 2.034801, 1.473751 and 0.835218, respectively, which are lower than other benchmark criteria, the market-expected criteria DA is 0.731621, which is second only to the 0.843725 of the NAGARCHSK-MLP model. It is worth noting that the NAGARCHSK-LSTM model, which has the advantage in fitting financial time series, has the worst performance in carbon prediction among all models. The error criteria RMSE, MAE and MAPE are 4.616283, 2.380641 and 3.237167, respectively, and the market-expected criteria DA is 0.505615. This shows that the NAGARCHSK-LSTM model's prediction ability and generalization ability are declining. The conclusion that the gap between the fitting curve and the real value is extremely obvious is also shown in Figure 5, and the correlation is poor.

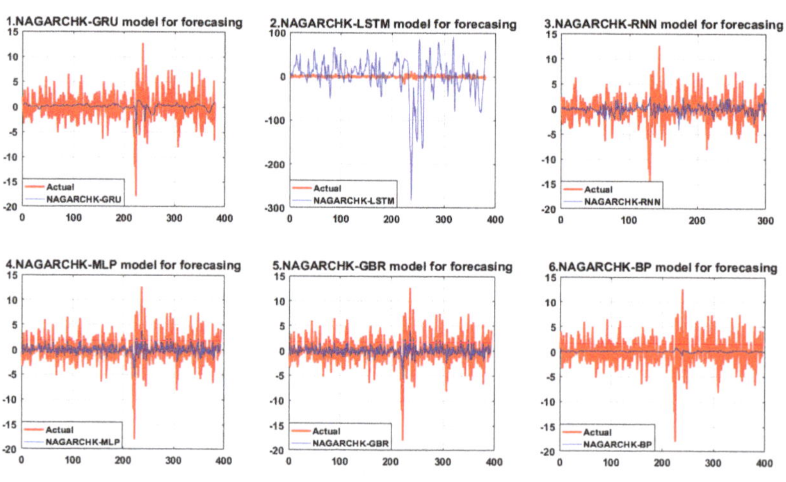

Figure 5. The forecasting performance of the proposed and benchmark models consider the feature of time-varying conditional variance and conditional kurtosis.

For the carbon pricing model without considering the shock of time-varying, high-order moment, the GRU network model has the smallest error evaluation criteria (as shown in Panel C), which denotes that the GRU model still has strong prediction accuracy and robustness even without considering the characteristics of time-varying, high-order moment. However, the market-expected criteria DA is 0.74368, which is only higher than the 0.536842 of the LSTM model. We can conclude that the prediction performance of the model makes it difficult to satisfy the investors' psychological expectations.

Although the error criteria of other benchmark models are lower than those of the GRU model, the difference is not significant (as shown in Figure 6), particularly the RMSE and MAE of all models are basically close, and the deviation is small. More obviously, the market-expected criteria of the GBR and MLP models are significantly higher than those of other models, with DA values of 0.875318 and 0.872774, respectively, indicating the carbon-prediction performance of those two models is relatively stable and has a certain robustness. The market prediction performance suggests a reliable reference for investors making investment decisions.

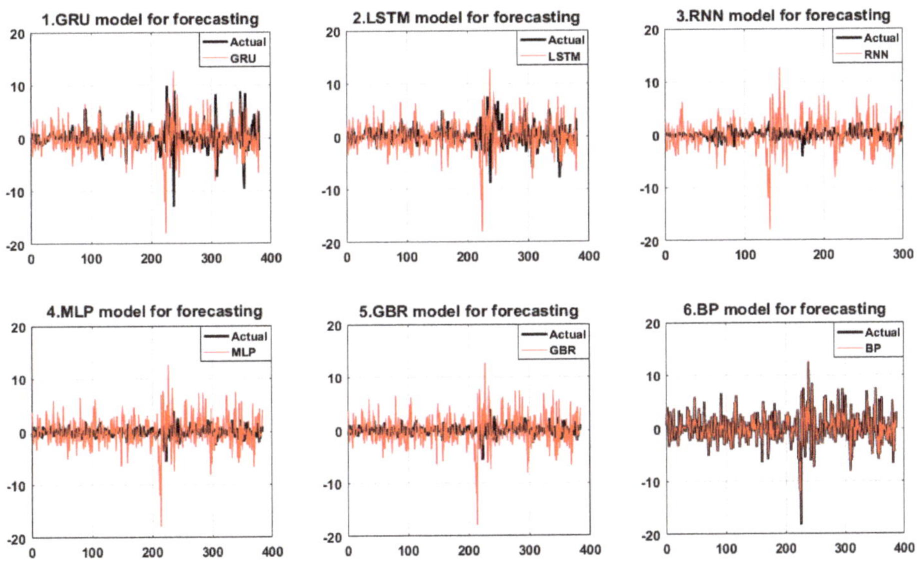

Figure 6. The forecasting performance of the proposed and benchmark models without considering the feature of time-varying, high-order moment.

Comparing the prediction results of all the pricing models of Panel A, Panel B and Panel C in Table 3, firstly, the carbon price-forecasting performance of the NAGARCHSK-GRU model is the best among all pricing models in Panel A, Panel B and Panel C. Secondly, the error criteria RMSE, MAE, and MAPE of the carbon pricing model in Panel A are significantly smaller than those of the error criteria in Panel B and Panel C, while the market-expected indicator DA is significantly higher than other models. Furthermore, the error criteria RMSE, MAE, and MAPE of the pricing model in Panel C are relatively high, while the market-expected index DA is relatively low.

The empirical results shown in Table 3 conclude that the deep-learning, carbon price forecasting model that considers the time-varying, high-order moment characteristics can provide more confident carbon premium evidence. This conclusion further proves that the carbon return is not only affected by the low-order moment attribute pricing factor, but also that the time-varying, high-order moment attribute that reflects the market asymmetric information and extreme shock is also an important explanatory factor for carbon return. The research results of this article can provide valuable reference for investors, commercial banks, and emission-reduction companies to judge market conditions and predict market trends.

4.3.3. Robustness of the NAGARCHSK-GRU Model

The significant advantage of the GRU model is that the parameter training structure of the long memory function can fit the finance time series, especially the time series over a long period time. Therefore, to prove the robustness of the NAGARCHSK-GRU model in different prediction period, this paper analyzes the performance of the proposed pricing model in the short-term, medium-term and long-term, respectively. Among them, the last 4 months, 10 months, and 15 months of the carbon returns are used as the prediction set in the short-term, medium-term, and long-term, respectively, and the rest of the data are used as the training set. The pricing model structure adopts the optimal network structure decided in the previous section. This part mainly describes the prediction performance of the carbon price-forecasting model that considers the time-varying, high-order moment characteristics, furthermore, the RMSE, MAE, MAPE error criteria are used to evaluate the model's pricing accuracy and stability.

For carbon price-forecasting performance in different periods (as shown in Table 4), the NAGARCHSK-GRU pricing model has significant superiority in the short-term, medium-term and long-term for all the error criteria, that is, the values of RMAE, MAE and MAPE are significantly lower than other benchmark models, and the proposed model has satisfactory robustness over all periods. The error distribution of the proposed and benchmark pricing models can be seen in Figures 7–9.

Table 4. Prediction performance of the pricing model considering the feature of time-varying, high-order moment.

	Proposed Model	Benchmark Model				
	NAGARCHSK-GRU	NAGARCHSK-LSTM	NAGARCHSK-RNN	NAGARCHSK-MLP	NAGARCHSK-GRB	NAGARCHSK-BP
Panel A: Long-term prediction performance (15 months)						
RMSE	**0.752385**	1.157871	2.996408	1.981738	1.985327	3.001473
MAE	**0.218883**	0.412011	2.352625	1.559398	1.560740	2.393809
MAPE	**0.354984**	0.299397	0.765891	0.692497	0.689473	6.234842
Panel B: Medium-term prediction performance (10 months)						
RMSE	0.825573	0.877745	3.699027	2.388462	2.388496	3.686389
MAE	0.246388	0.345985	2.629891	1.698282	1.698174	2.590053
MAPE	0.476214	1.997452	10.473157	0.809171	0.838451	11.833577
Panel C: Short-term prediction performance (4 months)						
RMSE	1.109585	0.761311	3.348173	2.160763	2.162539	3.209178
MAE	0.408066	0.258531	2.414090	1.560627	1.562554	2.334026
MAPE	0.264886	0.873019	7.400078	0.794943	0.807474	2.576337

Note: Bold numbers are the minimum MSE, RMSE, MAE, respectively. The network structure of the proposed and benchmark models adopts the optimal structure decided experimentally in the previous section, that is, the structure of 4-16-16-1.

Based on the estimation errors of the pricing models over different periods, it is found that as the forecasting period gradually increases from short-term to long-term, the forecasting errors of all pricing model gradually decrease, resulting the improvement of model accuracy and stability.

In particular, the NAGARCHSK-GRU model has the smallest prediction error and the best prediction performance, that is, the long-term prediction error RMSE, MAE and MAPE are 0.752385, 0.218883 and 0.354984, respectively, the medium-term prediction error RMSE, MAE and MAPE are 0.825573, 0.246388 and 0.476214, respectively, and the short-term prediction error RMSE, MAE and MAPE are 1.109585, 0.408066 and 0.264886, respectively. This evidence shows that the accuracy and stability of the NAGARCHSK-GRU model are gradually optimized with the extension of forecasting time, and it is significantly better than other benchmark models for forecasting the 15 month lagged returns. Since the advantage of the GRU model is fitting the longer finance time series, the findings of this article provide further evidence for this convinced conclusion and also the robust performance of the proposed model for different prediction periods.

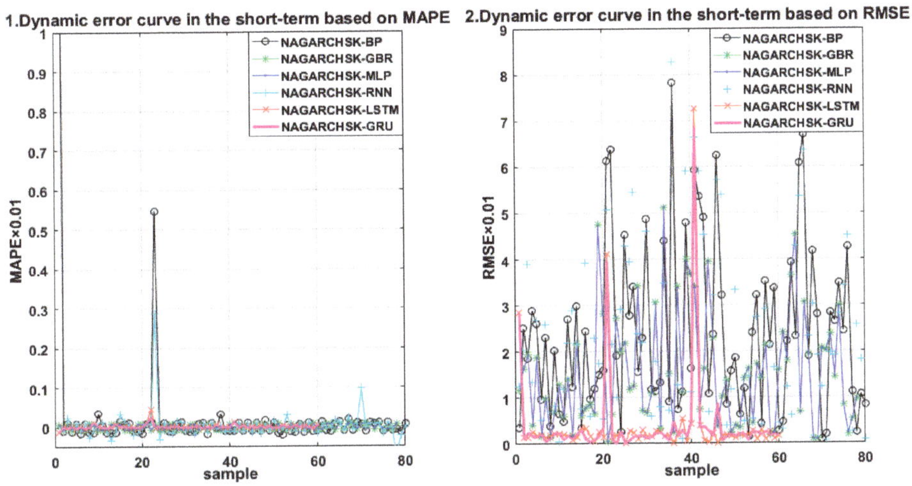

Figure 7. Error scatter distribution of the proposed and benchmark forecasting models considering the feature of time varying, high-order moment in the long term.

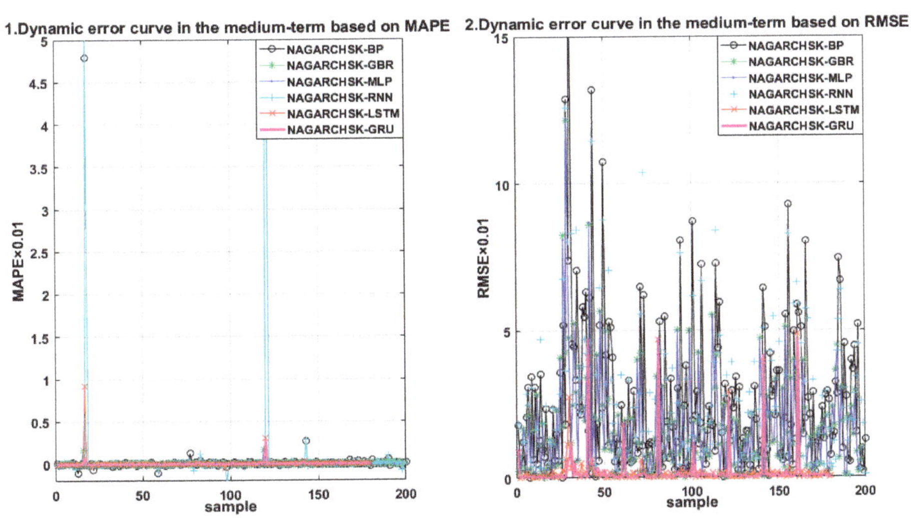

Figure 8. Error scatter distribution of the proposed and benchmark forecasting models considering the feature of time-varying, high-order moment in the medium term.

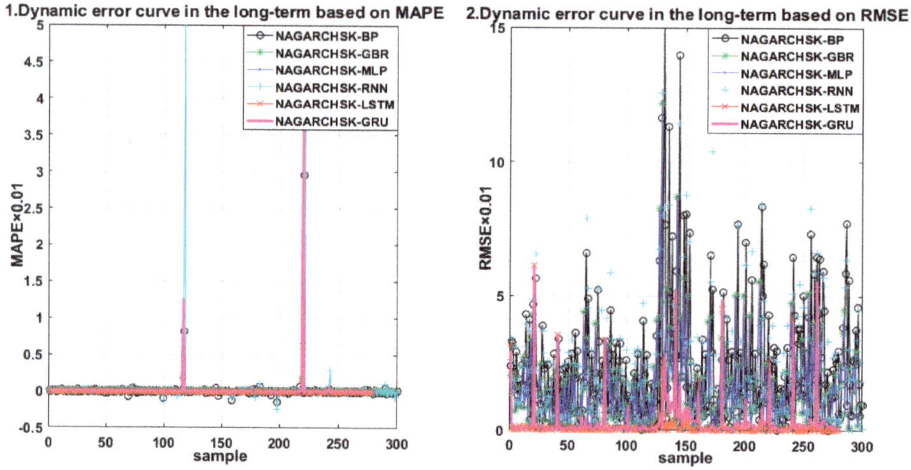

Figure 9. Error scatter distribution of the proposed and benchmark forecasting models considering the feature of time-varying, high-order moment in the short term.

5. Conclusions and Prospects

5.1. Conclusions

As a market-oriented mechanism for innovation to curb global climate issues, the carbon market is recognized as the most effective means to reducing the global carbon dioxide emissions and realize the sustainability of human society and economic growth. Compared with other financial markets, the carbon market has obvious market asymmetry, is sensitive to policy shocks and has time-varying volatility. However, the existing carbon price-forecasting research mainly focuses on the price information transmission and risk volatility spillover from the perspective of low-order moment of return, and ignores the time-varying impact of asymmetric information and extreme policies on carbon assets from the perspective of high-order moment attributes (market skewness and kurtosis). The explanation for carbon premium lacks sufficient evidential support.

The innovation and contribution of this article are constructing an integrated carbon price-forecasting model, NAGARCHSK-GRU, based on the special characteristics of carbon assets such as market asymmetry, strong policy-shock sensitivity, and time-varying volatility. The proposed forecasting model considers the time-varying impact of market asymmetric information and extreme factors on carbon prices from the perspective of high-order moment attributes, so as to provide new evidence to explain the carbon premiums. The main work and research conclusions of this paper are as follows:

Firstly, carbon assets have obvious time-varying, high-order moment volatility characteristics. Compared with constant high-order moment volatility models, the time-varying, high-order moment volatility NAGARCHSK model can reveal the time-varying impact of systemic risk, asymmetric information and extreme factors on carbon premium by the function of time-varying variance, time-varying skewness and time-varying kurtosis equations. Moreover, the time-varying, high-order moment characteristics estimated by the NAGARCHSK model can explain the volatility clustering and premium mechanism of carbon price.

Secondly, the proposed machine-learning pricing model has more accuracy and stability in predicting carbon price with time-varying, high-order moment volatility characteristics. The time-varying impact of asymmetric information and extreme factors on carbon price is also important evidence for explaining carbon premium. This conclusion shows that the carbon pricing model proposed in this paper can fit and forecast carbon return effectively, specifically, the NAGARCHSK-GRU model is significantly better than other deep-network models. Further research shows that the NAGARCHSK-GRU model has reliable advantages in long-term, medium-term and short-term carbon price fitting and forecasting. In particular, the long-term carbon price-forecasting ability is outstanding, that is, it has perfect stability and accuracy for 15 months of prices forecasting. This conclusion not only confirms the advantages of the NAGARCHSK-GRU model in fitting long-term financial data, but also proves that the carbon pricing model considering time-varying, high-order moment volatility can provide a strong explanation for carbon price.

The theoretical and practical implications of this paper are: firstly, as for the theoretical innovation, the findings show the rationality and effectiveness of incorporating the time-varying impact characteristics of asymmetric market information and extreme factors into the carbon price-forecasting model. The accuracy of carbon price forecasting can suggest a stronger reference for investors to judge market conditions, formulate investment strategies and may serve the implementation of carbon emission reduction. Secondly, as for the practical function, the maturity of carbon pricing mechanisms provide a decision-making basis for the government to speed up the construction of carbon market mechanisms and enhance the ability of the financial system to manage climate change. The conclusion of this paper also provides technical support for investors, emission-reduction entities and other market participants to capture price information and predict price changes.

5.2. Prospects

The focus of this paper is carbon premium explanation from the perspective of high-order moments. In the proposed high-order-moment pricing framework, each pricing term is a statistically structured factor, and the actual meaning behind the statistical indicators of high-order-moment attributes is not clear. Based on this, employing text-mining technology to obtain unstructured carbon pricing data that represent investor sentiment, policy impacts and other pricing factors, rather than the statistical moment attributes, is a valuable avenue for continuing relevant research in the future.

Author Contributions: Conceptualization, P.Y. and Y.Y.; methodology, P.Y. and Y.W.; model code analysis, P.Y. and C.Z.; formal analysis, Y.W. and C.Z.; writing—original draft preparation, P.Y. and Y.W.; writing—review and editing, P.Y. and Y.Y.; supervision, C.Z. All authors have read and agreed to the published version of the manuscript.

Funding: This work was funded by the following projects: the Youth Fund Project for Humanities and Social Sciences of the Ministry of education of China, the grant number is 21YJC790152; the University Humanities and Social Sciences Research Project of Anhui Province of China, the grant number is SK2021A0574; the Social Sciences innovation and development project of Anhui Province of China, the grant number is 2021CX028; the National Natural Science Foundation of China, the grant number are 72101006 and 71971071.

Institutional Review Board Statement: Not applicable.

Informed Consent Statement: Not applicable.

Data Availability Statement: The data in this paper are from the trading information of Environmental Markets in European Energy Exchange. The data can be obtained free of charge on the websites of Environmental Markets (eex.com). Of cause, the data are also available from the corresponding author upon reasonable request.

Conflicts of Interest: The authors declare no conflict of interest.

References

1. Aktar, M.A.; Alam, M.M.; Al-Amin, A.Q. Global economic crisis, energy use, CO_2 emissions, and policy roadmap amid COVID-19. *Sustain. Prod. Consump.* **2021**, *26*, 770–781. [CrossRef] [PubMed]
2. Sheng, P.; Li, J.; Zhai, M.; Majeed, M.U. Economic growth efficiency and carbon reduction efficiency in china: Coupling or decoupling. *Energy Rep.* **2021**, *7*, 289–299. [CrossRef]
3. Yang, X.; Zhang, C.; Yang, Y.; Wang, W.; Wagan, Z.A. A new risk measurement method for China's carbon market. *Int. J. Financ. Econ.* **2020**. [CrossRef]

4. Yuan, N.; Yang, L. Asymmetric risk spillover between financial market uncertainty and the carbon market: A GAS-DCS–Copula approach. *J. Clean. Prod.* **2020**, *259*, 120750. [CrossRef]
5. Zhu, B.; Ye, S.; Wang, P.; He, K.; Zhang, T.; Wei, Y.M. A novel multiscale nonlinear ensemble leaning paradigm for carbon price forecasting. *Energy Econ.* **2018**, *70*, 143–157. [CrossRef]
6. Yun, P.; Zhang, C.; Wu, Y.; Yang, X.; Wagan, Z.A. A Novel Extended Higher-Order Moment Multi-Factor Framework for Forecasting the Carbon Price: Testing on the Multilayer Long Short-Term Memory Network. *Sustainability* **2020**, *12*, 1869. [CrossRef]
7. Daskalakis, G.; Psychoyios, D.; Markellos, R.N. Modeling CO_2 emission allowance prices and derivatives: Evidence from the European trading scheme. *J. Bank Financ.* **2009**, *33*, 1230–1241. [CrossRef]
8. Seifert, J.; Uhrig-Homburg, M.; Wagner, M. Dynamic behavior of CO_2 spot prices. *J. Environ. Econ. Manag.* **2008**, *56*, 180–194. [CrossRef]
9. Taschini, L.; Paolella, M.S. An econometric analysis of emission trading allowances. *J. Bank Financ.* **2008**, *32*, 6–26.
10. Byun, S.J.; Cho, H. Forecasting carbon futures volatility using GARCH models with energy volatilities. *Energy Econ.* **2013**, *40*, 207–221. [CrossRef]
11. Chevallier, J. Detecting instability in the volatility of carbon prices. *Energy Econ.* **2011**, *33*, 99–110. [CrossRef]
12. Dhamija, A.K.; Yadav, S.S.; Jain, P.K. Forecasting volatility of carbon under EU ETS: A multi-phase study. *Env. Econ. Policy Stud.* **2017**, *19*, 299–335. [CrossRef]
13. Oberndorfer, U. EU emission allowances and the stock market: Evidence from the electricity industry. *Ecol. Econ.* **2009**, *68*, 1116–1126. [CrossRef]
14. Kim, H.S.; Koo, W.W. Factors affecting the carbon allowance market in the US. *Energy Policy* **2010**, *38*, 1879–1884. [CrossRef]
15. Ji, Q.; Zhang, D.; Geng, J. Information linkage, dynamic spillovers in prices and volatility between the carbon and energy markets. *J. Clean. Prod.* **2018**, *198*, 972–978. [CrossRef]
16. Chevallier, J. Carbon futures and macroeconomic risk factors: A view from the EU ETS. *Energy Econ.* **2009**, *31*, 614–625. [CrossRef]
17. Ren, X.; Zhu, L. Influence of allowance allocation events on prices in China's carbon market pilots–an AR-GARCH-based analysis. *Energy Source Part B* **2020**, *15*, 157–171. [CrossRef]
18. Koop, G.; Tole, L. Forecasting the European carbon market. *J. R. Stat. Soc. Ser. A-Stat. Soc.* **2013**, *176*, 723–741. [CrossRef]
19. Benz, E.; Stefan, T. Modeling the price dynamics of CO_2 emission allowances. *Energy Econ.* **2009**, *31*, 4–15. [CrossRef]
20. Zhang, C.; Yun, P.; Wagan, Z.A. Study on the wandering weekday effect of the international carbon market based on trend moderation effect. *Financ. Res. Lett.* **2019**, *28*, 319–327. [CrossRef]
21. Tiwari, A.K.; Abakah, E.J.A.; Le, T.L.; Leyva-de la Hiz, D.I. Markov-switching dependence between artificial intelligence and carbon price: The role of policy uncertainty in the era of the 4th industrial revolution and the effect of COVID-19 pandemic. *Technol. Forecast. Soc.* **2021**, *163*, 120434. [CrossRef]
22. Han, M.; Ding, L.; Zhao, X.; Kang, W. Forecasting carbon prices in the Shenzhen market, China: The role of mixed-frequency factors. *Energy* **2019**, *171*, 69–76. [CrossRef]
23. Zhou, J.; Wang, S. A Carbon Price Prediction Model Based on the Secondary Decomposition Algorithm and Influencing Factors. *Energies* **2021**, *14*, 1328. [CrossRef]
24. Zhang, W.; Wu, Z. Optimal hybrid framework for carbon price forecasting using time series analysis and least squares support vector machine. *J. Forecast.* **2021**. [CrossRef]
25. Sun, W.; Wang, Y. Factor analysis and carbon price prediction based on empirical mode decomposition and least squares support vector machine optimized by improved particle swarm optimization. *Carbon Manag.* **2020**, *11*, 315–329. [CrossRef]
26. Zhang, J.; Li, D.; Hao, Y.; Tan, Z. A hybrid model using signal processing technology, econometric models and neural network for carbon spot price forecasting. *J. Clean. Prod.* **2018**, *204*, 958–964. [CrossRef]
27. Qin, Q.; He, H.; Li, L.; He, L.Y. A novel Decomposition-Ensemble based carbon price forecasting model integrated with local polynomial prediction. *Comput. Econ.* **2020**, *55*, 1249–1273. [CrossRef]
28. Wang, J.; Sun, X.; Cheng, Q.; Cui, Q. An innovative random forest-based nonlinear ensemble paradigm of improved feature extraction and deep learning for carbon price forecasting. *Sci. Total Environ.* **2021**, *762*, 143099. [CrossRef]
29. Hao, Y.; Tian, C.; Wu, C. Modelling of carbon price in two real carbon trading markets. *J. Clean. Prod.* **2020**, *244*, 118556. [CrossRef]
30. Sun, W.; Xu, C. Carbon price prediction based on modified wavelet least square support vector machine. *Sci. Total Environ.* **2021**, *754*, 142052. [CrossRef]
31. Hochreiter, S.; Schmidhuber, J. Long short-term memory. *Neural Comput.* **1997**, *9*, 1735–1780. [CrossRef] [PubMed]
32. Liu, J.; Wang, P.; Chen, H.; Zhu, J. A combination forecasting model based on hybrid interval multi-scale decomposition: Application to interval-valued carbon price forecasting. *Expert Syst. Appl.* **2021**, *191*, 116267. [CrossRef]
33. Zhu, B.; Ye, S.; Wang, P.; Chevallier, J.; Wei, Y.M. Forecasting carbon price using a multi-objective least squares support vector machine with mixture kernels. *J. Forecast.* **2022**, *41*, 100–117. [CrossRef]
34. Muncharaz, J.O. Comparing classic time series models and the LSTM recurrent neural network: An application to S&P 500 stocks. *Financ. Mark. Valuat.* **2020**, *6*, 137–148. [CrossRef]
35. Ji, L.; Zou, Y.; He, K.; Zhu, B. Carbon futures price forecasting based with ARIMA-CNN-LSTM model. *Proce. Comput. Sci.* **2019**, *162*, 33–38. [CrossRef]

36. Sun, W.; Huang, C. A novel carbon price prediction model combines the secondary decomposition algorithm and the long short-term memory network. *Energy* **2020**, *207*, 118294. [CrossRef]
37. Yang, S.; Chen, D.; Li, S.; Wang, W. Carbon price forecasting based on modified ensemble empirical mode decomposition and long short-term memory optimized by improved whale optimization algorithm. *Sci. Total Environ.* **2020**, *716*, 137117. [CrossRef]
38. Huang, Y.; Dai, X.; Wang, Q.; Zhou, D. A hybrid model for carbon price forecasting using GARCH and long short-term memory network. *Appl. Energy* **2021**, *285*, 116485. [CrossRef]
39. Harvey, C.R.; Siddique, A. Time-varying conditional skewness and the market risk premium. *Res. Bus. Financ.* **2002**, *1*, 27–60. [CrossRef]
40. Cho, K.; Van, M.B.; Gulcehre, C.; Bahdanau, D.; Bougares, F.; Schwenk, H.; Bengio, Y. Learning phrase representations using RNN encoder-decoder for statistical machine translation. In Proceedings of the 2014 Conference on Empirical Methods in Natural Language Processing (EMNLP), Doha, Qatar, 25–29 October 2014; Volume 1406, p. 1078.
41. Dey, R.; Salem, F.M. Gate-variants of gated recurrent unit (GRU) neural networks. In Proceedings of the 2017 IEEE 60th International Midwest Symposium on Circuits and Systems (MWSCAS), Boston, MA, USA, 6–9 August 2017; pp. 1597–1600. [CrossRef]
42. Wen, F.; Zhao, H.; Zhao, L.; Yin, H. What drive carbon price dynamics in China? *Int. Rev. Financ. Anal.* **2022**, *79*, 101999. [CrossRef]
43. Engle, R.F.; Manganelli, S. CAViaR: Conditional autoregressive value at risk by regression quantiles. *J. Bus. Econ. Stat.* **2004**, *22*, 367–381. [CrossRef]
44. Le, R.N.; Bengio, Y. Representational power of restricted Boltzmann machines and deep belief networks. *Neural Comput.* **2008**, *20*, 1631–1649. [CrossRef]

International Journal of
Environmental Research and Public Health

MDPI

Article

Evaluating the Influences of Natural Resources and Ageing People on CO_2 Emissions in G-11 Nations: Application of CS-ARDL Approach

Usman Mehmood [1,2,*], Ephraim Bonah Agyekum [3], Solomon Eghosa Uhunamure [4], Karabo Shale [4] and Ayesha Mariam [2]

1 Department of Political Science, University of Management and Technology, Lahore 54770, Pakistan
2 Remote Sensing, GIS and Climatic Research Lab (National Center of GIS and Space Applications), Centre for Remote Sensing, University of the Punjab, Lahore 54590, Pakistan; Ayeshamariam60@gmail.com
3 Department of Nuclear and Renewable Energy, Ural Federal University Named after the First President of Russia Boris, 19 Mira Street, 620002 Ekaterinburg, Russia; agyekumephraim@yahoo.com
4 Faculty of Applied Sciences, Cape Peninsula University of Technology, P.O. Box 652, Cape Town 8000, South Africa; uhunamures@cput.ac.za (S.E.U.); shalek@cput.ac.za (K.S.)
* Correspondence: usmanmehmood.umt@gmail.com

Abstract: Globalization as well as the ratio of ageing people in the group of 11 (G-11) countries has seen a rapid increase in recent years. Therefore, this study aims to provide effective policy recommendations for sustainable development goals 13, 8, and 7, for the G-11 countries. This work estimates the impact of natural resources and the ageing population on the emission of carbon dioxide (CO_2) in G-11 countries using panel data from 1990–2020. For empirical results, second-generation methods were applied. The Westerlund co-integration test that assesses co-integration confirms the firm association among the parameters, and the values of coefficient of the cross-sectional autoregressive distributed lag (CS-ARDL) approach show that a 1% increase in the ageing population will lower the emissions of CO_2 by 13.41% among G-11 countries. Moreover, the findings show that there exists an environmental Kuznets curve (EKC) among natural resources, globalization, economic growth, ageing people, and the emission of CO_2. Based on the findings, this work presents some important policy implications for achieving sustainable growth in the G-11 countries. These countries need to lower the amount of energy obtained from fossil fuels to improve air quality.

Keywords: G-11 countries; ageing population; natural resources; globalization; CS-ARDL

Citation: Mehmood, U.; Agyekum, E.B.; Uhunamure, S.E.; Shale, K.; Mariam, A. Evaluating the Influences of Natural Resources and Ageing People on CO_2 Emissions in G-11 Nations: Application of CS-ARDL Approach. *Int. J. Environ. Res. Public Health* **2022**, *19*, 1449. https://doi.org/10.3390/ijerph19031449

Academic Editors: Qin Zhu and Taoyuan Wei

Received: 11 December 2021
Accepted: 20 January 2022
Published: 27 January 2022

1. Introduction

Rapid industrialization has created hurdles on the way to achieving sustainable development. According to United Nations (UN), developed and developing nations are striving to address climate problems but industrialization is making their efforts fruitless. To accomplish the economic targets of various countries, different resources are being shared across borders. These trade activities have been possible through globalization. Globalization affects the process of production, which ultimately affects environmental quality [1–4].

Today, different economies are creating economic targets by enhancing cross border trade. Most countries, however, overlooked the factors that could affect environmental quality when pursuing their economic goals [4]. The group of 11 (G-11) countries was formed on 20 September 2006, when most of the countries were in their developing stages. This group was established to enhance their economic progress by cooperation. Since its formation, the G-11 countries have shown remarkable economic progress [5]. After joining the Paris agreement, the (G-11) nations have shown a strong commitment to reduce environmental pollution and they are revising their current economic and demographic policies. This includes environmental actions, usage of clean energy, and improved living standards.

Due to improving health facilities, the ageing population is increasing and the rise in the ageing population may have environmental consequences. According to the World Bank, the ageing population in G-11 countries has seen a rapid increase. This population is projected to reach 923 million by 2050 [6], and the governments of these countries are not well prepared for this demographic change. In an economic context, a labour supply shortage might be created by an upsurge in ageing people [7]. However, in the context of environmental quality, ecological degradation is caused by moving ageing people, household pattern variations, and building of separate homes for such people. Ageing people have environmental awareness but their preferences for a cleaner environment may vary. Ageing people have less preference for the use of private vehicles to move from one place to another. At the same time, they may need additional energy in terms of health-related facilities and if the energy is coming from non-renewable energy resources, it will degrade the environment. Therefore, it is essential to probe the linkages of ageing people with emissions of CO_2.

The objective of the current study is to evaluate the influence of the ageing population on CO_2 emissions in G-11 nations. Furthermore, considering the importance of other socio-economic variables, this work includes energy use, globalization, economic growth, energy innovations, and natural resources on CO_2 emissions. To the best of the author's knowledge, there is a gap in literature, and very few studies have investigated the factors of environmental degradation in the context of SDGs for G-11 countries. Also, for the analytical framework, the environmental Kuznets curve (EKC) has not been addressed. EKC proposes that after reaching a threshold level, economic activity may reduce environmental pollution. This may be due to environmental awareness or the use of efficient means of energy. This work probes the EKC among globalization, natural resources, economic growth, and ageing populations in G-11 nations.

For effective policy instruments, this work utilizes the second-generation methods because the first-generation methods may not incorporate the cross-sectional heterogeneities, which the study seeks to address. Therefore, the present study uses the cross-sectional autoregressive distributed lag (CS-ARDL) technique. This method controls the structural similarities to provide the effects of independent parameters on the dependent parameters. Considering these advantages, this work uses the CS-ARDL instead of the traditional autoregressive distributed lag (ARDL) [8].

To capture the evolutionary impacts on CO_2 emissions, a suitable theoretical model is required. Hence, the EKC by Grossman and Krueger [9] has been employed to determine the evolutionary associations among the variables. The paper is structured as follows: a literature background is described in Section 2, the methodology and data are described in Section 3, the fourth section comprises an analysis of the results, and the last section outlines the conclusions of the study.

2. Literature Review

2.1. Energy and Air Pollution

The EKC hypothesis has been discussed widely by environmental economists [9,10]. This assumption presented that economic growth influences environmental pollution in three ways: technique, composition, and scale [11,12]. Currently, several studies have posited that innovations in energy are the key factors that lower global warming [7,13]. According to Torras et al. [14], technical novelties lower environmental pollution but recent literature also suggests that the scale effect can be lowered by using low carbon emissions technologies.

2.2. Natural Resources and Air Pollution

Several studies have shown that more natural resources are important to impact economic growth. For example, Auty [15] found that rich natural resources slow down the pace of economic growth. However, Bravo-Ortega and de Gregorio [16], observed that natural resources increase income but have a negative effect on the national growth rate.

Shahbaz et al. [17], validated the natural resource curse hypothesis. Brunnschweiler and Bulte [18], described the difference between natural resource dependence and abundance. They presented that natural resource abundance increases economic growth whereas gross domestic product (GDP) is unaffected by natural resource dependence. Balsalobre-Lorente et al. [7], argued that the abundance of natural resources reduces CO_2 emissions in European countries. They argued that countries with ample natural resources utilize them instead of fossil fuels and maintain economic growth. However, Danish et al. [19], presented contradictory evidence for Brazil, Russia, India, China and South Africa (the BRICS nations).

2.3. Globalization and Air Pollution

Globalization is increasing political, social, and economic integration across the globe. Dreher [20], reported that globalization put a positive impact on the growth of an economy. Dollar and Kraay [21] have observed the positive impact of globalization on economic growth. Similarly, Alam [22] found a positive nexus between environmental degradation and globalization. Kahuthu [23], investigated the association between CO_2 emissions and economic growth. They observed that globalization is playing a moderating role in this association by importing efficient technologies. Globalization is increasing GDP but lowering CO_2 emissions. Shahbaz et al. [24], found that globalization has degraded the environment. Shahbaz et al. [11], investigated the emissions of CO_2–globalization nexus in Indian economic growth. They also found that globalization is degrading the environment. Shahbaz et al. [25] suggested that globalization is triggering foreign direct investment, which enhances the reckless use of non-renewable energy, which contaminates the quality of the environment. However, for the Australian economy Shahbaz et al. [26], found that globalization is environmentally friendly. They argued that due to effective resource policy and administrative grip, globalization is a blessing to Australia.

2.4. Ageing and Air Pollution

Some studies have shown the association between ageing and air pollution [27]. York et al. [28] and Shi et al. [29] found that an ageing population can create more emissions of CO_2. Fan et al. [30] argued that the working class is lowering air quality in developing countries, but this class is improving air quality in developed nations. However, reference [31] argued that the elderly people use fewer resources and prefer public transportation and therefore they are environmentally friendly. Hassan and Salim [32], found that aged people are reducing emissions of CO_2 by 1.55%. O'Neill et al. [33], found that aged people do not participate in labour activities, and they slow economic growth with little to no emissions of CO_2.

Contrary to this, various studies have shown the adverse impacts of ageing people on the emission of CO_2. Farzin et al. [27] indicated that a society with more aged people will generate more CO_2 emissions. Menz and Welsch [34], presented that aged people use up more energy and increase CO_2 emissions. Menz et al. [35] found that the ageing people–CO_2 emissions linkage depends upon the country's position relative to development. Thalmann [36], presented that the wish for a cleaner environment decreases with age. They further highlighted that although elderly people are affected by environmental changes, they are not going to obtain the benefits of environmental regulations in the future. This thought further diminishes environmental awareness. According to Liddle and Lung [37], middle-age people require less energy requirements, but at an early and elderly age, they require more energy. Liddle [38], found the U-shaped linkages between ageing people and domestic energy consumption. It was observed that the youngest and elderly people positively affect energy demand.

Considering the potential impact of ageing people on the environment, this work attempts to enhance the current wave of knowledge.

2.5. Research Gap

The aforementioned literature above shows that different studies have contradictory results about the factors of environmental degradation. The contradictory results may be due to the level of development and the sample of variables collected from the countries. To the best of the authors' knowledge, there is a gap in literature, and very few studies have investigated the factors of environmental degradation in the context of SDGs for G-11 countries. Moreover, for the analytical framework, the EKC hypothesis has not been addressed. This gap in the literature is addressed by incorporating ageing people.

3. Data and Empirical Estimation

This work attempts to probe the impact of ageing people on CO_2 emissions by controlling the other socio-economic factors of globalization, natural resources, GDP, and energy use. In doing so, this work utilizes the annual data of 1990–2020 for G-11 nations of Croatia, Jordan, Ecuador, El Salvador, Georgia, Honduras, Morocco, Indonesia, Paraguay, Pakistan, and Sri Lanka. The data for GDP per capita (constant terms), CO_2 emissions (kilo tons), natural resources (% of GDP), energy use (kg of oil equivalent per capita), research, and development (number of patents), and the ageing population 65 and above were used. All the data were obtained from the World Bank [39] except the data for globalization, which was obtained from the KOF Economic Institute [40]. Figure 1a–e illustrates an increasing trend in globalization, GDP per capita, natural resource abundance, research and development, ageing population, and CO_2 emissions in G-11 countries. Croatia showed the highest globalization during 1990 to 2020. Similarly, Indonesia had the highest economic growth among the G-11 countries during the study period. The highest natural resource was in Ecuador. Indonesia had the highest number of ageing people during the study period.

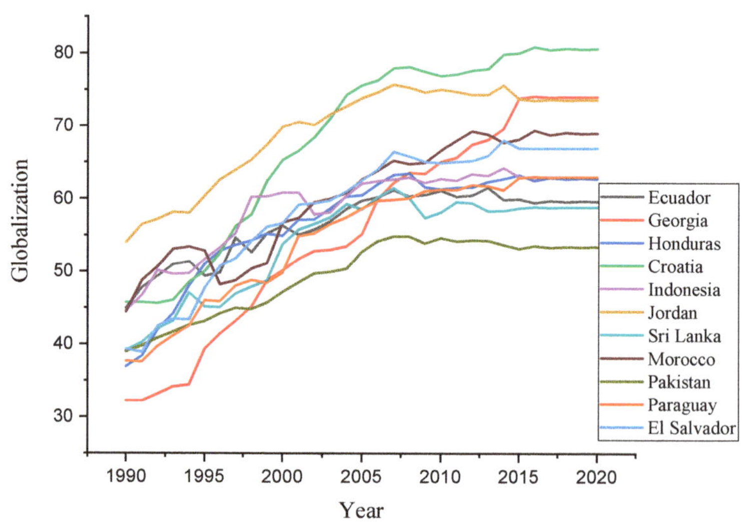

(a)

Figure 1. *Cont.*

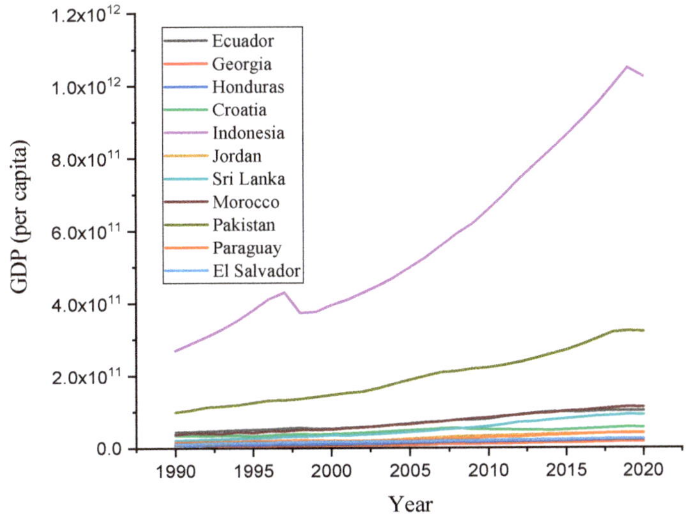

(b)

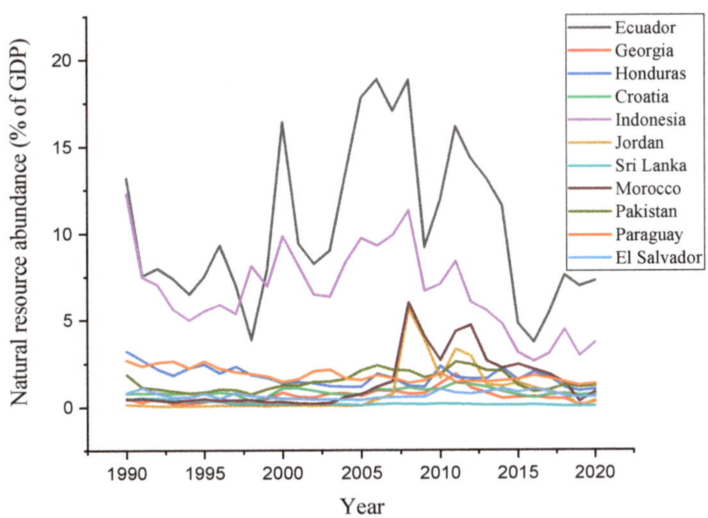

(c)

Figure 1. *Cont.*

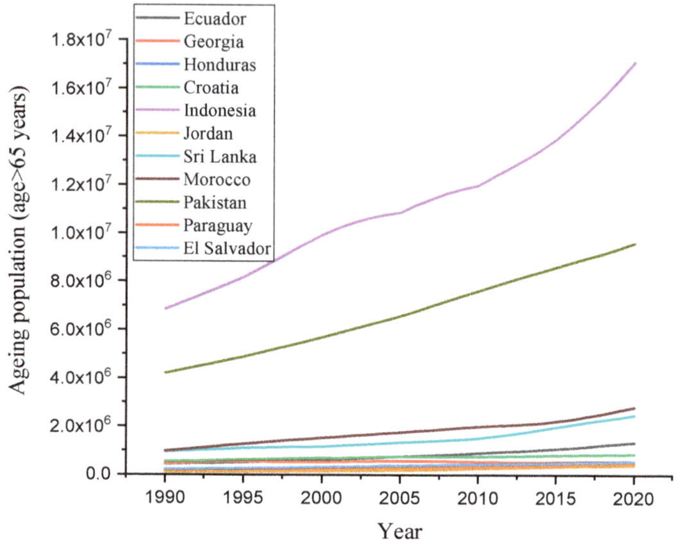

(**d**)

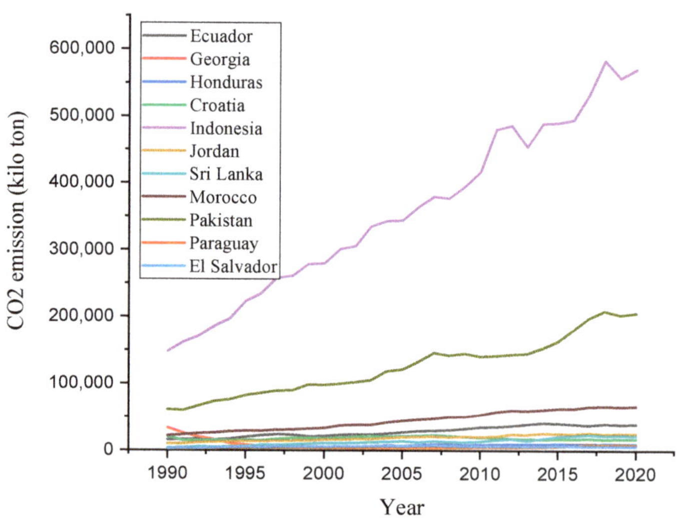

(**e**)

Figure 1. Graphical representation of variables used for the study for the various countries (**a**) globalization, (**b**) GDP per capita, (**c**) natural resource abundance, (**d**) ageing population, (**e**) CO_2 emissions.

Before the econometric analysis, all data were transformed into their natural logarithms. This form eliminates the problems of multicollinearity and provides robust findings [41–43]. This work follows the study of Balsalobre-Lorente et al. [44] in applying the empirical model, which is as follows:

$$\ln CO_{2t} = \beta_0 + \beta_1 \ln GL_t + \beta_2 \ln Gl_t^2 + \beta_3 \ln GDP_t + \beta_4 \ln GDP_t^2 + \beta_5 \ln NAT_t + \beta_6 \ln NAT_t^2 + \beta_7 \ln AG_t + \beta_8 \ln AG_t^2 + \beta_9 \ln EN + \beta_{10} \ln RD_t + ie_t \quad (1)$$

where, CO_2, GL, GDP, NAT, AG, EN, and RD represent the CO_2 emissions (kilo ton), the overall index of globalization, GDP per capita (constant term), natural resource abundance (% of GDP), ageing population 65 years and above, energy use (kg of oil equivalent per capita), and research and development (number of patents).

Table 1 presents the descriptive statistics of the variables, which shows that GDP, globalization, and number of ageing people in the population have the highest values.

Table 1. Descriptive statistics.

Parameters	Mean	Minimum	Maximum	Standard Deviation	Skewness	Kurtosis
CO_2e	57,553	2020	583,110	108,944.2	2.97	8.72
GDP	100,953,498,375.5	4,689,605,208.6	1,049,318,966,508.5	176,187,954,357.7	3.29	11.55
GL	58.47	32.23	80.89	10.28	−0.10	−0.26
NAT	2.44	0.03	18.85	3.44	2.50	6.65
AG	2,220,265.9	113,959	17,129,349	3,495,883.2	2.18	4.0

Table 2 shows the description and sources of data. For econometric analysis, this work adopts the second-generation methodology. There is a reason to use second-generation methods because the datasets obtained for South Asian countries may suffer from cross-section dependence (CD) due to common traditional methods, social norms, and economic policies. It might not be able to provide robust results. A second-generation unit root test is applied to find the order of integration among the panel data. The CS-generation technique is applied to present the values of long- and short-run coefficients.

Table 2. Description of the parameters under study.

Parameters	Symbol	Unit	Source
Carbon Dioxide emissions	CO_2	kilo ton (kt)	World Bank
Globalization	GL	Overall Index (Economic, political, and social globalization)	KOF institute
Gross Domestic Product	GDP	Constant 2015 US$	World Bank
Natural Resource abundance	NAT	Natural resource rents (%GDP)	World Bank
Research and Development	RD	Number of patents (residents)	World Bank
Ageing population	AG	Population more than 65 years	World Bank

Cross-Sectional Autoregressive Distributed Lag (CS-ARDL)

This work seeks to probe the linkages between economic growth, natural resources, energy use, globalization, ageing people, and CO_2 emissions for a panel of G-11 nations. Panel estimations can generate unreliable results because of the existence of cross-section dependence and slope heterogeneity issues. These issues are not considered by the traditional estimation techniques of FMOL and DOLS [45]. The issue of slope heterogeneity and CD is efficiently handled by the CS-ARDL approach, which is not catered for by the FMOL and DOLS techniques. Therefore, the current study used the CS-ARDL method to calculate the values of long- and short-run coefficients. This method caters for heterogeneity and CD problems by applying dynamic common correlated impact predictors [46,47]. Equation (1) represents the mathematical form of the CS-ARDL:

$$H_{i,t} = \sum_{I=0}^{p_w} \gamma_{I,i} W_{i,t-1} + \sum_{I=0}^{p_z} \beta_{I,i} Z_{i,t-I} + \varepsilon_{i,t} \quad (2)$$

Equation (1) represents the ARDL model; if we use Equation (5), by taking CD, it will produce uncertain results. Equation (4) is revised using averages of CS one by one repressor parameters. This will permit us to remove inappropriate interpretations concerning the existence of the threshold effect generated by CD [8].

$$H_{it} = \sum_{I=0}^{a_w} \gamma_{I,i}, H_{i,t-1} + \sum_{I=0}^{a_z} \beta_{I,i} Z_{i.t-I} + \sum_{I=0}^{a_x} \alpha'_i, I\overline{X}_{t-I} + \varepsilon_{i,t} \tag{3}$$

where the average value of dependent and independent parameters can be calculated by using the following equation:

$$\overline{X}_{t-I} = \overline{H}_{i,t-I}\overline{Z}_{i,t-I}$$

Existing lags among all the variables are denoted by a_w, a_x and a_z. H_{it} denotes emission of carbon per capita depending upon its utilization and $Z_{i.t}$ represents all the independent variables. Furthermore, X denote the average of CS (disregarding the trends) to overwhelm the spillover issues [48]. The CS-ARDL method estimates the long-run coefficients by using short-run coefficients as its input. Equations (4)–(6) represent the mean group (MG) predictor and the value of long-run and short-run coefficients, respectively:

$$\hat{\varphi}_{MG} = \frac{1}{N} \sum_{i=1}^{N} \hat{\varphi}_i \tag{4}$$

$$\hat{\varphi}_{CS-ARDL,i} = \frac{\sum_{I=0}^{p_z} \beta_{I,i}^{pw}}{1 - \sum_{I=0}} \hat{\gamma}_{I,i} \tag{5}$$

$$\Delta H_{i,t} = \vartheta_i [H_{i,t-1} - \varphi_i Z_{i,t}] - \sum_{I=1}^{a_{w-1}} \gamma_{I,i} \Delta_I H_{i,t-1} + \sum_{I=1}^{a_{ws-1}} \beta_{I,i} \Delta_I Z_{i,t} + \sum_{I=0}^{a_x} \alpha'_i, I\overline{X}_{t-I} + \varepsilon_{i,t} \tag{6}$$

where $\Delta_I = t - (t-1)$,

$$\hat{\delta}_i = -(1 - \sum_{I=1}^{a_w} \hat{\gamma}_{I,t}) \tag{7}$$

$$\varphi_i = \frac{\sum_{I=0}^{a_z} \beta_{I,i}^{aw}}{\hat{\delta}_i} \tag{8}$$

$$\hat{\overline{\varphi}}_{MG} = \frac{1}{N} \sum_{i=1}^{N} \hat{\varphi} \tag{9}$$

In the CS-ARDL approach, the economy achieves an equilibrium state as soon as the value of the error correction mechanism (ECM) approaches −1.

4. Results and Discussion

For panel data analysis, it is important to be sure of the CD. For this purpose, Table 3 shows the results that reject the null hypothesis of CD among the selected variables, i.e., CO_2, GL, GDP, AG, NAT, and RD, which confirms that the entire data have a CD at a 1% level. Thus, the results imply that a shock in one country will spill over to the other countries as well. These empirical findings agree with that of Mehmood et al. [49] and Musah at al. [50].

Table 3. Results obtained for cross-section dependence (CD) analysis.

Variable	Test Statistics (*p*-Values)
CO_2	20.45 *** (0.00)
GL	16.76 *** (0.00)
GDP	19.65 *** (0.00)
NAT	44.23 *** (0.00)
AG	27.67 *** (0.00)
RD	32.34 *** (0.00)

*** is significant at 1%.

Before the application of long-run analysis, it is required to know the integration order of the data. Therefore, this work applies the CIPS unit root test. Table 4 indicates the results of the unit root test which reveals that the panel data are integrated at the first difference. The results shows that in the CIPS unit root test almost all variables are integrated at the first difference except NAT, which is also integrated at this level. This indicates that except NAT, all remaining variables of interest acquired stationarity after the first difference indicating the integration sequence among the data. The findings are supported by the following studies of Musah et al. [50] and Adamu et al. [51].

Table 4. CIPS unit root test results from the study.

Variable	CIPS Test	
	At Level	1st Difference
CO_2	−2.94	−5.61 ***
GL	−2.78	−5.86 ***
GDP	−2.65	−6.90 ***
NAT	−3.52 ***	−6.16 ***
AG	−1.01	−3.45 **
RD	−3.12 ***	−6.10 ***

** and *** are significant at 5% and 1% levels, respectively.

After the CD and unit root test, the Hashem Pesaran and Yamagata [52] test was incorporated. A slope heterogeneity test was done to examine the slope heterogeneity between the selected variables. Table 5 depicts the analysis of heterogeneity of slope as measured by Pesaran and Yamagata [52]. This test was used to assess the coefficients of heterogeneous and homogenous slopes from the study. This test confirms the heterogeneity at the 1% significance level.

Table 5. Results obtained to show the slope heterogeneity.

Statistics	Test Value (*p*-Value)
Delta-tilde	23.46 *** (0.00)
Delta-tilde Adjusted	26.57 *** (0.00)

*** is significant at 1% level.

The findings of Westerlund and Edgerton [53] are presented in Table 6, which depicts the null hypotheses of no co-integration between the parameters in the existence of serial correlation, CD, and heterogeneity. The findings reject the null hypotheses with no mean shift and regime shift. This verifies the existence of a co-integrating association among the CO_2, GL, GDP, NAT, AG, EN, and RD at a 1% significance level. The results are consistent with Menz and Welsch [34].

Table 6. Westerlund and Edgerton [53] results obtained for panel co-integration test.

Test	No Shift	Mean Shift	Regime Shift
$Z_{\varphi}(N)$	−3.56 ***	−2.87 ***	−4.02 ***
P_{value}	0	0	0
$Z_{\tau}(N)$	−4.67 ***	−3.67 ***	−4.01 ***
P_{value}	0	0	0

*** is significant at 1% level

Table 7 shows the findings of the CS-ARDL, which shows different insights. The findings are presented sequentially considering the influence of globalization on the emissions of CO_2. In both short- and long-run estimations, the coefficient of globalization is positive and statistically significant. It is evident that globalization is exerting a positive impact on CO_2 emissions, but the square of globalization is negatively correlated with CO_2 emissions. This means that the evolutionary impact of globalization is inverted in the U-shape of CO_2 emissions in G-11 countries. Globalization increases economic opportunities and also makes room for importing efficient technologies to produce clean energy. This result is similar to the results of Sinha et al. [54]. The coefficient of GDP is positive having a value of 9.75% at a 5% significance level, whereas the square of the GDP is negatively associated with environmental degradation. This implies that the evolutionary impact of economic growth is also inverted in the U-shape, which means that recent economic growth is contaminating the environment. In the future, economic growth will improve air quality. Empirical findings by Balsalobre-Lorente et al. [7], Mehmood and Tariq [55] and Qayyum et al. [56] align with the findings of the study. The finding from the study also indicates that G-11 countries are spending on non-renewable resources in the energy sector but in the future, the ratio of renewable energy to the final energy output will increase which will lead to an improvement in air quality. The findings align with that of Mehmood et al. [55] and Abid et al. [57].

Table 7. Cross-sectional autoregressive distributed lag (CS-ARDL) results from the study.

Short Run	Coefficient	Std. Error	Significance Level
ΔCO_2	−0.95 ***	0.09	0.00
ΔGL	0.24 **	0.25	0.05
ΔGL^2	−2.32	2.24	0.78
ΔGDP	9.75 **	3.84	0.01
ΔGDP^2	−1.60 ***	0.70	0.02
ΔNAT	−0.05 ***	0.02	0.09
ΔNAT^2	0.02	0.03	0.53
ΔAG	−23.20 ***	25.40	0.36
ΔAG^2	1.60	1.84	0.35
ΔRD	−0.07 **	0.01	0.53
ΔEN	2.32 ***	0.87	0.00
	Long Run		
CO_2	−0.04 **	0.05	0.09
GL	0.09	0.13	0.09
GL^2	−2.32 **	2.19	0.03

Table 7. *Cont.*

Short Run	Coefficient	Std. Error	Significance Level
GDP	5.24 ***	2.11	0.00
GDP^2	−0.84 **	0.38	0.06
NAT	−0.07	0.01	0.53
NAT^2	0.01	0.02	0.54
AG	−13.41**	13.48	0.03
AG^2	0.91	0.99	0.02
RD	−0.46 ***	0.06	0.03
EN	1.65 **	0.74	0.01
ECT	−0.95 ***	0.09	0.00

** and *** are significant at 5% and 1% levels, respectively.

The evolutionary effect of natural resources on the emissions of CO_2 is an inverted U-shape. This means that dependence on natural resources is lowering air quality, but less consumption of natural resources is improving air quality in G-11 countries. The influence of natural resources on the emissions of CO_2 can be better explained if the joint effects of innovations in the energy sector and the use of energy on the emissions of CO_2 are examined. It can be observed that the impact of research and development on the emissions of CO_2 are negative. This means that research and development are lowering CO_2 emissions in G-11 countries.

Lastly, the effect of the ageing population on CO_2 emissions shows that, currently, elderly people positively improve the air quality but in the future this association becomes inverse. This outcome rejects the existence of EKC between ageing population and CO_2 emissions. This confirms that the current changing demographic patterns in the G-11 countries are environmentally friendly. However, in the future, due to the ageing population growth, they will negatively affect air quality. This outcome is similar to the findings of Hamza et al. [58].

5. Conclusions

During the last few years, the G-11 countries have made commitments to lower the concentration of CO_2 emissions and to improve air quality. These commitments require a comprehensive environmental policy. Therefore, considering the importance of SDGs in the G-11 countries, this work incorporates globalization and the ageing population to present some important recommendations. This work will be helpful for policymakers to achieve SDGs 13, 8, and 7. This study has proposed a comprehensive policy recommendation by analyzing the role of globalization, research and development, and ageing people. The study shows that the G-11 countries are spending on non-renewable resources in the energy sector but, in the future, the ratio of renewable energy to the final energy output will increase which will lead to an improvement in air quality. This work also revealed the need for policymakers to improve the ratio of renewable energy to the final energy output utilized in the industrial sectors. In increasing the ratio of renewable energy, the governments of these countries need to give special attention to employment opportunities because this aspect can be a hurdle in achieving sustainable development. The impacts of research and development on the emissions of CO_2 are negative. This means that research and development are lowering CO_2 emissions in the G-11 countries. Research and development will help invent renewable energy technologies. Currently, ageing people are environmentally friendly but in the future ageing people will start to contaminate air quality by increasing the CO_2 emissions. This result is important for policymakers, and they should divert their attention towards the environmental awareness of ageing people.

The role of natural resources is very important for achieving sustainable development. The evolutionary effect of natural resources on the emissions of CO_2 is not an inverted U-shape. Currently, natural resources are environmentally friendly but in future due to mismanagement resources will also contaminate the environmental quality by increasing CO_2 emissions. This result has also highlighted the policy instruments to preserve and use natural resources sustainably. The abundance of natural resources helps to reduce greenhouse gases and also serves as a catalyst for sustainable growth. Therefore, nations should be aware of the need to conserve natural resources.

This work validates the existence of EKC between globalization, GDP, and environmental quality. This means that globalization is currently creating environmental problems but, in the future, will start to improve air quality by reducing CO_2 emissions. This finding sheds light on the importance of globalization for the G-11 nations. It is expected that the G-11 nations should explore more markets to export their products, especially to the developed nations. This will provide the nations with the opportunities to import cleaner technologies to deal with CO_2 emissions.

Apart from the contribution of this study, future research can be applied to highly globalized and developed countries. Moreover, future works can be undertaken by utilizing the other panel data analysis.

Author Contributions: Conceptualization, U.M.; Data curation, U.M.; Formal analysis, U.M., E.B.A., S.E.U., K.S. and A.M.; Funding acquisition, S.E.U. and K.S.; Investigation, U.M., E.B.A. and A.M.; Methodology, U.M.; Project administration, U.M.; Software, U.M.; Validation, U.M., E.B.A. and A.M.; Writing—original draft, U.M., E.B.A., S.E.U., K.S. and A.M.; Writing—review and editing, U.M., E.B.A., S.E.U., K.S. and A.M. All authors have read and agreed to the published version of the manuscript.

Funding: This research received no external funding.

Institutional Review Board Statement: Not applicable.

Informed Consent Statement: Not applicable.

Data Availability Statement: Sources of data used are provided in the text.

Acknowledgments: We are thankful to World Bank for providing the data.

Conflicts of Interest: The authors declare no conflict of interest.

References

1. Mehmood, U.; Mansoor, A.; Tariq, S.; Ul-Haq, Z. The Interactional Role of Globalization in Tourism-CO2 Nexus in South Asian Countries. *Environ. Sci. Pollut. Res.* **2021**, *28*, 26441–26448. [CrossRef] [PubMed]
2. Mehmood, U. Globalization-Driven CO_2 Emissions in Singapore: An Application of ARDL Approach. *Environ. Sci. Pollut. Res.* **2020**, *28*, 11317–11322. [CrossRef] [PubMed]
3. Adebayo, T.S.; Awosusi, A.A.; Oladipupo, S.D.; Agyekum, E.B.; Jayakumar, A.; Kumar, N.M. Dominance of Fossil Fuels in Japan's National Energy Mix and Implications for Environmental Sustainability. *Int. J. Environ. Res. Public Health* **2021**, *18*, 7347. [CrossRef] [PubMed]
4. Adebayo, T.S.; Agboola, M.O.; Rjoub, H.; Adeshola, I.; Agyekum, E.B.; Kumar, N.M. Linking Economic Growth, Urbanization, and Environmental Degradation in China: What Is the Role of Hydroelectricity Consumption? *Int. J. Environ. Res. Public Health* **2021**, *18*, 6975. [CrossRef] [PubMed]
5. Mehmood, U. Contribution of Renewable Energy towards Environmental Quality: The Role of Education to Achieve Sustainable Development Goals in G11 Countries. *Renew. Energy* **2021**, *178*, 600–607. [CrossRef]
6. ADB Data and Research | Asian Development Bank. Available online: https://www.adb.org/data/main (accessed on 17 August 2019).
7. Balsalobre-Lorente, D.; Shahbaz, M.; Roubaud, D.; Farhani, S. How Economic Growth, Renewable Electricity and Natural Resources Contribute to CO2 Emissions? *Energy Policy* **2018**, *113*, 356–367. [CrossRef]
8. Chudik, A.; Pesaran, M.H. Common Correlated Effects Estimation of Heterogeneous Dynamic Panel Data Models with Weakly Exogenous Regressors. *J. Econom.* **2015**, *188*, 393–420. [CrossRef]
9. Grossman, G.; Krueger, A. *Environmental Impacts of a North American Free Trade Agreement*; MIT Press: Cambridge, MA, USA, 1991.
10. Sinha, A.; Gupta, M.; Shahbaz, M.; Sengupta, T. Impact of Corruption in Public Sector on Environmental Quality: Implications for Sustainability in BRICS and next 11 Countries. *J. Clean. Prod.* **2019**, *232*, 1379–1393. [CrossRef]

11. Shahbaz, M.; Mallick, H.; Mahalik, M.K.; Loganathan, N. Does Globalization Impede Environmental Quality in India? *Ecol. Indic.* **2015**, *52*, 379–393. [CrossRef]
12. Álvarez-Herránz, A.; Balsalobre, D.; Cantos, J.M.; Shahbaz, M. Energy Innovations-GHG Emissions Nexus: Fresh Empirical Evidence from OECD Countries. *Energy Policy* **2017**, *101*, 90–100. [CrossRef]
13. Chen, M.; Sinha, A.; Hu, K.; Shah, M.I. Impact of Technological Innovation on Energy Efficiency in Industry 4.0 Era: Moderation of Shadow Economy in Sustainable Development. *Technol. Forecast. Soc. Chang.* **2021**, *164*. [CrossRef]
14. Torras, M.; Boyce, J.K. Income, Inequality, and Pollution: A Reassessment of the Environmental Kuznets Curve. *Ecol. Econ.* **1998**, *25*, 147–160. [CrossRef]
15. Auty, R.M. *Resource Abundance and Economic Development*; Oxford University Press: Oxford, UK, 2005; ISBN 9780199275786.
16. Bravo-Ortega, C.; de Gregorio, J. *The Relative Richness of the Poor? Natural Resources, Human Capital, and Economic Growth*; Policy Research Working Papers; The World Bank: Washington, DC, USA, 2005.
17. Shahbaz, M.; Destek, M.A.; Okumus, I.; Sinha, A. An Empirical Note on Comparison between Resource Abundance and Resource Dependence in Resource Abundant Countries. *Resources Policy* **2019**, *60*, 47–55. [CrossRef]
18. Brunnschweiler, C.N.; Bulte, E.H. The Resource Curse Revisited and Revised: A Tale of Paradoxes and Red Herrings. *J. Environ. Econ. Manag.* **2008**, *55*, 248–264. [CrossRef]
19. Baloch, M.A.; Mahmood, N.; Zhang, J.W. Effect of Natural Resources, Renewable Energy and Economic Development on CO_2 Emissions in BRICS Countries. *Sci. Total Environ.* **2019**, *678*. [CrossRef]
20. Dreher, A. Does Globalization Affect Growth? Evidence from a New Index of Globalization. *Appl. Econ.* **2006**, *38*, 1091–1110. [CrossRef]
21. Dollar, D.; Kraay, A. Trade, Growth, and Poverty. *Econ. J.* **2004**, *114*, 22–49. [CrossRef]
22. Alam, S. Globalization, Poverty and Environmental Degradation: Sustainable Development in Pakistan. *J. Sustain. Dev.* **2010**, *3*, 103. [CrossRef]
23. Kahuthu, A. Economic Growth and Environmental Degradation in a Global Context. *Environ. Dev. Sustain.* **2006**, *8*, 55–68. [CrossRef]
24. Shahbaz, M.; Ozturk, I.; Afza, T.; Ali, A. Revisiting the Environmental Kuznets Curve in a Global Economy. *Renew. Sustain. Energy Rev.* **2013**, *25*, 494–502. [CrossRef]
25. Shahbaz, M.; Balsalobre, D.; Shahzad, S.J.H. The Influencing Factors of CO_2 Emissions and the Role of Biomass Energy Consumption: Statistical Experience from G-7 Countries. *Environ. Modeling Assess.* **2018**, *24*, 143–161. [CrossRef]
26. Shahbaz, M.; Bhattacharya, M.; Ahmed, K. CO_2 Emissions in Australia: Economic and Non-Economic Drivers in the Long-Run. *Appl. Econ.* **2017**, *49*, 1273–1286. [CrossRef]
27. Farzin, Y.H.; Bond, C.A.; Farzin, Y.H.; Bond, C.A. Democracy and Environmental Quality. *J. Dev. Econ.* **2006**, *81*, 213–235. [CrossRef]
28. York, R.; Rosa, E.A.; Dietz, T. STIRPAT, IPAT and ImPACT: Analytic Tools for Unpacking the Driving Forces of Environmental Impacts. *Ecol. Econ.* **2003**, *46*, 351–365. [CrossRef]
29. Shi, A. The Impact of Population Pressure on Global Carbon Dioxide Emissions, 1975–1996: Evidence from Pooled Cross-Country Data. *Ecol. Econ.* **2003**, *44*, 29–42. [CrossRef]
30. Fan, Y.; Liu, L.C.; Wu, G.; Wei, Y.M. Analyzing Impact Factors of CO_2 Emissions Using the STIRPAT Model. *Environ. Impact Assess. Rev.* **2006**, *26*, 377–395. [CrossRef]
31. Dalton, M.; O'Neill, B.; Fürnkranz-Prskawetz, A.; Jiang, L.; Pitkin, J.; Dalton, M.; O'Neill, B.; Fürnkranz-Prskawetz, A.; Jiang, L.; Pitkin, J. Population Aging and Future Carbon Emissions in the United States. *Energy Econ.* **2008**, *30*, 642–675. [CrossRef]
32. Hassan, K.; Salim, R. Population Ageing, Income Growth and CO_2 Emission: Empirical Evidence from High Income Oecd Countries. *J. Econ. Stud.* **2015**, *42*, 54–67. [CrossRef]
33. O'Neill, B.C.; Dalton, M.; Fuchs, R.; Jiang, L.; Pachauri, S.; Zigova, K. Global Demographic Trends and Future Carbon Emissions. *Proc. Natl. Acad. Sci. USA* **2010**, *107*, 17521–17526. [CrossRef]
34. Menz, T.; Welsch, H. Population Aging and Carbon Emissions in OECD Countries: Accounting for Life-Cycle and Cohort Effects. *Energy Econ.* **2012**, *34*, 842–849. [CrossRef]
35. Menz, T.; Welsch, H.; Menz, T.; Welsch, H. Population Aging and Environmental Preferences in OECD Countries: The Case of Air Pollution. *Ecol. Econ.* **2010**, *69*, 2582–2589. [CrossRef]
36. Thalmann, P. The Public Acceptance of Green Taxes: 2 Million Voters Express Their Opinion. *Public Choice* **2004**, *119*, 179–217. [CrossRef]
37. Liddle, B.; Lung, S. Age-Structure, Urbanization, and Climate Change in Developed Countries: Revisiting STIRPAT for Disaggregated Population and Consumption-Related Environmental Impacts. *Popul. Environ.* **2010**, *31*, 317–343. [CrossRef]
38. Liddle, B. Urban Transport Pollution: Revisiting the Environmental Kuznets Curve. *Int. J. Sustain. Transp.* **2015**, *9*, 502–508. [CrossRef]
39. World Bank Indicators | Data. Available online: https://data.worldbank.org/indicator?tab=all (accessed on 12 January 2022).
40. KOF. KOF Globalisation Index. Available online: https://kof.ethz.ch/en/forecasts-and-indicators/indicators/kof-globalisation-index.html (accessed on 12 January 2022).
41. Mehmood, U. Transport Energy Consumption and Carbon Emissions: The Role of Urbanization towards Environment in SAARC Region. *Integr. Environ. Assess. Manag.* **2021**. [CrossRef] [PubMed]

42. Mehmood, U. Examining the Role of Financial Inclusion towards CO_2 Emissions: Presenting the Role of Renewable Energy and Globalization in the Context of EKC. *Environ. Sci. Pollut. Res.* **2021**, 1–9. [CrossRef]
43. Mehmood, U.; Askari, M.U.; Saleem, M. The Assessment of Environmental Sustainability: The Role of Research and Development in ASEAN Countries. *Integr. Environ. Assess. Manag.* **2021**. [CrossRef]
44. Balsalobre-Lorente, D.; Sinha, A.; Driha, O.M.; Mubarik, M.S. Assessing the Impacts of Ageing and Natural Resource Extraction on Carbon Emissions: A Proposed Policy Framework for European Economies. *J. Clean. Prod.* **2021**, *296*, 126470. [CrossRef]
45. An, H.; Razzaq, A.; Haseeb, M.; Mihardjo, L.W. The Role of Technology Innovation and People's Connectivity in Testing Environmental Kuznets Curve and Pollution Heaven Hypotheses across the Belt and Road Host Countries: New Evidence from Method of Moments Quantile Regression. *Environ. Sci. Pollut. Res.* **2021**, *28*, 5254–5270. [CrossRef]
46. Yao, Y.; Ivanovski, K.; Inekwe, J.; Smyth, R. Human Capital and Energy Consumption: Evidence from OECD Countries. *Energy Econ.* **2019**, *84*, 104534. [CrossRef]
47. Çoban, S.; Topcu, M. The Nexus between Financial Development and Energy Consumption in the EU: A Dynamic Panel Data Analysis. *Energy Econ.* **2013**, *39*, 81–88. [CrossRef]
48. Liddle, B. Consumption-Based Accounting and the Trade-Carbon Emissions Nexus in Asia: A Heterogeneous, Common Factor Panel Analysis. *Sustainability* **2018**, *10*, 3627. [CrossRef]
49. Mehmood, U.; Tariq, S.; Haq, Z. Effects of Population Structure on CO_2 Emissions in South Asian Countries: Evidence from Panel Estimation. *Environ. Sci. Pollut. Res.* **2021**, *28*, 66858–66863. [CrossRef] [PubMed]
50. Musah, M.; Owusu-Akomeah, M.; Boateng, F.; Iddris, F.; Mensah, I.A.; Antwi, S.K.; Agyemang, J.K. Long-Run Equilibrium Relationship between Energy Consumption and CO_2 Emissions: A Dynamic Heterogeneous Analysis on North Africa. *Environ. Sci. Pollut. Res.* **2021**, *29*, 10416–10433. [CrossRef] [PubMed]
51. Adamu, T.M.; Shafiq, M. Analyzing the Impact of Energy, Export Variety, and FDI on Environmental Degradation in the Context of Environmental Kuznets Curve Hypothesis: A Case Study of India. *Energies* **2019**, *12*, 1076. [CrossRef]
52. Hashem Pesaran, M.; Yamagata, T. Testing Slope Homogeneity in Large Panels. *J. Econom.* **2008**, *142*, 50–93. [CrossRef]
53. Westerlund, J.; Edgerton, D.L. A Simple Test for Cointegration in Dependent Panels with Structural Breaks. *Oxf. Bull. Econ. Stat.* **2008**, *5*, 665–704. [CrossRef]
54. Sinha, A.; Shahbaz, M.; Balsalobre, D. Exploring the Relationship between Energy Usage Segregation and Environmental Degradation in N-11 Countries. *J. Clean. Prod.* **2017**, *168*, 1217–1229. [CrossRef]
55. Mehmood, U.; Tariq, S. Globalization and CO_2 Emissions Nexus: Evidence from the EKC Hypothesis in South Asian Countries. *Environ. Sci. Pollut. Res.* **2020**, *27*, 37044–37056. [CrossRef]
56. Qayyum, F.; Mehmood, U.; Tariq, S.; Nawaz, H. Particulate Matter (PM2. 5) and Diseases: An Autoregressive Distributed Lag (ARDL) Technique. *Environ. Sci. Pollut. Res.* **2021**, *28*, 67511–67518. [CrossRef]
57. Abid, A.; Mehmood, U.; Tariq, S.; Haq, Z.U. The Effect of Technological Innovation, FDI, and Financial Development on CO_2 Emission: Evidence from the G8 Countries. *Environ. Sci. Pollut. Res.* **2021**, 1–9. [CrossRef] [PubMed]
58. Hamza, N.; Gilroy, R. The Challenge to UK Energy Policy: An Ageing Population Perspective on Energy Saving Measures and Consumption. *Energy Policy* **2011**, *39*, 782–789. [CrossRef]

International Journal of
Environmental Research and Public Health

MDPI

Article

Impact of Resource on Green Growth and Threshold Effect of International Trade Levels: Evidence from China

Haiying Liu [1,2], Wenqi Guo [1,*], Yu Wang [3] and Dianwu Wang [2]

1 School of Business and Management, Jilin University, Changchun 130012, China; liuhaiying@jlu.edu.cn
2 School of Maritime Economics and Management, Dalian Maritime University, Dalian 116026, China; wdw20210051@dlmu.edu.cn
3 Corporate Strategy Research Department, State Grid Energy Research Institute, Beijing 102211, China; wangyu34120513@163.com
* Correspondence: jluguowq@163.com

Abstract: International trade levels can change the relationship between resource endowments and green economic growth. Therefore, this study tested the resource curse hypothesis from the perspective of green growth in China using provincial-level panel data for 2005–2017. Energy conservation and environmental improvement were considered under green growth to further analyze the regional mechanism of the resource curse. A panel threshold model was used to identify the impact of import and export threshold effects on the transformation of this mechanism. The resource curse hypothesis was found to be valid nationwide; it hindered green economic growth mainly by impeding energy conservation and curbing environmental improvement. In terms of regional differences in green growth, resource endowment had a positive impact on the eastern region, a negative impact on the central region, and no effect on the western region. When the levels of import and export trade exceeded the threshold values, the resource curse effect was enhanced by reducing energy conservation and weakened by promoting environmental improvement, respectively. Therefore, the Chinese government should establish a more reasonable import and export trade structure, promote changes to the energy structure and green technological innovation, and reduce the negative impact of resource endowment on green growth.

Keywords: resource curse; green growth; import; export; panel threshold model

Citation: Liu, H.; Guo, W.; Wang, Y.; Wang, D. Impact of Resource on Green Growth and Threshold Effect of International Trade Levels: Evidence from China. *Int. J. Environ. Res. Public Health* **2022**, *19*, 2505. https://doi.org/10.3390/ijerph19052505

Academic Editors: Taoyuan Wei and Qin Zhu

Received: 21 January 2022
Accepted: 18 February 2022
Published: 22 February 2022

1. Introduction

Abundant natural resources are important sources of national economic growth as they are indispensable input factors for production. Simultaneously, the comparative advantages from resource endowments can considerably crowd out other input factors [1] by weakening the impetus for technological innovation [2] and hindering high-quality economic development; this is termed as the resource curse [3]. This phenomenon is reflected at the regional level in China, where Shanxi, Shaanxi, Inner Mongolia, and Heilongjiang are rich in mineral resources; however, in recent years, the economic development level of these provinces has remained lower than that of the southeast coastal regions [4]. Therefore, the relationship between resource abundance and economic development should be re-examined. Economic development does not only refer to an increase in the economic aggregate, but also to the content it encompasses that changes according to different stages of social development. Since joining the World Trade Organization (WTO), China has been committed to economic and trade exchanges with other countries, pursuing a high-growth economic development model and producing numerous low-tech value-added primary products. Moreover, some regions have rapidly developed large industries, forming several industrial clusters with high energy consumption and high pollution characteristics, and resulting in an increasingly severe short-board effect on resources and the environment. The Chinese government has gradually recognized the importance of conserving resources and protecting the environment, and introduced a series of policies to eliminate ineffective production capacity and optimize the industrial structure. From proposing that "clear waters and green mountains are as valuable as golden and silver mountains" to advocating the five major development concepts of innovation, coordination, green, openness, and sharing, green growth has become the theme of development at present. The traditional resource curse theory only considers the relationship between resource abundance and economic growth. However, the concept of growth defined by this theory is different from that of green growth in the current social context in China. Under the theme of green growth, if the relationship between resources and development can be interpreted from multiple dimensions, such as resource conservation, environmental protection, and economic growth, green total factor productivity (TFP) can be used to characterize the level of green growth and expand the conceptual scope of the resource curse theory. This would be more conducive to formulating and implementing various ecological and environmental policies.

Although the resource curse hypothesis remains unconfirmed, irrespective of whether it is true or false, the importance of resources for economic and social development cannot be neglected. Resource input is the starting point of production, and natural resources themselves cannot act as a "curse" on the development of human society, but must be a "blessing." The resource curse is attributed to the scenario where economic development is excessively dependent on resources, and the production entities lack the incentive for technological innovation, resulting in a crowding-out effect on the other production factors. Owing to economic globalization, all countries participate in international trade with their respective advantages. For example, some countries in the Middle East and North Africa supply energy to the international market, and oil export trade increases the dependence of the economic development of the region on its resource endowments. However, an improved trade system can reduce the adverse impact of oil reserves on the performance of the real economy [5]. Import and export trade affect the input–output structure of resources; therefore, when the import and export levels are at different stages, their influence in changing the mechanism by which resource endowments affect green TFP should be explored. Therefore, we proposed the following hypothesis in this study: by adjusting the trade structure, the mechanism of the resource curse can be changed, and the negative impact of resource endowment on green growth can be reduced or even reversed. Thus, to test this hypothesis, we will investigate how the mechanism of the resource curse change under different levels of import and export trade, and through what route the changes occur.

In summary, this study first measures the green TFP of provinces in China through the super-efficiency data envelopment analysis (DEA) method and the Luenberger productivity index to characterize green growth, and uses a fixed-effect panel model to verify the existence and regional heterogeneity of the resource curse from the perspective of green growth. Then, green TFP is decomposed into energy conservation effect and environmental improvement effect to analyze the specific path of the impact of resource endowment on green growth. Finally, the threshold regression model is used to test whether international trade levels play a threshold role in the resource curse hypothesis, and through which route import and export trade levels may change the impact of resource endowments on green growth.

The study is structured as follows: Section 2 reviews the relevant literature on resource curse. Section 3 explains the adopted methods and data, including measurement and decomposition of green total factor productivity, fixed-effects model, and threshold regression model. Section 4 presents the results of models in Section 3, and their interpretation, and states the limitations of the study. Section 5 provides the main conclusions.

2. Literature Review

As a starting point of social production, natural resources should be a "blessing" for economic development [6,7]; however, at the national level, the economic growth performance of some resource-rich countries or regions is not outstanding, and even poor for some countries with scarce resources. Prebisch [8] first explored this distorted relationship between resource endowment and economic growth, and the resource curse hypothesis was formally proposed by Auty [3]. Many scholars have conducted extensive research on the existence of the resource curse hypothesis. Sachs et al. [9] conducted empirical studies using panel data from 95 countries between 1970 and 1990, and the results showed a negative correlation between resource abundance and economic growth. The SW model developed by them is known as the paradigm model of resource curse empirical research. They also found that in resource abundant countries, there is often a wage premium in the natural resource sector, crowding out entrepreneurial activity and curbing the country from upgrading its industrial structure, thus, inhibiting economic growth [10]. Numerous subsequent studies have also reached the same conclusion that the resource curse hypothesis holds at the national level [11–13]. Some scholars in China have confirmed the existence of a resource curse at the provincial or prefecture level in China, such as Xinjiang, Shanxi, and Inner Mongolia, where resource-rich provinces fall into resource traps [14–20]. Nevertheless, various studies have opposed the resource curse hypothesis [21]. Fang et al. [22] and Jing [23] conducted empirical tests using prefecture-level and provincial-level data in China, respectively, and did not observe any significant negative correlation between resource endowment and economic growth. In addition, based on Kuznets theory [24], various scholars have proposed that there is a nonlinear relationship between resource dependence and economic development [25,26].

Economic development is not the only criterion for evaluating social progress, and in the context of the country's active advocacy toward sustainable development, the quality of the ecological environment has also become an important indicator to measure economic growth. In recent years, numerous studies have expanded the theoretical scope of the resource curse and further explored the relationship between resources and development from the perspective of green growth [27–29]. Because green TFP can evaluate the quality of economic development based on resource input, environmental pollution, and economic growth, it is widely used in empirical research. Shao et al. [27] proposed the "conditional resource curse" hypothesis and reported that the dependence of the resource industry shows an inverted U-shaped curve relationship for both economic growth and green TFP growth. Li and Xu [28] used the nonradial directional distance function to measure green TFP in 275 prefecture-level cities in China and found that resource abundance is a "curse" to green economic growth. Cheng et al. [29] used the Malmquist–Luenberger index to measure green TFP at the provincial level in China and found that resource industry dependence negatively affects the green growth of the economy. The phenomenon mainly occurred through the extrusion of investment in innovation and human capital, hindering industrial development and reducing the quality of local systems.

The resource curse phenomenon occurs across the entire economic and social system, and its mechanism is affected by other external factors, and country's openness to international markets was proved to be one of the essential factors [30,31]. In recent years, China's economy has entered a new period. China has gradually lost its comparative advantage in labor due to the increase in labor prices. Owing to the global manufacturing shift to Southeast Asia, the Sino–US trade war has reached an unstable condition, and the traditional growth model of relying on exports to drive the economy has been severely challenged. Therefore, to investigate the existence of a resource curse in China, we could not ignore the moderating effect of the trade environment. Arezki and Ploeg [32] proved that natural resource endowments are negatively correlated with economic growth, but increasing trade openness can reduce this negative effect. Dong and Yan [33] used China's provincial panel data from 1997 to 2012 as a sample, and found that the level of expansion has a threshold effect on the resource curse phenomenon; the level of expansion can effectively improve the relationship between resource endowment and economic growth. When the level of expansion is higher than the threshold, the abundance of resources does not hinder economic growth. These studies have identified the moderating effect of trade level on the resource curse, but they have not specifically analyzed the route through which trade level changes the relationship between resource endowment and green growth. Therefore, in this study, we decomposed green TFP into the effects of energy conservation and environmental improvement to analyze the specific route through which the import and export threshold effects change the resource curse mechanism.

3. Materials and Methods

3.1. Measurement and Decomposition of Green Total Factor Productivity

3.1.1. Super-Efficiency Data Envelopment Analysis (DEA) Model

In this study, green growth is characterized by green total factor productivity (TFP). In order to calculate green TFP of each province in China, we first need to measure the level of inefficiency relevant to energy and the environment. Data envelopment analysis (DEA) is a commonly used relative efficiency evaluation model. Charnes et al. [34] proposed the first DEA model, termed the CCR-DEA model, which is an efficiency measurement method based on the assumption of constant returns to scale. Banker et al. [35] modified the CCR-DEA model and proposed a BCC-DEA model based on the assumption of variable returns to scale. When such traditional DEA models are used to evaluate the efficiency of decision-making units, multiple decision-making units may be at the forefront of input and output simultaneously, and the traditional DEA models cannot efficiently rank multiple effective units. To overcome this shortcoming, Andersen et al. [36] proposed a super-efficiency DEA model, which is based on the radial directional distance function for planning and solving, requiring input or output to approach the frontier with the same ratio. The nonradial directional distance function considers the relaxation of variables, allowing input and output to shrink and expand at different proportions. Therefore, this study improved upon the model proposed by Andersen et al. [36] and used the super-efficiency DEA model based on the nonradial directional distance function to measure the green TFP.

Equation (1) represents a super-efficiency DEA model based on a nonradial directional distance function, considering the efficiency evaluation of the ith province in year t as an example.

$$D_i^t(k^t, l^t, y^t, e^t, u_t; g^t) = \max_{\lambda, \beta_{i,e}^t, \beta_{i,u,j}^t, s_{i,k}^t, s_{i,l}^t} : \beta_i^t = \omega_e \beta_{i,e}^t + \sum_{j=1}^{3} \omega_{u,j} \beta_{i,u,j}^t + \varepsilon_k s_{i,k}^t + \varepsilon_l s_{i,l}^t$$

$$s.t. \begin{cases} \sum_{n=1,n\neq i}^{N} \lambda_n^t \times k_n^t + s_{i,k}^t \leq k_i^t \\ \sum_{n=1,n\neq i}^{N} \lambda_n^t \times l_n^t + s_{i,l}^t \leq l_i^t \\ \sum_{n=1,n\neq i}^{N} \lambda_n^t \times y_n^t \geq y_i^t \\ \sum_{n=1,n\neq i}^{N} \lambda_n^t \times e_n^t \leq (1-\beta_{i,e}^t) e_i^t \\ \sum_{n=1,n\neq i}^{N} \lambda_n^t \times u_{n,j}^t \leq (1-\beta_{i,u,j}^t) u_{i,j}^t \; j = 1,2,3 \\ \beta_{i,e}^t, \beta_{i,u,j}^t \leq 1 \\ \lambda_n^t \geq 0 \end{cases} \tag{1}$$

The main difference between the super-efficient and the traditional DEA models is that in the super-efficient DEA model, the efficiency of the ith province must be excluded from the set of decision-making units, that is, the ith province does not contribute to the process of building the frontier. In Equation (1), D_i^t represents the directional distance function of the ith province in year t, N represents the total number of provinces, and $\lambda_n^t \geq 0$ represents that the model satisfies the assumption of constant returns to scale; k_n^t, l_n^t, y_n^t, e_n^t, and $u_{n,j}^t$ denote the capital input, labor input, desired output, energy input, and undesired output of the nth $(n = 1, 2 \ldots, N; n \neq i)$ province in year t, respectively, in which $j (j = 1, 2, 3)$ indicates that there are three types of undesired outputs. The term g^t is the direction vector, indicating the directions of input and output optimization; in this study, $g^t = \left(0, 0, 0, -e_i^t, -u_{i,1}^t, -u_{i,2}^t, -u_{i,3}^t\right)$. $s_{i,k}^t$ and $s_{i,l}^t$ denote the slack variables of capital input and labor input, respectively; $\beta_{i,e}^t$ and $\beta_{i,u,j}^t$ denote the ratio of energy input and undesired output that need to be reduced to reach the production frontier level in the ith province. A positive value of $\beta_{i,e}^t$ or $\beta_{i,u,j}^t$ indicates the inefficiency level of energy input and undesired output, while a negative value indicates the super-efficiency level. Here, it is not required that $\beta_{i,e}^t$ and $\beta_{i,u,j}^t$ be equal, or that the energy input and undesired output change in the same proportion; β_i^t denotes the level of inefficiency in the ith province in the tth year, which equals the weighted average of the above four inefficiency values as well as $\varepsilon_k s_{i,k}^t$ and $\varepsilon_l s_{i,l}^t$. The optimized objective function was used to maximize the β_i^t. The weights of $\beta_{i,e}^t$ and $\beta_{i,u,j}^t$ are ω_e and $\omega_{u,j}$ ($\omega_e + \sum_{j=1}^3 \omega_{u,j} = 1$), respectively. Because the efficiency level is evaluated from the perspectives of energy conservation and environmental improvement, we assigned the weight ω_e of the energy inefficiency level $\beta_{i,e}^t$ to 1/2, and the weights $\omega_{u,j}$ (j = 1,2,3) of the three undesired output inefficiencies were 1/6. Variables ε_k and ε_l are the non-Archimedean infinitesimal quantities. In the objective function, the inefficiency level of capital and labor inputs are denoted by $\varepsilon_k s_{i,k}^t$ and $\varepsilon_l s_{i,l}^t$, respectively, which are the products of a finite constant and a non-Archimedean infinitesimal. Their values remain infinitesimal, and these do not have a significant effect on the objective function β_i^t.

3.1.2. Luenberger Green Total Factor Productivity Index and Decomposition

Equation (1) measures the inefficiency level using the nonradial directional distance function. Because of the additive form of the nonradial directional distance function, green TFP can be constructed by the results of inefficiency through the Luenberger productivity index [37]. Green TFP refers to the level of change in green efficiency in the current period, based on the previous period. A green TFP greater than zero indicates an increase in green efficiency, and a value less than zero indicates a decline in green efficiency. We assumed that the previous period is recorded as period 0, and the current period is recorded as period 1. The Luenberger green TFP ($L_{0,i}^1$) of the ith province follows.

$$L_{0,i}^1 = \frac{1}{2} \times [D_i^1(k^0, l^0, y^0, e^0, u^0; g^0) - D_i^1(k^1, l^1, y^1, e^1, u^1; g^1) + D_i^0(k^0, l^0, y^0, e^0, u^0; g^0) - D_i^0(k^1, l^1, y^1, e^1, u^1; g^1)] \quad (2)$$

Luenberger green TFP comprises four nonradial directional distance functions, of which the same-phase directional distance functions $D_i^0(k^0, l^0, y^0, e^0, u^0; g^0)$ and $D_i^1(k^1, l^1, y^1, e^1, u^1; g^1)$ are shown in Equations (3) and (4), respectively.

$$D_i^0(k^0, l^0, y^0, e^0, u^0; g^0) = max : \beta_i^{0,0} = \omega_e \beta_{i,e}^{0,0} + \sum_{j=1}^{3} \omega_{u,j} \beta_{i,u,j}^{0,0} + \varepsilon_k s_{i,k}^{0,0} + \varepsilon_l s_{i,l}^{0,0}$$

$$s.t. \begin{cases} \sum_{n=1,n\neq i}^{N} \lambda_n^{0,0} \times k_n^0 + s_{i,k}^{0,0} \leq k_i^0 \\ \sum_{n=1,n\neq i}^{N} \lambda_n^{0,0} \times l_n^0 + s_{i,l}^{0,0} \leq l_i^0 \\ \sum_{n=1,n\neq i}^{N} \lambda_n^{0,0} \times y_n^0 \geq y_i^0 \\ \sum_{n=1,n\neq i}^{N} \lambda_n^{0,0} \times e_n^0 \leq (1 - \beta_{i,e}^{0,0}) e_i^0 \\ \sum_{n=1,n\neq i}^{N} \lambda_n^{0,0} \times u_{n,j}^0 \leq (1 - \beta_{i,u,j}^{0,0}) u_{i,j}^0 \; j = 1,2,3 \\ \beta_{i,e}^{0,0}, \beta_{i,u,j}^{0,0} \leq 1 \\ \lambda_n^{0,0} \geq 0 \end{cases} \quad (3)$$

$$D_i^1(k^1, l^1, y^1, e^1, u^1; g^1) = max : \beta_i^{1,1} = \omega_e \beta_{i,e}^{1,1} + \sum_{j=1}^{3} \omega_{u,j} \beta_{i,u,j}^{1,1} + \varepsilon_k s_{i,k}^{1,1} + \varepsilon_l s_{i,l}^{1,1}$$

$$s.t.\begin{cases} \sum_{n=1,n\neq i}^{N}\lambda_n^{1,1}\times k_n^1+s_{i,k}^{1,1}\leq k_i^1 \\ \sum_{n=1,n\neq i}^{N}\lambda_n^{1,1}\times l_n^1+s_{i,l}^{1,1}\leq l_i^1 \\ \sum_{n=1,n\neq i}^{N}\lambda_n^{1,1}\times y_n^1\geq y_i^1 \\ \sum_{n=1,n\neq i}^{N}\lambda_n^{1,1}\times e_n^1\leq(1-\beta_{i,e}^{1,1})e_i^1 \\ \sum_{n=1,n\neq i}^{N}\lambda_n^{1,1}\times u_{n,j}^1\leq(1-\beta_{i,u,j}^{1,1})u_{i,j}^1\ j=1,2,3 \\ \beta_{i,e}^{1,1},\beta_{i,u,j}^{1,1}\leq 1 \\ \lambda_n^{1,1}\geq 0 \end{cases} \quad (4)$$

In the interperiod program, because the set of decision-making units and the decision-making unit being evaluated were not from the same period of data, the interperiod data of the ith province were not excluded from the set of decision-making units. The super-efficiency DEA model can not only sort multiple effective decision-making units, but also solve the problem of unsolvable intertemporal planning. The intertemporal directional distance functions $D_i^0(k^1,l^1,y^1,e^1,u^1;g^1)$ and $D_i^1(k^0,l^0,y^0,e^0,u^0;g^0)$ are shown in Equations (5) and (6), respectively.

$$D_i^0(k^1,l^1,y^1,e^1,u^1;g^1)=max\ :\ \beta_i^{0,1}=\omega_e\beta_{i,e}^{0,1}+\sum_{j=1}^{3}\omega_{u,j}\beta_{i,u,j}^{0,1}+\varepsilon_k s_{i,k}^{0,1}+\varepsilon_l s_{i,l}^{0,1}$$

$$s.t.\begin{cases} \sum_{n=1}^{N}\lambda_n^{0,1}\times k_n^0+s_{i,k}^{0,1}\leq k_i^1 \\ \sum_{n=1}^{N}\lambda_n^{0,1}\times l_n^0+s_{i,l}^{0,1}\leq l_i^1 \\ \sum_{n=1}^{N}\lambda_n^{0,1}\times y_n^0\geq y_i^1 \\ \sum_{n=1}^{N}\lambda_n^{0,1}\times e_n^0\leq(1-\beta_{i,e}^{0,1})e_i^1 \\ \sum_{n=1}^{N}\lambda_n^{0,1}\times u_{n,j}^0\leq(1-\beta_{i,u,j}^{0,1})u_{i,j}^1\ j=1,2,3 \\ \beta_{i,e}^{0,1},\beta_{i,u,j}^{0,1}\leq 1 \\ \lambda_n^{0,1}\geq 0 \end{cases} \quad (5)$$

$$D_i^1(k^0,l^0,y^0,e^0,u^0;g^0)=max\ :\ \beta_i^{1,0}=\omega_e\beta_{i,e}^{1,0}+\sum_{j=1}^{3}\omega_{u,j}\beta_{i,u,j}^{1,0}+\varepsilon_k s_{i,k}^{1,0}+\varepsilon_l s_{i,l}^{1,0}$$

$$s.t.\begin{cases} \sum_{n=1}^{N}\lambda_n^{1,0}\times k_n^1+s_{i,k}^{1,0}\leq k_i^0 \\ \sum_{n=1}^{N}\lambda_n^{1,0}\times l_n^1+s_{i,l}^{1,0}\leq l_i^0 \\ \sum_{n=1}^{N}\lambda_n^{1,0}\times y_n^1\geq y_i^0 \\ \sum_{n=1}^{N}\lambda_n^{1,0}\times e_n^1\leq(1-\beta_{i,e}^{1,0})e_i^0 \\ \sum_{n=1}^{N}\lambda_n^{1,0}\times u_{n,j}^1\leq(1-\beta_{i,u,j}^{1,0})u_{i,j}^0\ j=1,2,3 \\ \beta_{i,e}^{1,0},\beta_{i,u,j}^{1,0}\leq 1 \\ \lambda_n^{1,0}\geq 0 \end{cases} \quad (6)$$

Mahlberg et al. [37] and Chang et al. [38] proposed that the Luenberger productivity index based on the nonradial directional distance function can be decomposed into the sum of the productivity of each factor. The Luenberger green TFP in this study can be decomposed into energy conservation and environmental improvement effects (because $\varepsilon_k s_{i,k}^t$ and $\varepsilon_l s_{i,l}^t$ are infinitesimal, they can be ignored). $L_{0,i}^1$ represents the green TFP; $L_{0,i,e}^1$ and $L_{0,i,u}^1$ represent the efficiency changes of energy input and undesired output, that is, the energy conservation and environmental improvement effects, respectively.

$$L_{0,i,e}^1=\frac{1}{2}\times[(\beta_{i,e}^{1,0}-\beta_{i,e}^{1,1})+(\beta_{i,e}^{0,0}-\beta_{i,e}^{0,1})] \quad (7)$$

$$L_{0,i,u}^1=\frac{1}{2}\times[(\sum_{j=1}^{3}\frac{1}{3}\beta_{i,u,j}^{1,0}-\sum_{j=1}^{3}\frac{1}{3}\beta_{i,u}^{1,1})+(\sum_{j=1}^{3}\frac{1}{3}\beta_{i,u,j}^{0,0}-\sum_{j=1}^{3}\frac{1}{3}\beta_{i,u}^{0,1})] \quad (8)$$

$$L_{0,i}^{1} = \frac{1}{2} \times (L_{0,i,e}^{1} + L_{0,i,u}^{1}) \quad (9)$$

3.1.3. Input–Output Data in the Measurement of Green TFP

This study utilized the input–output data of 30 provinces and regions in China for 2005 to 2017. Because of the unavailability of data, the study did not include Tibet, Hong Kong, Macao, and Taiwan among the 34 provinces and regions of China. Moreover, Beijing, Tianjin, Shanghai, and Chongqing were not excluded from the calculation of TFP; however, because the functional positioning of municipalities is different from that of provinces and autonomous regions, the data of these four municipalities were excluded when calculating the threshold variable using the panel model. The data included capital input (k), labor input (l), energy input (e), expected output (y), and undesired output (u), whose sources and processing methods are explained as follows.

Capital input (k): We estimated the annual capital stock based on the perpetual inventory method proposed by Zhang et al. [39]. The earlier the selected base year, the lower the effect that the error of the initial capital stock estimated during the base year has in subsequent years. Therefore, 1952 was selected as the base year for estimation. The fixed asset depreciation rate of all provinces and regions was uniformly set to 9.6%, the total fixed capital formation was used as the current investment amount, and the regional fixed asset investment price index was used to convert the fixed asset investment price index into a constant price with 2005 as the base year.

Labor input (l): If the number of the employed population is used to represent labor input, the differences due to different levels of education can be ignored. Therefore, labor input was selected as the product of the total employed population in the primary, secondary, and tertiary industries and the average years of education in the region.

Energy input (e): Energy input represents the total energy consumption by region published in the *China Energy Statistical Yearbook*.

Desirable output (y): Desirable output is the gross domestic product (GDP), with 2005 as the base period, and the GDP index was used to account for deflation.

Undesirable output (u): Undesirable outputs include total SO_2 emissions, total wastewater emissions, and solid waste generation in each province.

The data were mainly obtained from the *China Statistical Yearbook*, *China Labor Statistics Yearbook*, *China Energy Statistics Yearbook*, statistical yearbooks of various provinces, and Wind Economic Database.

3.2. Methodology and Data

3.2.1. Model Settings

First, the linear relationship between resource endowment (*re*) and green TFP (*tfp*) should be examined. Because this study adopts panel data, and regions' individual fixed effects and time fixed effects need to be controlled in the regression process, a fixed-effect model was used to perform a basic regression analysis to test the existence and regional heterogeneity of the resource curse. Then, considering the existence of a nonlinear relationship between resource endowment (*re*) and green TFP (*tfp*) with certain variables as moderators and applying import and export trade levels as threshold variables, the panel threshold model was used to identify the changes in the mechanism of the impact of resource endowments on green growth before and after the threshold value.

Because green TFP can be decomposed into the energy conservation effect (*tfp_e*) and the environmental improvement effect (*tfp_u*), we also examined the influence of resource endowment (*re*) on the two effects in both models above to identify the route by which the resource curse phenomenon changes.

The fixed-effects model is shown in Equation (10):

$$gg_{i,t} = \alpha_0 + \alpha_1 re_{i,t} + \alpha_2 control_{i,t} + \mu_i + \lambda_t + \varepsilon_{i,t} \quad (10)$$

where i denotes the province and t denotes the year. The explanatory variable gg denotes the green growth effect, and the *tfp*, *tfp_e*, and *tfp_u* can be selected according to different study objectives. The core explanatory variable *re* is the resource endowment; *control* denotes the control variable, including imports, exports, environmental governance, research and development (R&D) investment, economic development level, industrial structure, urbanization level, and nationalization level; α_0 is a constant term; α_1 and α_2 are the regression coefficients for the explanatory and control variables, respectively; μ_i is a fixed effect in a region that does not change with time; λ_t denotes a fixed effect in time; and $\varepsilon_{i,t}$ is a random perturbation term.

As previously mentioned, import and export trade levels may play a threshold role in the resource curse hypothesis; however, it is difficult to determine the specific segmentation point. Therefore, the threshold regression model proposed by Hansen [40] was used for analysis. This model can accurately estimate the threshold value and perform a significant test of the threshold effect. The panel threshold model is shown in Equation (11):

$$gg_{i,t} = \alpha_0 + \alpha_{11} re_{i,t} \cdot I(q \leq \gamma) + \alpha_{12} re_{i,t} \cdot I(q > \gamma) + \alpha_2 control_{i,t} + \mu_i + \varepsilon_{i,t} \tag{11}$$

where q is the threshold variable, which represents the level of import and export trade, respectively, and γ is a threshold value. $I(*)$ is an indicative function; if the expression within the parentheses is true, the value is one, and the opposite is zero. When the threshold variable is lower than the threshold value $(q \leq \gamma)$, α_{11} is the regression coefficient of the resource endowment, and when the threshold variable is above the threshold value $(q > \gamma)$, α_{12} is the regression coefficient of the resource endowment. The meanings of the interpreted, explanatory, and control variables in Equation (11) are the same as those in Equation (10).

Equation (11) is the expression of a single threshold model, and if the threshold effect test proves the occurrence of a double threshold or triple threshold, a corresponding multithreshold model should also be established. Considering the double threshold model as an example, the corresponding expression is shown in Equation (12):

$$gg_{i,t} = \alpha_0 + \alpha_{11} re_{i,t} \cdot I(q \leq \gamma_1) + \alpha_{12} re_{i,t} \cdot I(\gamma_1 < q \leq \gamma_2) + \alpha_{13} re_{i,t} \cdot I(q > \gamma_2) + \alpha_2 control_{i,t} + \mu_i + \varepsilon_{i,t} \tag{12}$$

where α_{11}, α_{12}, and α_{13} are the regression coefficients of resource endowments under different threshold intervals, q is a threshold variable, and γ_1 and γ_2 are two different threshold values.

Stata 14.0 statistical analysis software was used to estimate the model.

3.2.2. Data Sources

The definitions and descriptions of the variables involved in the model are presented in Table 1.

Explained variables, including green total factor productivity (*tfp*), energy conservation effect (*tfp_e*), and environmental improvement effect (*tfp_u*), are derived from the model and data in 3.1.

Explanatory variable is resource endowment (*re*). The number of employees in the mining industry was chosen to be the proxy variable of resource endowment, because it can reflect the dependence of a region's economic development on resources and the abundance of resources [41]. The data are derived from the *China Labor Statistics Yearbook*. The impact of mineral resources on green growth was mainly considered in this study. Owing to the low-cost utilization of mineral resources, the regions rich in mineral resources lack the motivation for green production technology innovation, which may not be conducive to green growth. Although renewable energy is part of natural resource endowment, the development and utilization of renewable energy requires a high level of technology, which rarely causes a "curse" to the economy, and it accounts for a small proportion in energy consumption; thus, this study does not consider such resources.

Table 1. Definitions and descriptions of the variables.

Category	Symbol	Variables	Proxy Indicator
Explained variables	*tfp*	Green total factor productivity	Calculated by Equation (2)
	tfp_e	Energy conservation effect	Calculated by Equation (7)
	tfp_u	Environmental improvement effect	Calculated by Equation (8)
Explanatory variable	*re*	Resource endowment	Number of employees in the mining industry
Control variables	*import*	Import	Total import/Gross domestic product (GDP)
	export	Export	Total export/GDP
	govern	Environmental governance	Total investment in environmental pollution control/GDP
	rd	Research and development (R&D) investment	R&D capital stock/Gross domestic product
	pergdp	Economic development level	GDP per capita
	indus	Industrial structure	Secondary industry GDP/Total GDP
	urban	Urbanization level	Nonagricultural population/Total population
	own	Nationalization level	Number of employees in state-owned units/ Total number of employees

The calculation methods of control variables, including imports, exports, environmental governance, research and development (R&D) investment, economic development level, industrial structure, urbanization level, and nationalization level, are presented in Table 1. Except for the stock of R&D capital (*rd*), the data used to calculate other control variables are directly derived from the *China Statistical Yearbook*, *China Labor Statistics Yearbook*, *China Population and Employment Statistical Yearbook*, and statistical yearbooks of various provinces. Because the government has not released statistics on the stock of R&D capital, we used the perpetual inventory method to estimate the calculation equation as follows:

$$S_{i,t} = (1-\delta)S_{i,t-1} + RD_{i,t} \tag{13}$$

where $S_{i,t}$ and $S_{i,t-1}$ are the R&D capital stocks of province i in year t and $t-1$, respectively; and $RD_{i,t}$ is the internal R&D expenditure of province i in year t. The term δ is the depreciation rate, which is consistent with the previous estimate of capital stock, and is also set to 9.6%. Considering 2000 as the initial year, the calculation method for capital stock in 2000 follows:

$$S_{i,2000} = RD_{i,2000}/(\delta + g) \tag{14}$$

where $S_{i,2000}$ is the R&D capital stock of province i in 2000, $RD_{i,2000}$ is the internal expenditure of R&D expenditure in province i in 2000, δ is the depreciation rate (9.6%), and g represents the average growth rate of internal R&D expenditure from 2000 to 2017.

4. Results and Discussion

4.1. Green Total Factor Productivity Levels in Each Province and Region

This study used the Linprog function in MATLAB to calculate the green TFP. Equation (9) shows that green TFP can be decomposed into energy conservation and environmental improvement effects. Table 2 shows the average values (2005–2017) of the three indicators—*tfp*, *tfp_e*, and *tfp_u*—in each province. From the national average result, the green TFP is 0.062%, of which the negative energy conservation effect leads to an average annual decline of 0.409% in green TFP, but the environmental improvement effect contributes 0.472% of the increase in green TFP. According to the specific conditions of each province, the green TFP of Beijing (7.783%) and Shanghai (2.510%) were significantly higher than those of other provinces, while those of Heilongjiang, Hainan, and Xinjiang were all less than −1.000%. However, the growth effect was negative. For most provinces, environmental improvement was the main reason for the increase in green TFP, while the decline in energy use efficiency hindered green growth. However, the energy conservation effects of Beijing, Shanxi, and Jilin were positive, indicating that the energy use efficiencies of these three provinces have increased, which in turn increased the green TFP. The environmental improvement effects of Heilongjiang, Qinghai, and Ningxia were negative. For these provinces, the deterioration of environmental efficiency was the main reason for the decline in TFP.

Table 2. Green total factor productivity and its decomposition results for various provinces.

Regions	Green Total Factor Productivity *tfp* (%)	Energy Conservation Effect *tfp_e* (%)	Environmental Improvement Effect *tfp_u* (%)	Regions	Green total Factor Productivity *tfp* (%)	Energy Conservation Effect *tfp_e* (%)	Environmental Improvement Effect *tfp_u* (%)
Beijing	7.783	1.411	6.372	Henan	−0.176	−0.237	0.061
Tianjin	0.583	−0.228	0.811	Hubei	0.365	−0.162	0.527
Hebei	−0.449	−0.518	0.069	Hunan	0.424	−0.289	0.713
Shanxi	0.058	0.084	−0.026	Guangdong	−0.032	−0.587	0.555
Inner Mongolia Mongolia	−0.381	−0.303	−0.079	Guangxi	−0.153	−0.791	0.638
Liaoning	−0.808	−0.587	−0.222	Hainan	−1.359	−1.285	−0.074
Jilin	0.330	0.215	0.115	Chongqing	0.296	−0.291	0.587
Heilongjiang	−1.756	−1.073	−0.683	Sichuan	0.100	−0.204	0.304
Shanghai	2.510	−0.698	3.208	Guizhou	0.743	0.465	0.278
Jiangsu	−0.367	−0.965	0.598	Yunnan	−0.638	−0.369	−0.269
Zhejiang	−0.707	−0.913	0.205	Shaanxi	0.087	−0.165	0.252
Anhui	−0.386	−0.517	0.130	Gansu	−0.240	−0.291	0.051
Fujian	−0.621	−1.159	0.538	Qinghai	−0.881	−0.497	−0.384
Jiangxi	−0.474	−0.735	0.261	Ningxia	0.276	−0.019	0.295
Shandong	−0.979	−0.702	−0.278	Xinjiang	−1.274	−0.865	−0.409
				Mean	0.062	−0.409	0.472

4.2. Descriptive Statistics of the Variables in Fixed-Effects Model

To understand the variables more intuitively, Table 3 lists the descriptive statistics for each. Because the functional positioning of municipalities is different from that of provinces and autonomous regions, the sample data of Beijing, Tianjin, Shanghai, and Chongqing were excluded from the follow-up empirical research.

Table 3. Variable descriptive statistics.

Variables	Observations	Mean	Standard Deviation	Minimum	Minimum
Green total factor productivity (*tfp*)	338	−0.358	2.377	−12.391	7.010
Energy conservation effect (*tfp_e*)	338	−0.959	2.810	−15.879	7.746
Environmental improvement effect (*tfp_u*)	338	0.244	2.794	−10.774	10.576
Resource endowment (*re*)	338	20.485	20.455	0.471	103.014
Import (*import*)	338	11.079	12.005	0.417	72.594
Export (*export*)	338	13.338	16.937	0.728	92.927
Environmental governance (*govern*)	338	1.325	0.671	0.402	4.111
R&D investment (*rd*)	338	7.684	7.220	0.075	46.362
Economic development level (*pergdp*)	338	1.083	0.417	0.333	2.357
Industrial structure (*indus*)	338	47.526	6.942	22.327	61.478
Urbanization level (*urban*)	338	49.191	9.538	26.870	69.850
Nationalization level (*own*)	338	9.535	3.805	4.203	23.617

4.3. Analysis of the Existence and Regional Differences of the Resource Curse

First, regardless of the influence of threshold variables on the mechanism of the resource curse, a fixed-effect model (Equation (10)) was used to examine the linear relationship between resource endowment and green growth. Model 1, Model 2, and Model 3 describe the impact of resource endowment (*re*) on *tfp*, *tfp_e*, and *tfp_u* under the full sample, respectively, and the existence of the resource curse hypothesis was tested from these three aspects. Models 4, 5, and 6 are the estimation models of the impact of resource endowments in the eastern, central, and western regions on *tfp*, respectively, testing the regional heterogeneity of the resource curse. The eastern region includes Hebei and Liaoning, Jiangsu, Zhejiang, Fujian, Shandong, Guangdong, and Hainan; the central region includes Shanxi, Jilin, Heilongjiang, Anhui, Jiangxi, Henan, Hubei, and Hunan; and the western region includes Inner Mongolia, Guangxi, Sichuan, Guizhou, Yunnan, Shaanxi, Gansu, Qinghai, Ningxia, and Xinjiang. Table 4 presents the regression results of the fixed-effects model.

Table 4. Regression results of the fixed-effects model.

Variables	Model 1 (Full Sample)	Model 2 (Full Sample)	Model 3 (Full Sample)	Model 4 (Eastern)	Model 5 (Central)	Model 6 (Western)
	Green Total Factor Productivity (*tfp*)	Energy Conservation Effect (*tfp_e*)	Environmental Improvement Effect (*tfp_u*)	Green Total Factor Productivity (*tfp*)	Green Total Factor Productivity (*tfp*)	Green Total Factor Productivity (*tfp*)
Resource endowment (*re*)	−0.095 *** (−3.16)	−0.098 ** (−2.39)	−0.093 *** (−3.21)	0.151 ** (2.78)	−0.091 * (−2.37)	0.009 (0.10)
Import (*import*)	0.011 (0.17)	−0.031 (−0.54)	0.014 (0.18)	−0.052 (−0.46)	0.053 (0.29)	0.019 (0.37)
Export (*export*)	−0.072 (−1.01)	−0.025 (−0.40)	−0.119 (−1.47)	0.056 (0.43)	0.189 (1.62)	−0.186 ** (−2.32)
Environmental governance (*govern*)	−0.665 ** (−2.14)	−0.648 ** (−2.08)	−0.644 * (−1.93)	−0.498 (−1.21)	−1.161 (−1.36)	−0.564 * (−1.90)
R&D investment (*rd*)	0.121 (0.87)	0.147 (0.94)	0.057 (0.42)	0.711 * (2.16)	0.664 * (2.03)	0.058 (0.73)
Economic development level (*pergdp*)	3.018 (1.04)	4.074 (1.46)	1.039 (0.34)	6.705 * (2.08)	8.041 (1.76)	−2.927 (−1.55)
Industrial structure (*indus*)	0.042 (0.83)	−0.011 (−0.19)	0.111 ** (2.01)	0.347 ** (2.68)	−0.154 * (−2.10)	0.258 ** (3.09)
Urbanization level (*urban*)	0.194 * (1.72)	0.294 ** (2.50)	0.085 (0.71)	0.319 (1.18)	0.117 (0.61)	−0.002 (−0.01)
Nationalization level (*own*)	−0.244 (−1.30)	−0.212 (−1.14)	−0.250 (−1.17)	−0.484 (−1.76)	0.115 (0.41)	−0.001 (−0.00)
constant	Y	Y	Y	Y	Y	Y
year	Y	Y	Y	Y	Y	Y
province	Y	Y	Y	Y	Y	Y
Prob (F)	0.000	0.000	0.000	0.000	0.000	0.000
Observations	338	338	338	104	104	130

Notes: Robust t statistics aNotes: Robust t statistics are shown in parentheses. *** $p < 0.01$, ** $p < 0.05$, and * $p < 0.1$. The values in parentheses are T values. Y means yes, indicating that the constant term is included in the model, and the fixed effects for the year and province are controlled.

According to the regression results of the full sample in Table 4, a significant negative correlation exists between resource endowment and green TFP, energy conservation effect, and environmental improvement effect. From a national perspective, the resource curse hypothesis was found to be valid, and abundant natural resources hinder energy conservation and environmental improvement, while negatively affecting green growth through the effects of energy and the environment. However, regional heterogeneity was observed in the phenomenon of the resource curse. The regression results in the eastern region showed that resource endowment had a significant positive impact on green TFP. In the eastern region, abundant natural resources are conducive to increasing the level of green economic growth. The results for the central region are similar to those of the full sample. The central region also exhibited the resource curse phenomenon, but the negative correlation between resource endowments and green TFP in the central region was less than that in the full sample. Moreover, the severity was lower than that at the national level. The resource endowment in the western region did not significantly affect the green TFP; thus, the resource curse hypothesis was not true for this region. In summary, resource endowment is a "blessing" for the eastern region and a "curse" for the central region, but they do not affect the green growth effect in the western region.

Regression results from other control variables showed that imports would not have a significant impact on green growth at the national and regional levels, but exports would negatively affect green TFP in the western region. Environmental governance would have a significant negative impact on green TFP, energy conservation effect, and environmental improvement effect, and the increase in environmental governance is not conducive to green economic growth. Environmental governance can reflect the strength of local environmental regulations and environmental access standards to a certain extent. If the environmental regulations of a region are too strict or environmental access standards are too high, some polluting enterprises cannot enter the local market, resulting in damage to the output structure. Therefore, environmental regulation in China has not yet demonstrated an innovative compensation effect according to the Porter hypothesis [42].

The level of urbanization had a significant positive impact on the TFP and energy conservation effects. Increasing the level of urbanization helps achieve green economic growth and energy conservation. The industrial structure was positively related to the effect of environmental improvement; thus, the higher the proportion of the secondary industry, the greater the degree of improvement in environmental quality. However, this is contrary to people's traditional cognition, but it can be explained reasonably from two perspectives.①Even if the country has been emphasizing the adjustment of industrial structure, it cannot ignore the role of the secondary industry as a pillar of China's economic development. Increasing the proportion of secondary industries can increase the level of green growth by increasing output.②Environmental quality in areas with heavy industries is generally low, leading to greater opportunities for environmental improvement.

4.4. Transformation of the Resource Curse Mechanism and Analysis of the Mechanism under the Import Level Threshold

Import trade is not only a supplementary means to improve the structure of domestic consumer goods supply, but it is also an important method to determine technology spillovers. However, excessive dependence on imported products reduces domestic manufacturing. Therefore, import trade has two opposite effects on the economy: the technology spillover and the product crowding-out effects. Import trade not only affects the domestic product structure but may also indirectly affect the energy structure and environmental quality. Therefore, "import" was used as a threshold variable to further analyze the nonlinear relationship between resource endowments and green growth. In the following, Models 7, 8, and 9 used imports as the threshold variable. The explained variables of the three models are *tfp*, *tfp_e*, and *tfp_u*. Among them, Model 7 was used to identify the mechanism change of the resource curse under different import levels, and Models 8 and 9 were used to identify the route through which the import trade promotes the mechanism change of the resource curse.

Table 5 shows the analysis results for the import threshold effect. Both Models 7 and 8 had significant single threshold effects, but no double threshold effect was observed. Therefore, for both Models 7 and 8, a single threshold model was adopted with import as the threshold variable (Equation (11)). However, the single threshold effect of Model 9 did not pass the 10% significance level test; therefore, there was no threshold effect, indicating that there was no nonlinear relationship between resource endowment and environmental improvement effects when "import" was the threshold variable. Therefore, Model 9 became equivalent to Model 3 (the fixed-effects model), and it is not discussed further.

Table 5. Analysis of the import threshold effect.

Models	Threshold Type	F-Statistic	p	Critical Value		
				1%	5%	10%
Model 7 (*tfp*) Green total factor productivity	Single threshold	13.67 **	0.046	23.978	13.345	11.238
	Double threshold	8.98	0.132	53.122	19.254	10.539
Model 8 (*tfp_e*) Energy conservation effect	Single threshold	23.04 **	0.012	23.096	14.148	11.909
	Double threshold	10.35	0.158	90.519	58.523	20.983
Model 9 (*tfp_u*) Environmental improvement effect	Single threshold	4.73	0.620	16.913	13.163	10.895

Notes: Robust t statistics are shown in parentheses. ** $p < 0.05$.

The F-statistic and the critical value of P were simulated by repeated sampling (500 times) using the bootstrap method.

Table 6 shows the estimated value and confidence interval of the import threshold in Models 7 and 8. Because Model 9 did not have a threshold effect, there is no corresponding estimated value or confidence interval. Figure 1a–c shows the images of the likelihood ratio functions of the import trade threshold variables in the three models. The threshold values of Models 7 and 8 were the same (50.110), which implies that the impact of resource endowment on green TFP and energy conservation effects both undergo a mechanism change at approximately 50.110; that is, when the total imports accounted for more than 50.110% of GDP, the resource curse changed its mechanism. However, the impact of resource endowments on the mechanism of environmental improvement effects did not change under different import levels. In summary, the mechanism of resource curse changes when the import level is at different ranges; however, imports can only change the impact of resource endowment on green growth through the route of energy conservation, and the behavior of resource curse on the mechanism of environmental improvement has not changed.

Table 6. Estimated import threshold and confidence interval.

	Model 7 (*tfp*) Green Total Factor Productivity		Model 8 (*tfp_e*) Energy Conservation Effect		Model 9 (*tfp_u*) Environmental Improvement Effect	
	Estimated Value	95% Confidence Interval	Estimated Value	95% Confidence Interval	Estimated Value	95% Confidence Interval
Threshold γ	50.110	/47.632, 50.960/	50.110	/42.446, 50.960/	—	—

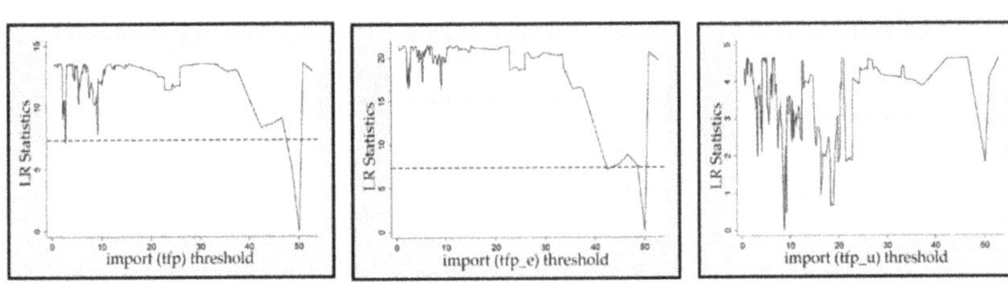

(a) (b) (c)

Figure 1. Likelihood ratio (LR) function graph of the import threshold variables: (**a**) green TFP (*tfp*), (**b**) energy conservation effect (*tfp_e*), and (**c**) environmental improvement effect (*tfp_u*).

Table 7 shows the regression results for Models 7 and 8. The regression results of Model 7 revealed that when the ratio of total imports to GDP was less than 50.110%, the regression coefficient of resource endowment to green TFP was −0.084, and it passed the 1% level of significance test; however, when the import level exceeded the threshold value, the regression coefficient of resource endowment was −0.636, and the negative impact of resource endowment on green TFP increased significantly. In the regression results of Model 8, when the import level did not exceed the threshold value, that is, when the ratio of total imports to GDP was less than 50.110%, no significant correlation was found between resource endowment and the energy conservation effect; however, when the ratio exceeded 50.110%, resource endowment had a negative impact on the energy conservation effect. In summary, an increase in the level of import trade intensifies the adverse impact of resource endowment on green growth and promotes the deterioration of the resource curse. When the import level exceeded the threshold value, the resource curse phenomenon occurred along the energy route. This shows that import trade hinders energy conservation, which in turn leads to the deterioration of the resource curse, while imports do not change the relationship between resources and development through the route of environmental improvement. From the perspective of the import commodity structure in China, the proportion of primary product imports in 2015, 2016, and 2017 accounted for 28.11%, 27.78%, and 31.44%, respectively. Raw materials and fossil fuels are the main imported primary products, in which raw oil imports account for approximately 8% of the total import value (the import ratio of primary products and raw oil is manually calculated based on data from the *China Statistical Yearbook* [43]). The high proportion of imports of primary products leads to weaker technology spillover effects of import trade, hindering the increase in green TFP in China through import trade. The higher the import trade level of a province, the higher the dependence of the province's economic growth on the resources of other countries. International trade has solved the scarcity of resources in the region to a certain extent, but the cost of importing raw materials and energy also has a crowding-out effect on R&D investment in production technology. Owing to the large uncertainty and positive externalities in green technological innovation, when the supply of raw materials and energy in the international market is sufficient, most production entities attempt to solve the problem of scarcity of production materials through imports rather than through technological innovation to save more energy. Therefore, imports intensify the adverse effects of the resource curse by hindering technological innovation, and the high proportion of energy imports leads to imports that can impede energy conservation to promote the mechanism of the resource curse. Table 7 also shows that the regression coefficient of imports in Model 8 is 0.095, and it is significant at the 10% level. This shows that even though import trade had a direct positive effect on TFP, the absolute value of this positive effect was lower than the absolute value of the negative effect from resource curse (the effect of resource endowment on TFP) when the import trade level exceeded the threshold value. To effectively reflect the positive role of import trade on TFP and avoid the occurrence of the resource curse, the level of import trade should be controlled below the threshold value.

Table 7. Regression results of the threshold model with import as the threshold variable.

Variables	Model 7 Green Total Factor Productivity (*tfp*)	Model 8 Energy Conservation Effect (*tfp_e*)
Resource endowment (*import* < 50.110) (*re_0*)	−0.084 *** (−2.72)	−0.043 (−1.24)
Resource endowment (*import* ≥ 50.110) (*re_1*)	−0.636 *** (−4.03)	−0.846 *** (−4.78)
Import (*import*)	−0.484 (−1.58)	−0.433 (−1.26)
Export (*export*)	0.073 (1.43)	0.095 * (1.66)
Environmental governance (*govern*)	−0.078 * (−1.64)	−0.056 (−1.10)
R&D investment (*rd*)	0.135 (1.18)	0.188 (1.46)
Economic development level (*pergdp*)	2.525 (1.53)	5.172 *** (2.80)
Industrial structure (*indus*)	0.136 *** (2.83)	0.146 *** (2.72)
Urbanization level (*urban*)	0.048 (0.75)	−0.004 (−0.05)
Nationalization level (*own*)	−0.220 * (−1.74)	−0.172 (−1.21)
constant	−8.250 *** (−2.82)	−11.844 *** (−3.61)
Prob (F)	0.000	0.000
observation	338	338

Notes: Robust t statistics are shown in parentheses. *** $p < 0.01$, and * $p < 0.1$. The values in parentheses are T values.

4.5. Transformation of the Resource Curse Mechanism and Analysis of Mechanism under the Export Level Threshold

Export trade is an important method for a country to participate in international trade and exert its comparative advantages. Export trade can not only directly affect green growth, but also indirectly by affecting the relationship between resources and green growth. We used export trade as the threshold variable to find the different impacts of resource endowment on green TFP under different export levels, and the specific mechanism for this difference. Export trade was used as the threshold variable in Models 10, 11, and 12. The explained variables of the three models are *tfp*, *tfp_e*, and *tfp_u*. The role of Model 10 was to identify the change of the resource curse mechanism under different export levels, and the roles of Models 11 and 12 were to identify the route through which the export leads to the mechanism change of the resource curse.

Table 8 presents the analysis results for the export threshold effect. Models 10 and 12 had a significant single threshold effect, and both passed the 10% significance level test. In contrast, the double threshold effect of the two models did not pass the significance test; therefore, a single threshold model with export trade as the threshold variable should be used (Equation (11)). However, Model 11 did not pass the single threshold test, indicating that there is no nonlinear relationship between resource endowment and the energy conservation effect with export trade as the threshold variable. Therefore, Model 11 became equivalent to Model 2 (the fixed-effects model), and it is not discussed further.

Table 8. Analysis of the export threshold effect.

Models	Threshold Type	F-Statistic	p	Critical Value		
				1%	5%	10%
Model 10(*tfp*) Green total factor productivity	Single threshold	11.64 *	0.068	17.065	12.264	10.293
	Double threshold	7.36	0.260	14.014	10.725	9.427
Model 11(*tfp_e*) Energy conservation effect	Single threshold	7.77	0.304	21.419	12.834	10.811
Model 12(*tfp_u*) Environmental improvement effect	Single threshold	9.06 *	0.096	13.446	10.381	9.025
	Double threshold	4.85	0.488	19.726	11.517	9.718

Notes: Robust t statistics are shown in parentheses. * $p < 0.1$.

The F-statistic and the critical value of p were simulated by repeated sampling (500 times) using the bootstrap method.

Table 9 shows the estimated value and confidence interval of the export threshold in Models 10 and 12. Because Model 11 does not have a threshold effect, there is no corresponding estimated value or confidence interval. Figure 2a–c shows the likelihood ratio (LR) functions of the exit threshold variables in the three models. The threshold values of Models 10 and 12 were 3.232 and 3.076, respectively. When the ratio of the total export trade to GDP exceeded 3.076%, the effect of resource endowment on environmental improvement changed, and when the ratio exceeded 3.232%, the effect of resource endowment on green TFP changed. Because there is no threshold effect in Model 11, the impact of resource endowments on energy conservation effects did not change at different export levels. In summary, when the export level is at different ranges, the mechanism of the resource curse changes. However, exports can only change the impact of resource endowment on green growth by hindering the mechanism of environmental improvement, and the resource curse changes, but exports can only change the impact of resource endowment on green growth by hindering the mechanism of environmental improvement.

Table 9. Estimated export threshold and confidence interval.

	Model 10 (*tfp*) Green Total Factor Productivity		Model 11 (*tfp_e*) Energy Conservation Effect		Model 12 (*tfp_u*) Environmental Improvement Effect	
	Estimated Value	**95% Confidence Interval**	**Estimated Value**	**95% Confidence Interval**	**Estimated Value**	**95% Confidence Interval**
Threshold γ	3.232	/2.781, 3.263/	——	——	3.076	/2.683, 3.141/

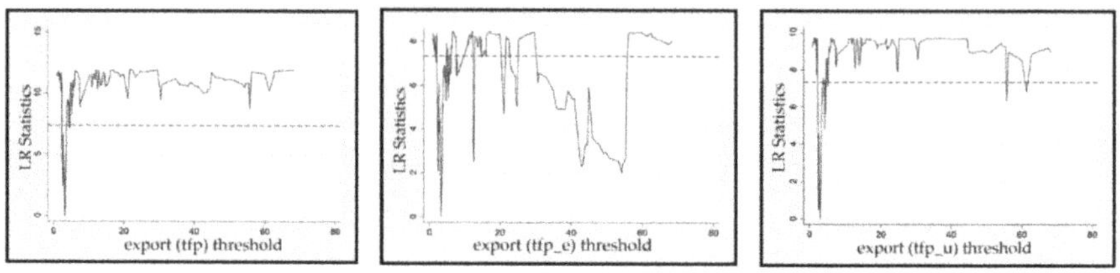

(a) **(b)** **(c)**

Figure 2. LR function graph of export threshold variables: (**a**) *tfp*, (**b**) *tfp_e*, and (**c**) *tfp_u*.

Table 10 shows the regression results of Models 10 and 12. From the results of Model 10, when the export level did not exceed the threshold value, that is, when the ratio of total exports to GDP was lower than 3.232%, a significant negative correlation existed between resource endowment and green TFP, with a regression coefficient of −0.184. However, when the export level exceeds the threshold of 3.232%, the regression coefficient of resource endowment was −0.094, and passed the 1% significance level test. Moreover, when the export level increased to the threshold value, the negative impact of resource endowment on green TFP was weakened. The regression results of Model 12 showed that when the ratio of total exports to GDP was lower than 3.076%, the correlation coefficient between resource endowment and environmental improvement effect was −0.233; however, when the export level exceeded the threshold value, the regression coefficient of resource endowment was −0.136. Furthermore, with the increase in export level, the negative effect of resource endowment on the environmental improvement was also be weakened. In summary, export trade can reduce the adverse impact of resource endowment on green growth and alleviate the severity of the resource curse phenomenon. However, export trade can only change the relationship between resources and development through the route of environmental improvement, but not that of energy conservation. The structure of China's export commodities in 2015, 2016, and 2017 revealed that the exports of industrial finished products accounted for 95.43%, 94.99%, and 94.80%, respectively, among which the export of machinery and transportation equipment accounted for a large proportion (the export ratio of industrial finished products is manually calculated based on data from the *China Statistical Yearbook* [34]). The continuous increase in the proportion of exports of heavy industrial products, such as machinery and equipment, indicates that the technological level of China's export commodities is constantly improving. Environmental barriers and international market demand in export trade have caused Chinese companies to undergo technological innovation. China relies on industrial products to obtain export trade income, and the commodity structure of resource-dependent provinces is mostly based on raw materials and fossil fuels; hence, the resource-dependent provinces in China do not have evident trade advantages. However, owing to the large demand for such commodities in the domestic market, resource-based provinces and regions can obtain a comfortable living space even if they only serve the domestic market without export trade. The geographical distribution of China's natural resources is uneven; the central and western regions are the areas with resource advantages, while the eastern regions rely on convenient transportation and trade conditions to ensure technological advantages. The separation of the resource and technological advantages has also led to differences in the division of labor between provinces and regions. The high level of export trade in the eastern region has intensified the demand for raw materials such as energy, which also increases the exploitation of natural resources by the central and western regions, deepens the dependence of the central and western regions on resources, and reduces the possibility of technological innovation in resource-dependent provinces. Unlike international trade, domestic trade cannot bring incentives for green technology innovation. Based on this analysis, the higher the export trade level of a province, the higher the industrial technology level of the region, and the less dependent the region is on resource endowments for economic development and green growth. The provinces with lower export levels mostly exhibited comparative advantages in terms of resources and lacked motivation for green technological innovation. Therefore, when the export level was lower than the threshold value, resource endowment had a severely negative impact on green growth. With the increase in the level of export trade, restrictions on environmental barriers have also continued to increase. High-exporting provinces give more attention to reducing the negative environmental externalities of the production process, while environmental barriers have less impact on provinces with low export levels; therefore, export trade can improve the resource curse phenomenon through the environmental improvement route. Table 10 also shows that even though the regression coefficient of exports is not significant, a negative correlation exists between exports and green TFP and environmental improvement effects. Therefore, the relationship between export trade and green growth should be adequately considered, and the Chinese government should allow export trade to play its role in improving the resource curse and adopt appropriate measures to eliminate its hindrance to green growth.

Table 10. Regression results of threshold model with export as the threshold variable.

Variables	Model 10 Green Total Factor Productivity (*tfp*)	Model 12 Environmental Improvement Effect (*tfp_u*)
Resource endowment (*export* < 3.232) (*re_0*)	−0.184 *** (−4.24)	−0.233 *** (−4.40)
Resource endowment (*export* ≥ 3.232) (*re_1*)	−0.094 *** (−3.04)	−0.136 *** (−3.63)
Import (*import*)	−0.519 * (−1.70)	−0.506 (−1.37)
Export (*export*)	0.026 (0.52)	−0.012 (−0.20)
Environmental governance (*govern*)	−0.066 (−1.37)	−0.087 (−1.52)
R&D investment (*rd*)	0.124 (1.07)	0.060 (0.43)
Economic development level (*pergdp*)	4.167 ** (2.44)	0.777 (0.38)
Industrial structure (*indus*)	0.105 ** (2.14)	0.115 ** (1.94)
Urbanization level (*urban*)	0.015 (0.24)	0.073 (0.94)
Nationalization level (*own*)	−0.235 * (−1.85)	−0.300 * (−1.96)
constant	−6.005 ** (−2.01)	−2.286 (−0.63)
Prob (F)	0.000	0.000
observation	338	338

Notes: Robust t statistics are shown in parentheses. *** $p < 0.01$, ** $p < 0.05$, and * $p < 0.1$. The values in parentheses are T values.

4.6. Limitations and Future Research

Although the study provided a useful conclusion concerning resource curse, the study still has some limitations. (1) Because of the availability of data, we choose the number of employees in the mining industry as the proxy variable of resource endowment, which may not fully represent the output scale of the mining industry. (2) In this study, green growth is characterized by green TFP. Although TFP has been widely used as an explained variable in the regression model [44], true TFP is unobserved and DEA estimates of productivity have their own limitations. (3) This study focuses on the threshold effect of trade level on the resource curse; thus, whether there is a relationship between green growth and the quadratic term of resources is not discussed.

Based on the results of the above analysis, we recommend the following future actions for Chinese government: (1) Because the resource curse phenomenon exists in China, the government should increase its efforts to promote economic green transformation and reduce the dependence of economic growth on natural resources. Because of the regional heterogeneity in the resource curse, the "one size fits all" approach should be avoided when implementing policies and regulations. For the eastern region, a policy promoting resource development should be implemented, while for the central region, a policy restricting resource development should be implemented. Although the mechanism of resource endowment in western China is not evident, the government should handle resource development activities cautiously and attempt to optimize the input factor structure from the supply side of resources. (2) Although increasing import trade level intensifies the resource curse, import trade also has a positive effect on environmental improvement. Therefore, the import trade level should be controlled below the threshold value and it must be ensured that the environmental dividend generated by import trade is fully utilized. (3) Although export trade could reduce the negative impact of the resource curse, its hindrance to green growth cannot be ignored. Hence, a reasonable level of export trade must be conducted in combination with economic development goals to alleviate resource dependence and mitigate its crowding-out effect on the output.

5. Conclusions

According to the resource curse hypothesis, abundant natural resources would become an obstacle to economic growth. Therefore, based on the concept of the resource curse phenomenon, this study attempts to interpret the relationship between resources and development from the perspective of green growth. However, the concept of development in this study is not limited to economic growth, but it evaluates the quality of economic development from multiple dimensions, such as energy conservation, environmental improvement, and economic growth. We re-examined the impact of resource endowments on green growth under the theoretical framework of the resource curse, and the level of green growth was indicated by the green TFP. Because there may be a nonlinear relationship between resource endowments and green growth with certain variables as moderators, the impact of import and export threshold effects on the transformation of the resource curse mechanism was further investigated, and the transformation route for the resource curse mechanism was identified from the perspectives of energy conservation and environmental improvement.

We provided evidence to support the resource curse hypothesis using a unique dataset of 26 provinces in China for 2005 to 2017 and applying them to a fixed-effects and a panel threshold model. The following are the analysis results. (1) The resource curse hypothesis was valid nationwide, and a significant negative correlation existed between resource endowments and green TFP, energy conservation effects, and environmental improvement effects. Resource endowments negatively affect green growth by hindering both energy conservation and environmental improvement. The phenomenon of resource curse exhibited regional heterogeneity; resource endowment was found to be a "blessing" for the eastern region, a "curse" for the central region, and did not affect the western region. (2) Import trade increased the adverse impact of resource endowment on green growth and promoted the deterioration of the resource curse situation. When the import level exceeded the threshold value, the resource curse phenomenon changed along the energy route. Import trade deteriorated the resource curse by impeding energy conservation, but it did not change the relationship between resources and development through the route of environmental improvement. (3) Export trade reduced the adverse impact of resource endowment on green growth and alleviated the severity of the resource curse phenomenon. However, export trade could only change the relationship between resources and development through environmental improvement, and not through energy conservation.

In conclusion, improving the import and export trade structure can reduce resource dependence to a certain degree; however, their roles are limited; the route to fundamentally alleviating the resource curse is through energy structure adjustment and green technological innovation. According to Vuong [45], investing in science, especially for research and development (R&D), will benefit society in the long run. China has invested a significant portion of its GDP in R&D, around 2.4 percent of GDP per year [46]; however, R&D investment in the ecological area still needs to be increased to improve energy structure and environmental quality. Additionally, in the context of the COVID-19 pandemic, promoting international trade in high-ecological technology value-added products is conducive to economic recovery and healthy economic development [47]. The Chinese government should take this as an opportunity, by shaping ecological values, promoting ecosurplus culture [48], and reducing the international trade of primary products, so as to reshape the innovative ecosystem, lessen the effects of the resource curse, and move toward a more sustainably green economy.

Author Contributions: Conceptualization, H.L., W.G., Y.W. and D.W.; methodology, W.G.; software, W.G.; validation, H.L., W.G. and Y.W.; formal analysis, Y.W. and D.W.; investigation, W.G. and Y.W.; resources, W.G.; data curation, W.G.; writing—original draft preparation, W.G. and Y.W.; writing—review and editing, H.L., W.G., Y.W. and D.W.; supervision, H.L.; project administration, H.L.; funding acquisition, H.L. All authors have read and agreed to the published version of the manuscript.

Funding: This study was financially supported by the research on "Research on the transmission mechanism of environmental regulation to promote the green transformation of China's industry" National Social Science Foundation of China, Chinese Academy of Social Sciences, China (20BJY102).

Institutional Review Board Statement: Not applicable.

Informed Consent Statement: Not applicable.

Data Availability Statement: Publicly available dataset *China Statistical Yearbook 2018* was analyzed in this study. These data can be found at: https://data.stats.gov.cn accessed on 21 January 2022. Other raw/processed data required to reproduce these findings cannot be shared at this time, as the data also form part of an ongoing study.

Conflicts of Interest: The authors declare no conflict of interest.

References

1. Gylfason, T. Natural Resources, Education, and Economic Development. *Eur. Econ. Rev.* **2001**, *45*, 847–859. [CrossRef]
2. Papyrakis, E.; Gerlagh, R. Resource Abundance and Economic Growth in the United States. *Eur. Econ. Rev.* **2007**, *51*, 1011–1039. [CrossRef]
3. Auty, R.M. *Sustaining Development in Mineral Economies: The Resource Curse Thesis*; Routledge: London, UK, 1993.
4. Li, J.; Zhang, J.; Gong, L.; Miao, P. The Spatial and Temporal Distribution of Coal Resource and Its Utilization in China—Based on Space Exploration Analysis Technique ESDA. *Energy Environ.* **2015**, *26*, 1099–1113. [CrossRef]
5. Apergis, N.; Payne, J.E. The Oil Curse, Institutional Quality, and Growth in MENA Countries: Evidence from Time-Varying cointegration. *Energy Econ.* **2014**, *46*, 1–9. [CrossRef]
6. Rosenstein-Rodan, P.N. Problems of Industrialization of Eastern and South-Eastern Europe. *Econ. J.* **1943**, *53*, 202–211. [CrossRef]
7. Matsuyama, K. Agricultural Productivity, Comparative Advantage and Economic Growth. *J. Econ. Theory* **1992**, *58*, 317–334. [CrossRef]
8. Prebisch, R. *The Economic Development of Latin America and Its Principal Problem*; United Nations Department of Economic Affairs, Economic Commission for Latin America (ECLA): New York, NY, USA, 1950; Available online: http://archivo.cepal.org/pdfs/cdPrebisch/002.pdf (accessed on 21 January 2022).
9. Sachs, J.D.; Warner, A.M. *Natural Resources Abundance and Economic Growth*; National Bureau of Economic Research: Cambridge, MA, USA, 1995. [CrossRef]
10. Sachs, J.D.; Warner, A.M. The curse of natural resources. *Eur. Econ. Rev.* **2001**, *45*, 827–838. [CrossRef]
11. Cockx, L.; Francken, N. Natural Resources: A Curse on Education Spending? *Energy Policy* **2016**, *92*, 394–408. [CrossRef]
12. Apergis, N.; Katsaiti, M. Poverty and the Resource Curse: Evidence from a Global Panel of Countries. *Res. Econ.* **2018**, *72*, 211–223. [CrossRef]
13. Song, D.; Yang, Q. Does Environmental Regulation Break the Resource Curse. *China Popul. Resour. Environ.* **2019**, *29*, 61–69.
14. Zhang, X.; Xing, L.; Fan, S.; Luo, X. Resource Abundance and Regional Development in China. *Econ. Transit.* **2008**, *16*, 7–29. [CrossRef]
15. Shao, S.; Zhang, Y.; Tian, Z.; Li, D.; Yang, L. The Regional Dutch Disease Effect Within China: A Spatial Econometric Investigation. *Energy Econ.* **2020**, *88*, 104766. [CrossRef]
16. Zhang, H.; Shen, L.; Zhong, S.; Elshkaki, A. Coal Resource and Industrial Structure Nexus in Energy-Rich Area: The Case of the Contiguous Area of Shanxi and Shaanxi Provinces, and Inner Mongolia Autonomous Region of China. *Resour. Policy* **2020**, *66*, 101646. [CrossRef]
17. Qiu, Q.; Chen, J. Natural Resource Endowment, Institutional Quality and China's Regional Economic Growth. *Resour. Policy* **2020**, *66*, 101644.

18. Ma, Y.; Cheng, D. 'Resource Gospel' or "Resource Curse"? On Panel Threshold. *Fin. Trade Res.* **2017**, *1*, 13–25.
19. Zhang, P.; Wu, J. Government Intervention, Resource Curse and Regional Innovation: An Empirical Research Based on the Provincial Panel Data from Mainland China. *Sci. Res. Manag.* **2017**, *38*, 61–69.
20. Yu, X.; Li, Y.; Chen, H.; Li, C. Study on the Impact of Environmental Regulation and Energy Endowment on Regional Carbon Emissions from the Perspective of Resource Curse. *China Popul. Resour. Environ.* **2019**, *29*, 52–60.
21. Hu, Y.; Yan, T. Resource Dependency: Curse on Growth or Poverty? *China Popul. Resour. Environ.* **2019**, *29*, 137–146.
22. Fang, Y.; Ji, K.; Zhao, Y. Does "Resource Curse" Exist in China? *J. World Econ.* **2011**, *4*, 144–160.
23. Jing, X. Natural Resource Exploitation and Economic Growth in China: A Disproof of "Resource Curse" Hypothesis. *Econ. Probl.* **2012**, *3*, 4–8.
24. Kuznets, M. Economy growth and income inequality. *Am. Econ. Rev.* **1955**, *45*, 1–28. [CrossRef]
25. Brock, W.A.; Taylor, M.S. Economic Growth and the Environment: A Review of Theory and Empirics. *Handb. Econ. Growth* **2005**, *1*, 1749–1821. [CrossRef]
26. Mosconi, E.M.; Colantoni, A.; Gambella, F.; Cudlinová, E.; Salvati, L.; Rodrigo-Comino, J. Revisiting the Environmental Kuznets Curve: The Spatial Interaction Between Economy and Territory. *Economies* **2020**, *8*, 74. [CrossRef]
27. Shao, S.; Fan, M.; Yang, L. How Resource Industry Dependence Affects the Efficiency of Economic Development? Test and Explanation of the Conditional Resource Curse Hypothesis. *Manag. World* **2013**, *2*, 32–63.
28. Li, J.; Xu, B. Curse or Blessing: How Does Natural Resource Abundance Affect Green Economic Growth in China? *Econ. Res. J.* **2018**, *9*, 151–167.
29. Cheng, Z.; Li, L.; Liu, J. Natural Resource Abundance, Resource Industry Dependence and Economic Green Growth in China. *Resour. Policy* **2020**, *68*, 101734. [CrossRef]
30. Sachs, J.D.; Warner, A.M. Sources of slow growth in African economies. *J. Afr. Econ.* **1997**, *6*, 335–376. [CrossRef]
31. Sachs, J.D.; Warner, A.M. The big push, natural resource booms and growth. *J. Dev. Econ.* **1999**, *59*, 43–76. [CrossRef]
32. Arezki, R.; Ploeg, F. *Can the Natural Resource Curse Be Turned into a Blessing? The Role of Trade Policies and Institutions*; Oxcarre Working Papers; University of Oxford: Oxford, UK, 2008; Volume 7.
33. Dong, L.; Yan, T. Technology Input, Openness Degree and "Resource Curse": An Examination of Terms of Trade with China's Provincial Panel Data. *J. Int. Trade* **2015**, *9*, 55–65.
34. Charnes, A.; Cooper, W.W.; Rhodes, E. Measuring the Efficiency of Decision Making Units. *Eur. J. Oper. Res.* **1978**, *2*, 429–444. [CrossRef]
35. Banker, R.D.; Charnes, A.; Cooper, W.W. Some Models for Estimating Technical and Scale Inefficiencies in Data Envelopment Analysis. *Manag. Sci.* **1984**, *30*, 1078–1092. [CrossRef]
36. Andersen, P.; Petersen, N.C. A Procedure for Ranking Efficient Units in Data Envelopment Analysis. *Manag. Sci.* **1993**, *39*, 1261–1264. [CrossRef]
37. Mahlberg, B.; Sahoo, B.K. Radial and Non-Radial Decompositions of Luenberger Productivity Indicator with an Illustrative Application. *Int. J. Prod. Econ.* **2011**, *131*, 721–726. [CrossRef]
38. Chang, T.P.; Hu, J.L.; Chou, R.Y.; Sun, L. The Sources of Bank Productivity Growth in China During 2002–2009: A Disaggregation View. *J. Bank. Financ.* **2012**, *36*, 1997–2006. [CrossRef]
39. Zhang, J.; Wu, G.; Zhang, J. The Estimation of China's Provincial Capital Stock: 1952–2000. *Econ. Res. J.* **2004**, *10*, 23–44.
40. Hansen, B.E. Threshold Effects in Non-Dynamic Panels: Estimation, Testing, and Inference. *J. Econom.* **1999**, *93*, 345–368. [CrossRef]
41. Shao, S.; Yang, L. Abundance of Natural Resources, Resource Industry Dependence and China's Regional Economic Growth. *Manag. World* **2010**, *9*, 26–44.
42. Porter, M.E.; Van der Linde, C. Toward a New Conception of the Environment-Competitiveness Relationship. *J. Econ. Perspect.* **1995**, *9*, 97–118. [CrossRef]
43. National Bureau of Statistics of China. *China Statistical Yearbook 2018*; China Statistics Press: Beijing, China, 2018.
44. Wu, H.; Hao, Y.; Ren, S.; Yang, X.; Xie, G. Does internet development improve green total factor energy efficiency? Evidence from China. *Energy Policy* **2021**, *153*, 112247. [CrossRef]
45. Vuong, Q.H. The (Ir)rational Consideration of the Cost of Science in Transition Economies. *Nat. Hum. Behav.* **2018**, *2*, 5. [CrossRef]
46. China Spending on Research and Development to Rise 7% per Year in Push for Major Tech Breakthroughs. Available online: https://www.cnbc.com/2021/03/05/china-to-boost-research-and-development-spend-in-push-for-tech-breakthroughs.html (accessed on 16 February 2022).
47. Vuong, Q.H.; Le, T.T.; La, V.P.; Nguyen, H.T.T.; Ho, M.T.; Khuc, Q.V.; Nguyen, M.H. Covid-19 Vaccines Production and Societal Immunization Under the Serendipity-mindsponge-3D Knowledge Management Theory and Conceptual Framework. *Humanit. Soc. Sci. Commun.* **2022**, *9*, 22. [CrossRef]
48. Vuong, Q.H. The Semiconducting Principle of Monetary and Environmental Values Exchange. *Econ. Bus. Lett.* **2021**, *10*, 284–290. [CrossRef]

International Journal of *Environmental Research and Public Health*

MDPI

Article

Does Human Capital Matter for China's Green Growth?—Examination Based on Econometric Model and Machine Learning Methods

Xiaoxue Liu [1], Fuzhen Cao [1] and Shuangshuang Fan [2,*]

1 School of Economics, Beijing Technology and Business University, Beijing 100048, China
2 School of Management, China University of Mining and Technology-Beijing, Beijing 100086, China
* Correspondence: bqt1900502011@student.cumtb.edu.cn

Abstract: To tackle the increasingly severe environmental challenges, including climate change, we should pay more attention to green growth (GG), a path to realize sustainability. Human capital (HC) has considered a crucial driving factor for developing countries to move towards GG, but the impact and mechanisms for emerging economies to achieve GG need to be further discussed. To bridge this gap, this paper investigates the relation between HC and GG in theory and demonstration perspective. It constructs a systematic theoretical framework for their relationship. Then, it uses a data envelopment analysis (DEA) model based on the non-radial direction distance function (NDDF) to measure the GG performance of China's 281 prefecture level cities from 2011 to 2019. Ultimately, it empirically tests the hypothesis by using econometric model and LightGBM machine learning (ML) algorithm. The empirical results indicate that: (1) There is a U-shaped relationship between China's HC and GG. Green innovation and industrial upgrading are transmission channels in the process of HC affecting GG. (2) Given other factors affecting GG, HC and economic growth contribute equally to GG (17%), second only to city size (21%). (3) China's HC's impact on GG is regionally imbalanced and has city size heterogeneity.

Keywords: human capital; green economy efficiency; green innovation; LightGBM machine learning; green growth; industrial upgrading

Citation: Liu, X.; Cao, F.; Fan, S. Does Human Capital Matter for China's Green Growth?—Examination Based on Econometric Model and Machine Learning Methods. *Int. J. Environ. Res. Public Health* **2022**, *19*, 11347. https://doi.org/10.3390/ijerph191811347

Academic Editors: Taoyuan Wei and Qin Zhu

Received: 10 June 2022
Accepted: 6 September 2022
Published: 9 September 2022

1. Introduction

The global industrialization and urbanization have disturbed the earth's natural balance. The critical imbalance in the carbon cycle between carbon sources and carbon sinks has forced the world to focus on issues of global warming and frequent natural disasters. The increasingly severe climate change has significantly impacted ecosystems and economics, as well as social development [1–3]. As the world's leading developing economy, China has become the world's largest carbon emitter [4] in recent years. Therefore, it is China's duty as a major power to transfer its development model to reduce energy consumption and carbon emissions while maintaining economic growth. To this end, at the 2015 Paris Climate Conference, the Chinese government made a commitment to hitting peak carbon emissions by 2030 [5]. Since then, the green growth model based on harmonious coexistence of humans and nature has become the core value orientation in China.

The concept of green growth was first proposed in the United Nations Economic and Social Commission for Asia and the Pacific (UNESCAP) in 2005. Green growth emphasizes that, when reducing poverty and improving human well-being through economic growth, countries should focus on transforming economic growth and consumption patterns, improving the ecological efficiency of economic growth, and coordinating environmental and economic development [6], so as to achieve sustainable development goals. The concept of green growth takes green economy, low-carbon economy, and circular economy as the main economic forms [7]. Its primary goal is to promote the transformation and upgrading of industrial civilization to ecological civilization [8]. However, the transition from brown economy to green economy will involve government policies [9,10], economic growth target pressure [11,12], technological progress [13], socio-cultural contexts [14], transition costs [15–17], and many other factors. Among many potential factors affecting green growth, human capital, known as a kind of living capital, refers to the knowledge and skill sets that workers have [18–20]. It is characterized by creativity, innovation, and subjective initiative, so it contributes greatly to green growth patterns [21,22].

Since China's reform and opening up, the human capital level has continued to improve [23]. According to the *China Human Capital Report 2021* (http://news.cufe.edu.cn/info/1002/52212.htm, accessed on 14 December 2021) released by China Center for Human Capital and Labor Market Research of Central University of Finance and Economics, the average education years for China's labour force increased from 6.1 years in 1985 to 10.5 years in 2019. The national workforce population with a college degree or above increased from 10% to 20.6%. This naturally arouses a series of questions worthy of attention: what is the relationship between China's human capital and green growth? What is the internal mechanism driving the formation of such relationship? What is the transmission path? Among many factors influencing green growth, how much does human capital contribute? Although academics and policy makers pay more and more attention to these issues, few of them can provide theoretical and empirical findings that systematically answer those questions.

Research conclusions in existing literature on the relationship between human capital and green growth are inconsistent. A lot of literature believes that there is a linear relationship between human capital and green growth. They propose two kinds of distinct conclusions: one is "promotion viewpoint", indicating that human capital promotes green growth (or reduces carbon emissions) [24–29]; the other is "inhibition viewpoint", arguing that human capital inhibits green growth (or increases carbon emissions) [30,31]. Other literature believes that the relationship between human capital and green growth is uncertain, changing with different time periods, different industries, macroeconomic variables and human capital regime level in various regions [32–34]. A few studies confirm the possible non-linear relationship between human capital and green growth [35–37]. In existing studies on the nonlinear relationship, both the measurement index and nonlinear shape of green growth and human capital are quite different. Reviewing conclusions about human capital and green growth, such as the promotion viewpoint, inhibition viewpoint and non-linear relationship viewpoint, we find it necessary to further clarify the relationship between human capital and green growth. Specifically, for emerging economies such as China, it will help them promote green growth more efficiently.

To answer the previous questions, by using the panel data of China's 281 prefecture-level cities (including municipalities directly under the Central Government) from 2011 to 2019, this paper examines the relationship between China's human capital and green growth from theoretical and empirical perspectives. Hence, we can precisely classify and implement specific policies. First, from a theoretical perspective, we propose the hypotheses that there is the nonlinear relationship between human capital and green growth, and that green innovation and industrial upgrading are transmission paths. Meanwhile, in order to reveal the time series trend and spatial distribution characteristics of China's green growth, this paper uses the data envelopment analysis (DEA, Appendix A) model based on the non-radial direction distance function (NDDF) to measure the sample cities' green economic efficiency (GEE). Furthermore, this paper empirically tests the previous hypotheses. Specifically, it uses econometric models to investigate whether human capital and green growth have the nonlinear relationship and the main transmission path. It also applies machine learning algorithms to measure the human capital's contribution weight among many influencing factors. The research findings are as follows: first, China's human capital and green growth have a U-shaped relationship rather than a simple linear relationship. That is, when human capital development cannot reach a certain threshold, it will inhibit green growth; when it exceeds a certain threshold, it will promote green growth. This conclusion is still reliable after the robustness test. Green innovation and industrial upgrading are transmission channels in the process of human capital affecting green growth. Second, the result of the machine learning algorithm reveals that among many factors influencing green growth, the human capital's contribution weight is higher, at about 17%. It is as important as the economic growth level, second only to city size (21%). In addition, the heterogeneity analysis indicates that human capital has exceeded the U-shaped threshold in southern regions. In the eastern region, it has been near the U-shaped threshold and is about to promote green growth. In other regions, human capital has not yet been able to promote green growth. In large cities, human capital has already exceeded the U-shaped threshold. While in small and medium-sized cities, it is still on the left side of the U-shaped threshold, indicating that city size can speed up crossing a threshold between human capital and green growth so that human capital can positively promote green growth.

This study provides the following three contributions: first, it illustrates the theoretical root of the nonlinear relationship between human capital and green growth from the production and consumption perspective. Meanwhile, the inner mechanism of human capital influencing green growth is analyzed in detail. It proposes two transmission paths of green innovation and industrial upgrading. This work directly proves that there is a U-shaped relationship between human capital and green growth, which enriches and expands the research results of nonlinear relations between them [35–37]. This means that the linear relationship assumptions between human capital and green growth, i.e., the promotion or inhibition viewpoints, are not suitable for China. Second, this study uses an econometric model and machine learning (ML) algorithms to test theoretical hypotheses, which not only clarifies the transmission mechanism of human capital affecting green growth, but also introduces advanced ML algorithms into economics field to study human capital's contribution to green growth. However, existing studies on nonlinear relationship [35–37] only use econometric models for empirical research, which cannot accurately reflect human capital's contribution to green growth. Third, this study conducts a series of grouping heterogeneity tests based on the city's location and size, respectively, and uses the U test econometric model to examine whether the human capital development level in different groups exceeds the U-shaped threshold. Hence, we can adjust measures to local conditions and implement the classified policies to ensure that human capital will positively impact green growth policy.

The remainder of this paper is organized as follows. Section 2 provides theoretical basis and research hypothesis. Section 3 discusses research design and data selection. Section 4 reveals empirical results of econometric models and machine learning algorithms. Section 5 concludes by proposing main conclusions and policy implications.

2. Literature Review and Hypothesis Proposal

2.1. Human Capital and Green Growth

Human capital refers to labour's ability composed of knowledge, skills, and physical ability. It is formed through human investment (such as education investment), takes workers as a carrier, and indicates worker's skills, intelligence and talents [18,20]. Schultz, the "father of human capital theory", believes that human capital is the source of driving economic growth. He also emphasized the important role played by the "quality" of human capital [38]. In the green transformation of the economy, human capital also contributes important value. Recently, some scholars have studied the relationship between human capital and green economy, but their conclusions are inconsistent. These conclusions include "promotion viewpoint", "inhibition viewpoint" and "non-linear relationship". The "promotion viewpoint" holds that human capital can improve natural resource conservation [24,28,39], reduce energy consumption intensity [25], and reduce pollutant emission [26,27,29]. The "inhibition viewpoint", on the contrary, emphasizes the positive correlation between human capital and carbon emissions [30,31]. Based on "promotion viewpoint" and "inhibition viewpoint", some scholars argue that the relationship between human capital and carbon emissions is time-varying. It varies in the short-term and long-term, in different industries as well as financial development and human capital at different regime levels [2,32,33]. Li and Ou Yang [33] argue that human capital increases CO_2 emissions in the short term and reduces CO_2 emissions in the long term. Çakar et al. [34] find that financial development and the development level of human capital affect whether human capital increases or suppresses carbon emissions. Human capital increases carbon emissions in both low regimes of financial development and human capital, and decreases in high regimes. In addition, some studies believe that human capital has a significant threshold effect on the green economy, resulting in nonlinear effect under different levels of complex variables of economic and social development [35–37]. Liu and Lv [36] test the non-linear relationship between rural human capital and agricultural green total factor productivity (AGTFP) in China. Maranzano et al. [37] test the nonlinear relationship between education and emissions, reflecting the dynamic change in OECD and European economic and social development. Chen et al. [35] believe that green R&D activities and sulfur dioxide emissions are in a nonlinear relationship, but are affected by technology absorption capacity. At present, the view that human capital and green growth have a nonlinear relationship comes from indirect evidence rather than direct evidence. There are great differences in the measurement of human capital and green growth. The existing green growth indicators include AGTFP, CO_2 emissions, SO_2 emissions and other measurement indicators. Human capital includes rural human capital, average years of education (population 15–64 years), green R&D and other measurement indicators; furthermore, the nonlinear shape is also inconsistent, and it is considered as an "N-shaped" relationship [36] or an inverted U-shaped relationship [37]. However, the human capital formed through education investment needs to be accumulated for a relatively long time before population endowment improves [40], which in turn positively affects the green growth. Therefore, we believe that human capital and green growth are not a simple linear relationship, but have different impacts on the green economy at different human capital development stages.

First, from the production sector perspective, human capital is closely related to productivity [41,42]. When human capital is at a low development level and employees have low education level and professional skills, the industry will absorb a large number of low-skilled labours and have very few high-skilled labourers [43]. Under such circumstances, the marginal contribution rate of talents to production is low, and the output improvement mainly depends on the large-scale investment of physical capital, which leads to "high energy consumption and high pollution emissions" that hinder the green economy. As human capital continues to accumulate and enters a higher development stage, the labour skill structure changes; the proportion of high-skilled labour increases significantly, and the complementarity between capital and skills begins to strengthen [44]. Hence, the individual production department's efficiency is significantly improved at first, and generates a positive spillover effect through the demonstration effect [45], driving the entire production department to reshape the production process to reduce physical capital input and improve production through technological iteration. It then further reduces energy consumption, pollution levels, and promotes green economy development. Second, from the consumption perspective, the human capital level is closely related to the consumption structure [46,47]. The low-level human capital development stage corresponds to the relatively low consumers' income and affordability [48]. Under such circumstances, as human capital improves, consumers often pay attention to related consumer goods to meet basic "material needs", such as purchasing household appliances, automobiles, and other large commodities. However, if such consumer demand continues to grow, it will increase carbon dioxide emissions and inhibit green growth [49]. As human capital development exceeds a certain level, on the one hand, after the basic "material needs" are fully satisfied, the consumption structure will undergo a "qualitative leap"; that is, "spiritual consumer goods" related to entertainment and health will take the lead. On the other hand, high-level human capital with good environmental awareness [50] will increase the consumption proportion of environmentally friendly, green, and low-pollution consumer goods that are conducive to green growth. In addition, consumption structure upgrade will also force the production sector to improve and iterate products [6,49,51], which is conducive to producing more environmentally friendly products [52]. Therefore, human capital at this stage will positively promote green economy development.

To sum up, the relationship between human capital and green growth is not a simple linear one. If the human capital does not reach a certain threshold, it will inhibit green growth. On the contrary, it will promote green growth. The human capital will first inhibit green growth and then promote it. Accordingly, we put forward the following hypothesis.

Hypothesis 1. *There is a U-shaped relationship between human capital and green growth.*

2.2. Human Capital, Green Innovation and Green Economic Efficiency

Romer's endogenous growth theory believes that human capital is an important source of driving total factor productivity improvement and technological progress. Human capital promotes innovation from both micro and macro perspectives. At the micro level, human capital represents a high level of human resources and can directly affect R&D activities within a company [53,54]. High-level human resources can promote a company's technology R&D through integrating both internal knowledge and external knowledge [21,55]. This integration mainly includes knowledge creation, knowledge dissemination, knowledge diffusion, and companies' internal R&D activities transformation [56]. Therefore, high-level human capital can directly affect companies' R&D activities, thereby enhancing their innovation level. On the other hand, at a macro level, when a city's human capital is at a high level, it can bring about knowledge spillover effect through the agglomeration, flow, and imitation of talents [57,58]. That is, companies with high-level human capital can share and transfer their tacit knowledge and resources to other companies in the industry chain to drive the entire industry chain and city to innovate and develop [59,60]. In short, human capital will promote innovation and development.

However, existing studies have shown that the relationship between green innovation and GEE is often nonlinear. Chen and Huo [61] and Shi et al. [62] argue that there is an inverted U-shaped relationship between innovation and carbon emissions. Hu et al. [63] finds that there is a U-shaped relationship between green innovation and green development. First, corporate innovation requires enterprises to expand their investment through years of operation and accumulation. In the process of realizing its technological innovation, enterprises will spare no effort to increase R&D investment in the early stage [64]. However, due to the long cycle and high risk of scientific and technological innovation, early R&D investment may not be able to be converted into R&D results in time to play the role of driving the city's green growth [35]. Moreover, since now company scale expansion and capital recycling have brought certain negative externality to the environmental system, the "rebound effect" of such negative externality is greater than the energy-saving effect brought by technological innovation [65,66], which will increase energy consumption and carbon emissions to some extent, and is not conducive to improving green economic performance [61,62]. Second, when the innovation level exceeds a certain threshold and reaches a high level, the early R&D investment is transformed into a real force to promote companies' technological improvement and product iteration. Hence, the innovation results can be transformed and resource use efficiency is improved, thus reducing carbon emissions [67,68] and achieving high-level green development [63,69]. Therefore, we propose the second hypothesis.

Hypothesis 2. *Green innovation is the intermediate variable between human capital and green growth.*

2.3. Human Capital, Industrial Upgrading, and Green Growth

British economist Crick was the first to interpret the connotation of industrial upgrading. That is, when the labour force transfers from the primary industry to the secondary and tertiary industries, a country's economy gradually evolves from the primary industry as the leading industry to the secondary and tertiary industries [70]. During industrial upgrading process, labour, capital, natural endowment, and technological progress have become important driving factors [71–73]. As a "living capital", human capital, with labour as its carrier, has greater value-added potential than hard capital such as capital and material, and is more innovative and creative [21,53]. Therefore, it has a non-negligible contribution rate to industrial upgrading [72,74,75]. Schultz believes that education is the most important form of human capital investment [76]. The improvement of labour education level has accelerated its transfer from the primary industry to the secondary and tertiary industries [73]. The accumulation of human capital stock also will help to break the original industrial chain and accelerating the process of forming a new economic and technological industrial chain, which will lead to changes in the industrial and market environment and help to create a new industrial chain [77,78]. In addition, the higher the level of human capital stock, the stronger the efficiency of knowledge dissemination and spillover, that is, the better the effect of "learning by doing", which is conducive to transforming and absorbing advanced technology, thus boosting the industrial structure leap [79].

However, the industrial upgrading process requires a leap from the accumulation of quantitative changes to qualitative changes, which is not achieved overnight. The relationship between industrial upgrading and green growth is not a simple linear one. The "accumulation" stage and "leap" stage of industrial upgrading may have different impacts on green growth. Existing studies, such as Wei and Zhang [80], Liang et al. [81], Yang et al. [82], and Zhang et al. [83], demonstrate the nonlinear relationship between the two. In the initial stage of industrial upgrading, since a large amount of capital and labour flow into the secondary and tertiary industries, and market-driven industrial changes often lack scientific policy supporting facilities [84], this type of industrial upgrading is relatively extensive. The profit-seeking nature of capital makes the industry focus on the return on investment measured in currency, while ignoring the governance of externalities such as environmental pollution [85,86]. As the industrial upgrading reaches a certain level, the industry gradually transforms from a low value-added, extensive, low-tech one to a high value-added, intensive, and high-tech one [87,88]. Meanwhile, with the implementation of a series of high-quality development strategies, innovation-driven, green development, and other initiatives drive the industrial upgrading process and the green development to run simultaneously, which actively promotes the GEE [89–91]. Based on this, we propose the third hypothesis:

Hypothesis 3. *Industrial upgrading is the intermediate variable between human capital and green growth.*

3. Methodology and Data

First, this section explains the variables selected in this study. Second, we apply the NDDF-DEA model to measure cities' green growth level during the statistical period. Then, we employ the econometric models and LightGBM machine learning model to explore the impact and mechanism of human capital on green growth in this study. Finally, the data source used in this study is briefly introduced.

3.1. Variable Measurement and Selection

The variables selected in this study are shown in Table 1.

Table 1. Variable definition and calculation method.

Variable Type	Definition	Code	Calculation Method
Dependent variable	Green economic efficiency	*GEE*	Measured by the NDDF-DEA model
Independent variable	Human capital	*HC*	Logarithm of financial education expenditures in prefecture-level cities at the end of each year
Intermediary variable	Green innovation	*GIN*	Logarithm of green invention patent applications in prefecture-level cities
	Industrial upgrading	*IU*	The output value of secondary industry plus output value of tertiary industry, divided by GDP
Control variable	Free trade zone	*FTA*	The variable is equal to one if the city is a free trade zone; otherwise, it is zero
	Level of economic development	*LED*	Logarithm of per capita GDP
	Government intervention	*GI*	Public budget expenditure divided by GDP
	City scale	*CS*	Logarithm of the total population of each city at the end of the year
	Foreign direct investment	*FDI*	The total amount of foreign capital divided by GDP
	Fiscal decentralization	*FD*	The ratio of the fiscal revenue in the municipal budget to the fiscal expenditure in the municipal budget
Other variables	Years of education	HC_1	(number of university students in city/number of university students in province) × ln(6 × the proportion of labour force in the sample with no higher than primary school education +9* the proportion of labour force with no higher than junior middle school education +12 × the proportion of labour force with no higher than senior high school education +16* the proportion of labour force with college education) (Wang et al. 2021) [22]
	Carbon dioxide emissions	CO_2	Logarithm of carbon dioxide emissions
	Year	*Year*	a dummy variable
	City	*City*	a dummy variable according to China Urban Statistical Yearbook

3.1.1. Dependent Variable: Green Economic Efficiency (GEE)

Green growth is the dependent variable in this study. Referring to Cheng et al. [92], as well as Wang and Chen [93], this study uses green economic efficiency (*GEE*) as the proxy variable of green growth. This study originally uses distance functions [94], including the Shephard distance function (SDF) and directional distance function (DDF), when measuring *GEE*. However, SDF cannot achieve pollutant emission reduction when ensuring an ideal output [94]. Although DDF overcomes this problem, it leads to an overestimation of efficiency [95]. On this basis, Zhou et al. [96] proposed NDDF.

This study introduces the DEA model to measure the *GEE* of sample cities. This model has the advantage of comprehensively considering the desirable outputs and undesirable outputs in the economic system from the aspects of input and output. In addition, Zhang and Li [97] and Li and Ji [98] both use NDDF of the DEA model to measure *GEE*. The input variables include energy (E), labour, and capital. In terms of output variables, the desirable output is GDP, and the undesirable outputs are industrial wastewater (WW), industrial sulfur dioxide gas (WG), and industrial soot and dust (SD), as well as carbon dioxide (CD). In this process, the weights of the energy input (E), GDP, WW, WG, and SD are set to 1/3, 1/3, 1/9, 1/9, and 1/9, respectively. The proportion of these five weights, which can be increased or decreased, is calculated by the super-efficiency DEA model. Finally, the GEE of the *i*-th city in the *t*-th period is constructed as the dependent variable of this article.

$$DDF_{it} = \frac{1}{2}\left[\frac{(E_{it} - \beta_{E,it} * E_{it})/(G_{it} - \beta_{G,it} * G_{it})}{E_{it}/G_{it}}\right] + \frac{1}{2}\left[\frac{1}{3}\sum_{N=WW,WG,SD} \frac{(N_{it} - \beta_{N,it} * N_{it})/(G_{it} - \beta_{G,it} * G_{it})}{N_{it}/G_{it}}\right]$$

where $\beta_E, \beta_G, \beta_{WW}, \beta_{wG}, \beta_{SD}$ are the optimal solutions of the DEA model. Capital stock data are calculated using the perpetual inventory method. The raw data required include the fixed asset investments of cities at the prefecture level and above; these data are obtained from the CEIC China Economy Database. Based on the estimation reported by Xiang [99], we can obtain the capital stock data of each city in the base year (2000) and the capital stock depreciation rate of each city [99]. We agree that the capital stock depreciation rates of cities in the same province are the same.

3.1.2. Independent Variable and Intermediate Variables

(1) Core Independent Variable: Measurement of Human Capital

The core variable considered in this article is the human capital. Human capital has many measurement dimensions, but mainly focuses on education [22,100]. Existing studies include the number of college students [22] and the average years of education of the population [100,101]. In addition, some studies have pointed out that government education expenditure is highly correlated with human capital formation and development [101,102]. Therefore, this paper uses the logarithm of urban government education expenditure to measure city human capital level. It also uses the average education years of the population [22] as the proxy variable of human capital for robustness test.

(2) Intermediate Variable: Green Innovation (GIN)

High-level human capital promotes green innovation [59]. Green innovation (*GIN*) will largely affect carbon emissions [61], affecting green economic development [63]. In this paper, green innovation is measured by the number of green invention patent applications in prefecture level cities.

(3) Intermediate Variable: Industrial Upgrading

Existing literature reveals that human capital will lead industrial upgrading [73]. Industrial structure in turn will certainly impact energy efficiency [103,104]. Industrial upgrading is now the main form of industrial structure change. Referring to Yao et al., (2019) [105], this study uses the ratio of the secondary and tertiary industries' total output value to GDP to measure industrial upgrading.

3.1.3. Control Variables

We select a series of control variables that affect urban GEE from the two aspects of urban development and government factors to better study the impact of the human capital on urban green economy development. First, the urban factors include level of economic development (*LED*), city scale (*CS*), and foreign direct investment (*FDI*). Second, the government factors include Free Trade Zone (FTZ), government intervention (*GI*) and fiscal decentralization (*FD*). The specific meaning of each variable is provided below.

(1) Free Trade Zone (FTZ)

Free trade zone, China's special functional area enjoying opening to the outside world, has greatly impacted the GTFP of China's manufacturing industry (Liu et al., 2019) [106]. This paper uses virtual variables to measure whether a city is a free trade zone (FTZ). The variable is equal to one if the city is a free trade zone; otherwise, it is zero.

(2) Level of Economic Development (*LED*)

The LED of city is the basis for a city to achieve green growth. According to the research findings, the scale of production and consumption changes with an increase in income level, and this affects energy consumption and environmental quality [107,108]. In this study, a city's economic development level is expressed by the logarithm of the ratio of the urban GDP value to the total population at the end of the year, i.e., the logarithm of per capita GDP.

(3) Government Intervention (*GI*)

Droste et al. [109] state that *GI* is key to urban green development. Some studies on *GI* and green economy have shown that *GI* can improve environmental performance [110] and affect the efficiency of urban pollutant emission [111]. In this article, *GI* is measured as the ratio of a city's public budget expenditure to *GDP*.

(4) City Size (*CS*)

Theoretically, a city with a larger population has more capital for green economy development. Islam and Ghani [112] believe that population size is a key factor affecting the environment. In this article, the city scale is measured by the logarithm of the total population of each city at the end of the year.

(5) Foreign Direct Investment (*FDI*)

Foreign direct investment is an inseparable and important factor affecting China's green economic development efficiency [113]. This factor is measured by the ratio of the total amount of foreign capital used by each city to the regional GDP.

(6) Fiscal Decentralization (*FD*)

Fiscal decentralization has a certain impact on carbon emissions, enterprise ecological innovation, and *GEE* [114,115]. This study uses fiscal autonomy to represent fiscal decentralization, i.e., the ratio of the fiscal revenue in the municipal budget to the fiscal expenditure in the municipal budget.

3.2. Research Methods and Model Resign

3.2.1. Combines Econometric Model and LightGBM Machine Learning Algorithm

This paper empirically tests the U-shaped relationship between human capital and green growth, the transmission channel, and the contribution weight of human capital on green growth by combining econometric model and ML algorithm. The econometric model includes the benchmark model and the intermediary effect model, which can explain the direction and transmission mechanism between variables. However, it is impossible to measure the contribution of the core explanatory variable to the explained variable, and there may be some potential problems, such as the inverse causality between the independent variable and the dependent variable, or the multicollinearity between the independent variable and control variables; the machine learning algorithm can well overcome the endogenous problems and multicollinearity problems that may exist in econometric models, predict the dependent variables according to multiple explanatory variables, and accurately measure the interpretation degree of the core explanatory variables to the dependent variables, but it is difficult to explain the mechanism of the independent variables and the dependent variables. Therefore, combining the two methods can give full play to their advantages and clarify the relationship between human capital and green growth and its importance to green growth.

3.2.2. Benchmark Model and Intermediary Effect Model

First, the benchmark model of human capital and green growth is as shown in Formula (1):

$$GEE_{it} = \alpha_{it} + \beta HC_{it} + \gamma_1 HC_{it}^2 + \sigma X_{it} + Year_i + City_t + \varepsilon_{it} \tag{1}$$

where GEE_{it} is the green economic efficiency of city i in t year, HC_{it} is the human capital of city i in t year, and X_{it} is the control variable, mainly including *FTA*, *LED*, *GI*, *CS*, *FDI* and *FD*. β and γ_1 are used to investigate whether there is a nonlinear relationship between HC_{it} and GEE_{it}. When $\beta > 0$ and $\gamma_1 < 0$, it means that there is an inverted U-shaped relationship between *GEE* and *HC*; when $\beta < 0$ and $\gamma_1 > 0$, there is a U-shape relationship between *GEE* and *HC*. After the regression coefficient is determined, it needs to be further determined in combination with the U test results to determine whether it is a U-shaped or inverted U-shaped relationship. ε_{it} is the residual. Year and city refer to control year and city effect, respectively.

Secondly, the intermediary effect model of human capital and green economic efficiency is as follows.

$$GIN_{it} = \alpha_{it} + \beta_1 HC_{it} + +\sigma X_{it} + \varepsilon_{it} \quad (2)$$

$$GEE_{it} = \alpha_{it} + \beta_2 GIN_{it} + \gamma_2 GIN_{it}^2 + \sigma X_{it} + \varepsilon_{it} \quad (3)$$

where GIN_{it} is the green innovation of city i in t year, HC_{it} is the human capital of city i in t year, GEE_{it} is the green economic efficiency of city i in t year, which is the measure of green growth, and X_{it} is the control variable, including *FTA*, *LED*, *GI*, *CS*, *FDI* and *FD*. The regression coefficient β_1 reflects the relationship between *HC* and *GIN*; β_2 and γ_2 are used to investigate whether there is a nonlinear relationship between GIN_{it} and GEE_{it}. When $\beta_2 > 0$ and $\gamma_2 < 0$, it means that there is an inverted U-shaped relationship between *GIN* and *GEE*; when $\beta_2 < 0$ and $\gamma_2 > 0$, there is a U-shape relationship between *GIN* and *GEE*. After the regression coefficient is determined, it is also necessary to be in combination with U test results to determine whether it is a U-shaped or inverted U-shaped relationship. φ_{it} is the residual.

Then this paper takes industrial upgrading (IU_{it}) as an intermediary variable, and uses IU_{it} to replace GIN_{it} in the above Equations (2) and (3), that is, to test the intermediary effect of industrial upgrading.

3.2.3. LightGBM Algorithm

When considering other factors affecting green growth, we further used the LightGBM algorithm to measure the contribution of human capital to green growth. The processing of the LightGBM algorithm is according to Fan and Liu [116]. LightGBM is an efficient implementation of XGBoost. The commonly used GBDT machine learning algorithm has limitations when processing massive data. The main reason for the birth of LightGBM is to solve the problems encountered by GBDT in massive data, so that GBDT can be better and faster used in industrial practice. Its idea is to discretize continuous floating-point features into k discrete values and construct a histogram with a width of k. Then, traverse the training data and calculate the cumulative statistics of each discrete value in the histogram. In the feature selection, we only need to traverse to find the optimal segmentation point according to the discrete value of histogram. In addition, the use of leaf wire strategy with a depth limit saves a lot of time and space consumption. Its features are: optimizing speed and memory usage; sparse optimization; optimizing accuracy; using leaf-wise growth mode, to process categorical variables; and optimizing network communication. We build a machine learning model with the help of python software. The ratio of data training set to test set is 8:2. See Appendix B for the specific hyperparametric settings of the model.

3.3. Data Source

This study takes the panel data of China's 281 prefecture-level cities from 2011 to 2019 as the sample to empirically measure human capital's impact on green growth and its internal mechanism. The data are obtained from the China Economy Database (CEIC), China City Statistical Yearbook, China Population and Employment Statistics Yearbook, and China Statistical Yearbook. When measuring the *GEE*, we obtain the data of the capital, labour, energy consumption, and GDP from CEIC; data of the SD from the China City Statistical Yearbook; and data of the two pollutants of WW and WG from CEIC. The data of human capital are obtained from China Population and Employment Statistics Yearbook and China Statistical Yearbook. The data of the intermediate variable and control variables are obtained from China City Statistical Yearbook.

Table 2 shows the sample descriptive statistics of each variable, including sample size, mean, standard deviation, minimum, maximum, Skewness, Kurtosis. The mean value of GEE is 0.334, the maximum value is 1, and the minimum value is 0.11. That is, the overall GEE is low and there is obvious regional imbalance. The difference in the HC of the different cities is relatively large. The maximum value is 16.2456, the minimum value is only 9.9059 and the mean is 13.1288. All variables are right biased except the *CS* and *IU*. In addition to *LED*, *FD* and *GIN*, the kurtosis of other variables is greater than 3, which does not obey the standard normal distribution and shows obvious characteristics of "fat-tail distribution". The variance inflation factor (VIF) of all explanatory variables is less than 10, which means that there is no serious multicollinearity.

Table 2. The statistics summary of variables.

Variable	Obs	Mean	Std. Dev.	Min	Max	Skewness	Kurtosis	VIF
GEE	2297	0.3341	0.1623	0.1107	1.0000	2.6018	10.8540	
HC	2297	13.1288	0.7805	9.9059	16.2456	2.4250	9.0127	8.6100
FTA	2297	0.2037	0.4029	0	1.0000	1.5530	3.4119	7.8900
LED	2297	10.7171	0.5790	8.8416	13.0557	0.2192	2.8621	5.5300
GI	2297	0.0793	0.0281	0.0234	0.2273	1.1885	5.6820	4.5200
CS	2297	5.9025	0.6963	2.9704	8.1362	−0.5567	4.0910	2.3700
FDI	2297	0.0027	0.0027	0	0.0299	2.2460	13.5210	1.3500
FD	2297	0.4790	0.2255	0.0680	1.5413	0.5302	2.6254	1.2400
GIN	2297	4.3325	1.7641	0	10.1825	0.4849	2.9046	4.2300
IU	2297	4.4730	0.1035	3.6618	4.6049	−2.1419	11.1377	2.0000

4. Empirical Results

4.1. Spatiotemporal Characteristics of GEE

We reveal the spatiotemporal characteristics of Chinese cities' *GEE* and describe it using a geographic distribution map before empirically analysing the relationship between human capital and *GEE*. Chinese cities' geographic distribution map of *GEE* (Figures 1 and 2) indicates that the overall level of *GEE* is not high, and *GEE* in most cities is between 0 and 0.3341. The development level of GEE in different regions is uneven. The *GEE* level in the eastern is higher than that in the central and western regions, and the *GEE* of cities in the northeast regions has not been continuously optimized after the phased improvement. Specifically, from 2011 to 2016, some cities in the northeast regions became national new industrialization comprehensive reform pilot areas, with high overall *GEE*. However, Liaoning Province is dominated by heavy industry with high energy consumption and pollution. This industrial structure is not conducive to the continuous improvement of *GEE*. In 2019, the overall *GEE* in the northeast region decreased. Among them, the areas with the fastest improvement in *GEE* are the Yangtze River Delta and the eastern coastal areas of the Pearl River Delta, which is mainly related to the national green planning for rapid urban development during the 12th and 13th Five-Year Plans.

The average value change trend of Chinese cities' *GEE* and *HC* from 2011 to 2019 (Figures 2 and 3) indicates that Chinese cities' *GEE* generally shows a U-shaped change, and HC is approximately linear. From the change trend of both, it is likely that HC and GEE have a U-shaped relationship. Moreover, *GEE* has been significantly improved since the 13th Five-Year Plan. This indicates that the improvement of GEE is related to government policy guidance.

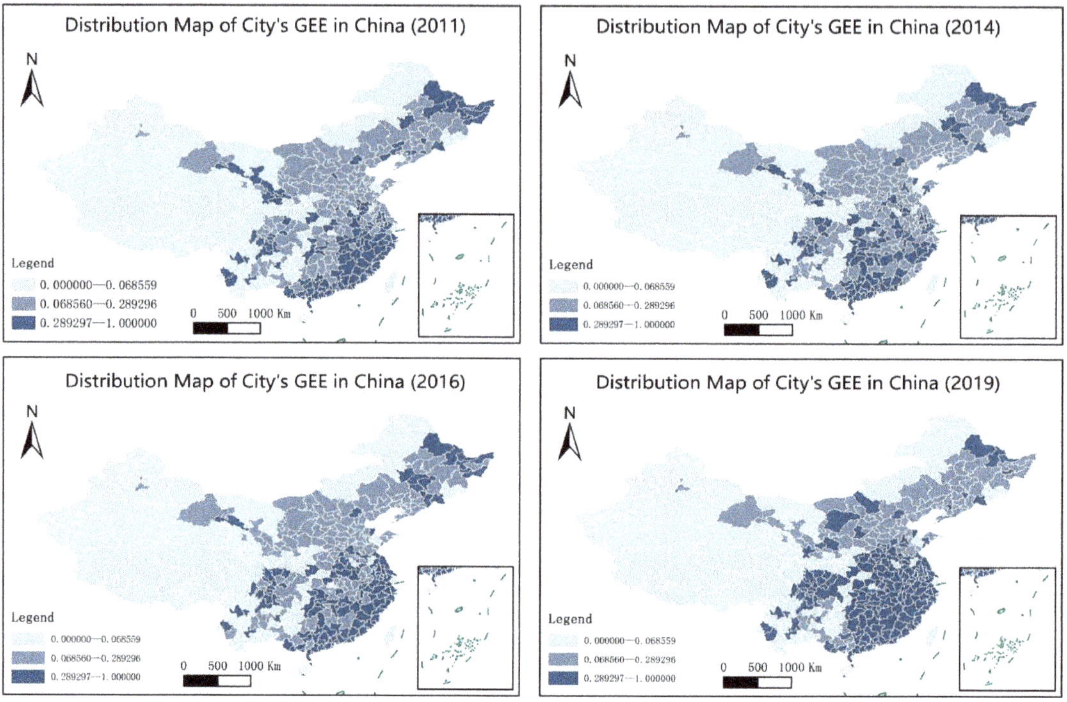

Figure 1. Geographical distribution map of city's GEE in China (2011–2019).

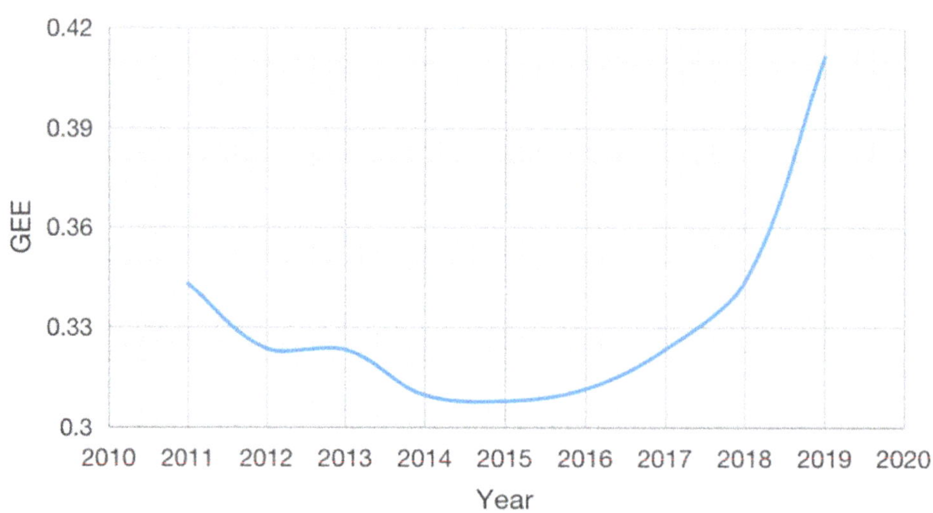

Figure 2. The average value change trend of GEE from 2011 to 2019.

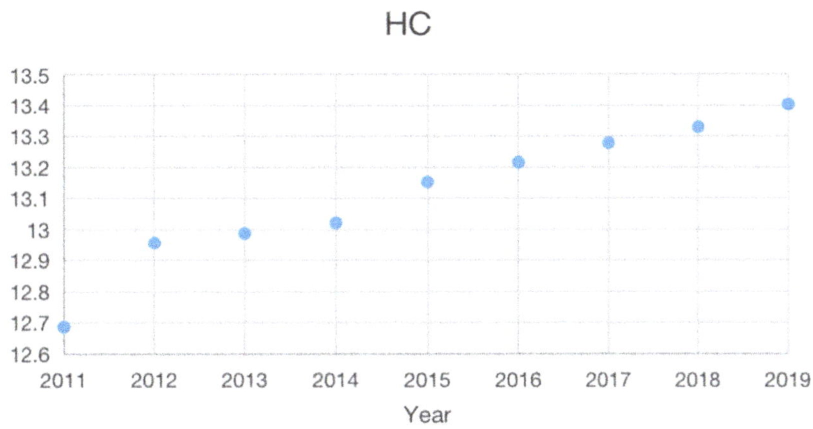

Figure 3. The change trend of HC from 2011 to 2019.

4.2. Test of Nonlinear Relationship between HC and GG

To explore the nonlinear relationship between *HC* and *GG*, under the control of other variables, first, we use the OLS model to regress *HC* and *GEE*, and then examine with the U test. The regression results of *HC* and *GEE* are shown in Table 3. The results are as follows:

First, there is a U-shaped relationship between *HC* and *GEE*. Specifically, the regression coefficient of *HC* is −0.932, while the regression coefficient of HC^2 is 0.037, both of which are significant at the level of 1%, indicating a U-shaped relationship between *HC* and *GEE*. On the left side of the U-shape, with the improvement of *HC*, the green growth is suppressed; when the level of *HC* exceeds a certain threshold, it will promote green growth. The results of U test show that there is a U-shaped relationship between *HC* and *GEE* at the significance level of 1%.

Second, the current *HC* level is on the left side of the U-shape, which has not reached the threshold of *HC* promoting *GEE*. The current *HC* level is 12.469, which has not reached the threshold value of *HC* promoting *GEE* development (12.595). It is on the left side of the U-shaped fitting diagram of *HC* and *GEE* (Figure 4a).

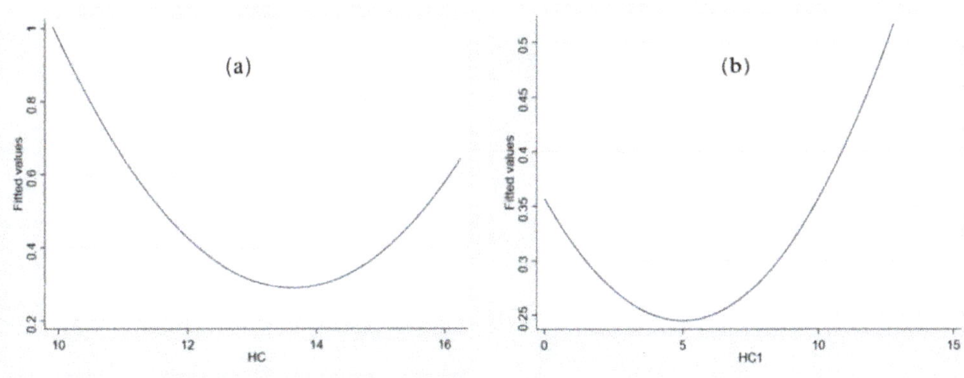

Figure 4. Fitting diagram of U-shaped relationship between HC and GEE, (**a**) fitting diagram of HC and GEE (HC = education expenditure) (**b**) fitting diagram of HC_1 and GEE (HC_1 = years of education).

The result that HC and green growth have a U-shaped relationship in this study supports the view of Maranzano et al. [37] to a certain extent, but the green growth measurement index is different. We adopt the NDDF-DEA model to measure *GEE* more comprehensively, while Maranzano et al. [37] adopt a carbon emission index. The result is different from Wang et al. [22] and Xiao and You [88]. They all support that total *HC* can improve green growth, and Wang et al. [22] further conclude that different HC levels have different effects on GTFP.

Table 3. Regression results of U-shaped relationship between HC and GEE.

Variables	GEE
HC	−0.932 ***
	(−7.71)
HC^2	0.0370 ***
	(8.03)
FTA	0.0010
	(0.14)
LED	0.0260
	(1.57)
GI	−0.8730 ***
	(−3.67)
CS	0.1150 ***
	(2.85)
FDI	0.1610
	(0.14)
FD	0.1280 ***
	(2.72)
U test	12.469 ***
	(5.95)
U test lower bound interval	9.9060
U test upper bound interval	16.2460
_cons	4.892 ***
	(5.57)
Year	controlled
City	controlled
N	2493
R^2	0.7640

Note: (1) *t* statistics in parentheses; (2) *** represent significance levels of 1.

4.3. The Mechanism Test Results Analysis

4.3.1. Human Capital, Green Innovation, and Green Growth

We use the intermediary effect model to examine whether green innovation acts as an intermediary variable between *HC* and *GEE*. The empirical results are shown in Table 4.

The results in column (1) of Table 4 shows that there is a U-shaped relationship between *HC* and *GEE*; the results in column (2) show that *HC* significantly and positively promotes the development of green innovation at the significance level of 1%, and the regression coefficient is 0.332, that is, every 1% increase in *HC* increases green innovation by 0.332%. The results in column (3) shows that the regression coefficient of *GIN* is significant at the level of 1%, which is −0.051, while the coefficient of GIN^2 is positive at the significance level of 1%. According to *GIN* and GIN^2 coefficients, there may be a U-shaped relationship between green innovation and *GEE*. Before the green innovation level reaches the threshold value, green innovation suppresses *GEE*. Once the green innovation level reaches the threshold value, the high utilization rate of resources promotes the development of *GEE*. The U-shaped relationship between green innovation and *GEE* supports the views of Hu et al. [63] and Liu et al. [117]. The U test results also show that the U-shaped relationship between green innovation and *GEE* is significant at the level of 1%. This means that human capital is positively promoting green innovation, and there is a U-shaped relationship between green innovation and *GEE*. This verifies hypothesis 2 that green innovation acts as transmission channel between *HC* and *GEE*.

Table 4. Empirical results of the relationship between HC, green innovation and GEE.

Variables	(1)	(2)	(3)
	GEE	*GIN*	*GEE*
HC	−0.932 ***	0.332 ***	
	(−7.71)	(3.94)	
HC^2	0.037 ***		
	(8.03)		
GIN			−0.051 ***
			(−6.97)
GIN^2			0.008 ***
			(10.17)
FTA	0.001	−0.021	0.001
	(0.14)	(−0.63)	(0.21)
LED	0.026	0.467 ***	0.034 **
	(1.57)	(5.67)	(2.27)
GI	−0.873 ***	2.168 *	−0.788 ***
	(−3.67)	(1.82)	(−3.46)
CS	0.115 ***	0.576 ***	0.097 **
	(2.85)	(2.87)	(2.56)
FDI	0.161	−1.049	0.805
	(0.14)	(−0.18)	(0.70)
FD	0.128 ***	−0.134	0.123 ***
	(2.72)	(−0.57)	(2.66)
U test	12.469 ***		3.073 ***
	(5.95)		(6.97)
U test lower bound interval	9.906		0
U test upper bound interval	16.246		10.182
_cons	4.892 ***	−6.294 ***	−0.726 **
	(5.57)	(−3.88)	(−2.26)
N	2493.000	2493.000	2493.000
R^2	0.764	0.950	0.768

Note: (1) t statistics in parentheses; (2) * $p < 0.1$, ** $p < 0.05$, *** $p < 0.01$

4.3.2. Human Capital, Industrial Upgrading and Green Growth

We use the intermediary effect model to test whether industrial upgrading acts as an intermediary variable between *HC* and *GEE*. The empirical results are shown in Table 5.

The results in column (2) of Table 5 show that *HC* significantly promotes industrial upgrading. The regression coefficient is 0.039, which means that every 1% increase in *HC* will improve industrial upgrading by 0.039%; the results in column (3) of Table 5 show that the regression coefficient of *IU* is negative and that of IU^2 is positive, indicating that there is a U-shaped relationship between industrial upgrading and *GEE*. On the left side of the U-shaped turning point, that is, the "accumulation" stage of industrial upgrading, with the development of industrial upgrading, industrial upgrading inhibits *GEE*; once the industrial upgrading exceeds the threshold and enters the "leap" stage of industrial upgrading, the industrial upgrading is dominated by the development of high-tech and digital industries, which improves the utilization rate of resources and promotes *GEE*. The U test results further verify that the U-shaped relationship between industrial upgrading and *GEE* is significant. On the whole, *HC* is positively promoting industrial upgrading. There is a U-shaped relationship between industrial upgrading and GEE. This verifies hypothesis 3 that industrial upgrading is the transmission channel between *HC* and *GEE*.

Table 5. Empirical results of the relationship between HC, industrial upgrading and GEE.

	(1)	(2)	(3)
	GEE	*IU*	*GEE*
HC	−0.932 ***	0.039 ***	
	(−7.71)	(4.65)	
HC^2	0.037 ***		
	(8.03)		
IU			−5.770 ***
			(−3.90)
IU^2			0.708 ***
			(4.01)
FTA	0.001	0.008 **	0.002
	(0.14)	(2.42)	(0.37)
LED	0.026	0.059 ***	0.033 ***
	(1.57)	(7.19)	(1.98)
GI	−0.873 ***	−0.228 *	−0.478 **
	(−3.67)	(−1.92)	(−2.02)
CS	0.115 ***	−0.065 ***	0.216 ***
	(2.85)	(−3.27)	(5.70)
FDI	0.161	0.012	−0.598
	(0.14)	(0.02)	(−0.46)
FD	0.128 ***	0.119 ***	0.042
	(2.72)	(5.04)	(0.87)
U test	12.469 ***		4.074 **
	(5.95)		(3.07)
U test lower bound interval	9.906		3.66
U test upper bound interval	16.246		4.60
_cons	4.892 ***	3.702 ***	10.279 **
	(5.57)	(22.90)	(3.28)
N	2493.000	2417.000	2417.000
R^2	0.764	0.862	0.757

Note: (1) t statistics in parentheses; (2) * $p < 0.1$, ** $p < 0.05$, *** $p < 0.01$.

4.4. Robustness Test

To prove that the conclusion is reliable, we examine the robustness of the benchmark model. The robustness test can be carried out by replacing either dependent variables or independent variables.

4.4.1. Using Substitute Variables of Human Capital

To further test the U-shaped relationship between human capital and green growth, we use the human capital measured by education years in each prefecture-level city [118,119] to replace current human capital measured by education expenditure to verify the relationship between human capital and green growth. The regression results in Column (1) of Table 6 and Figure 4b show that the relationship between human capital and *GEE* is U-shaped and has passed the U test. On the left side of the U-shaped turning point, when human capital increases, it inhibits green growth; when human capital level exceeds a certain threshold, it promotes green growth.

4.4.2. Using Substitute Variables of GEE

We use CO_2 emissions to replace the dependent variable, i.e., *GEE*, and verify the relationship between human capital and green growth through the relationship between human capital and CO_2 emissions. The regression results in column (2) of Table 6 reveal that the coefficients of *HC* and HC^2 are 1.294 and −0.046, respectively, and are significant at the level of 1%, indicating that the relationship between human capital and CO_2 emissions is an inverted U-shape and has passed the U test. Since lower CO_2 emissions mean higher

GEE, this demonstrates the U-shaped relationship between human capital and GEE. On the left side of the U-shape, CO_2 emissions are increasing as human capital enhances, indicating that human capital inhibits green growth. When human capital level exceeds a certain threshold, an increase in human capital will reduce CO_2 emissions, thus promoting green growth.

Table 6. Robustness test of the U-shaped relationship between human capital and *GEE*.

Variables	*GEE*	CO_2
	(1)	**(2)**
HC_1	−0.028	1.294 ***
	(−1.25)	(3.39)
$HC_1{}^2$	0.007 ***	−0.046 ***
	(3.29)	(−3.31)
FTZ	0.005	0.038 *
	(0.76)	(1.86)
LED	0.065 ***	0.192 ***
	(4.09)	(3.66)
GI	−0.543 **	−0.035
	(−2.24)	(−0.05)
CS	0.207 ***	0.162
	(5.33)	(1.27)
FDI	−0.207	−10.045 ***
	(−0.16)	(−2.74)
FD	0.074	−0.105
	(1.52)	(−0.71)
U test	2.018 *	14.051 **
	(1.25)	(1.80)
U test lower bound interval	0	9.906
U test upper bound interval	12.782	16.256
_cons	−2.334 ***	−2.852
	(−5.93)	(−1.03)
N	2297.000	2482.000
R^2	0.758	0.952

Note: (1) t statistics in parentheses; (2) ***, ** and * represent significance levels of 1%, 5% and 10%, respectively; (3) column (1) is the regression result of HC_1 and *GEE*; column (2) is the regression result of *HC* and CO_2 emissions.

4.5. Heterogeneity Analysis

4.5.1. Heterogeneity Analysis Based on Different Location of Cities

Since human capital development level differs in different regions, its relationship with green growth may also be different. According to the location of cities, we divide the samples into eastern, central, and western regions, and examine the relationship between human capital and *GEE* in different regions. The results are shown in Table 7 and Figure 5. The regression results indicate that the relationship between human capital and *GEE* in eastern, central, and western regions is U-shaped. The U test results in the eastern and western regions are significant, but insignificant in the central region. We draw a conclusion that eastern cities' human capital level (13.635) is higher than that of central cities (12.9394); and that in central cities is higher than that of western cities (12.457). Both the regression results and U-test demonstrates that there is a significant U-shaped relationship between *HC* and *GEE* in eastern and western cities, but not in central cities. Both eastern and western cities' human capital level is on the left side of the threshold (Figure 5). The development of human capital still inhibits *GEE*. Compared with the western region, the eastern region is closer to the U-shaped threshold.

Table 7. Test of the relationship between HC and GEE in different regions.

Variables	East	Centre	West	South	North
	(1)	(2)	(3)	(4)	(5)
	GEE	*GEE*	*GEE*	*GEE*	*GEE*
HC	−0.749 ***	−0.609 **	−1.026 ***	−0.933 ***	−1.115 ***
	(−4.37)	(−2.18)	(−4.06)	(−6.44)	(−4.73)
HC^2	0.027 ***	0.030 ***	0.041 ***	0.033 ***	0.046 ***
	(4.23)	(2.75)	(4.13)	(6.07)	(4.95)
FTZ	−0.001	−0.026 **	0.048 ***	−0.011	0.007
	(−0.11)	(−1.98)	(3.24)	(−1.47)	(0.55)
LED	0.066 ***	0.078 **	−0.161 ***	0.025	−0.087 ***
	(3.28)	(2.23)	(−4.25)	(1.16)	(−2.75)
GI	−0.642 **	−2.309 ***	−1.491 ***	−0.500 *	−0.899 **
	(−2.01)	(−5.05)	(−2.87)	(−1.68)	(−2.11)
CS	0.460 ***	0.086	−0.234 **	0.145 ***	−0.089
	(5.35)	(1.54)	(−2.55)	(3.15)	(−0.94)
FDI	−0.189	4.157 *	0.162	−4.781 ***	3.683 *
	(−0.13)	(1.77)	(0.04)	(−2.90)	(1.79)
FD	0.048	0.405 ***	0.188	0.112 **	0.087
	(0.74)	(4.68)	(1.53)	(2.00)	(0.94)
U test	13.635 **	10.238	12.457 ***	14.176 ***	12.066 ***
	(2.98)	(0.28)	(3.44)	(3.22)	(3.61)
U test lower bound interval	9.906	9.906	9.906	9.906	9.906
U test upper bound interval	16.246	16.246	16.246	16.246	16.246
_cons	1.536	1.622	9.716 ***	5.561 ***	8.399 ***
	(1.08)	(0.86)	(5.20)	(5.23)	(4.76)
N	956.000	759.000	702.000	1607.000	810.000
R^2	0.839	0.759	0.771	0.737	0.804

Note: (1) t statistics in parentheses; (2) * $p < 0.1$, ** $p < 0.05$, *** $p < 0.01$.

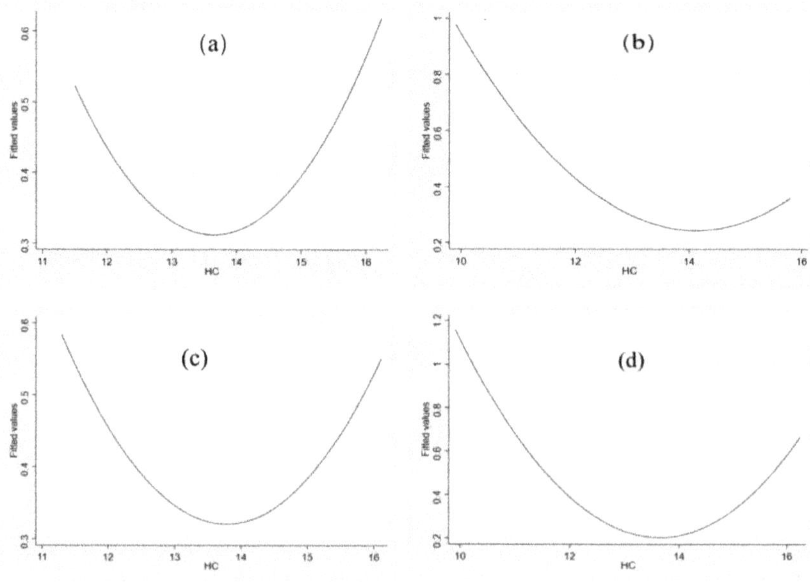

Figure 5. Fitting diagram of city's HC and GEE in different locations. Note: (**a**) fitting diagram of HC and GEE for eastern cities; (**b**) fitting diagram of HC and GEE for western cities; (**c**) fitting diagram of HC and GEE for southern cities; (**d**) fitting diagram of HC and GEE for northern cities.

Moreover, we further divide the samples into southern and northern regions, and examines the relationship between human capital and *GEE* in different regions. The results are shown in Table 7 and Figure 5. Both the regression and the U test results indicate that there is a significant U-shaped relationship between human capital and *GEE* in the southern and northern regions.

Specifically, the human capital level in southern cities (14.176) is higher than that in northern cities (12.066). The southern cities' human capital level is already on the right side of the threshold (14.136), indicating that it will promote *GEE* development as it improves. Human capital level in northern cities is 12.066, which is still on the left side of the threshold for northern cities (12.12, Figure 5d). That means human capital in northern cities still inhibits *GEE*. Southern cities' human capital promotes *GEE*, while northern cities' human capital inhibits *GEE*. This result is also consistent with the fact that human capital and city development levels in the southern cities are higher than those in the northern cities.

4.5.2. Heterogeneity Analysis of Different Size of Cities

Influenced by resource endowment, cities of different size have different human capital development level. According to cities' development size, we divided the samples into large cities and small and medium-sized cities for heterogeneity analysis. If the urban population in that year is larger than the sample average level, it is considered as a big city; otherwise, it is regarded as a small and medium-sized city. The regression and U test results reveal that (Table 8) the human capital level in large cities (13.665) is higher than that in small and medium-sized cities (11.556). There is a significant U-shaped relationship between *HC* and *GEE* in both large and small and medium-sized cities. The human capital level in large cities (13.665) is on the right side of the threshold (13.558) (Figure 6a). The relationship between *HC* and *GEE* exceeds the U-shaped turning point, indicating that *HC* will promote *GEE*. Small and medium-sized cities' human capital level (11.556) is still on the left side of the threshold (11.591) (Figure 6b), which has not yet reached the U-shaped threshold. It is still in the state of *HC* inhibiting *GEE*, which means that city scale development level will speed up crossing the threshold between *HC* and *GEE*, helping *HC* promote *GEE*.

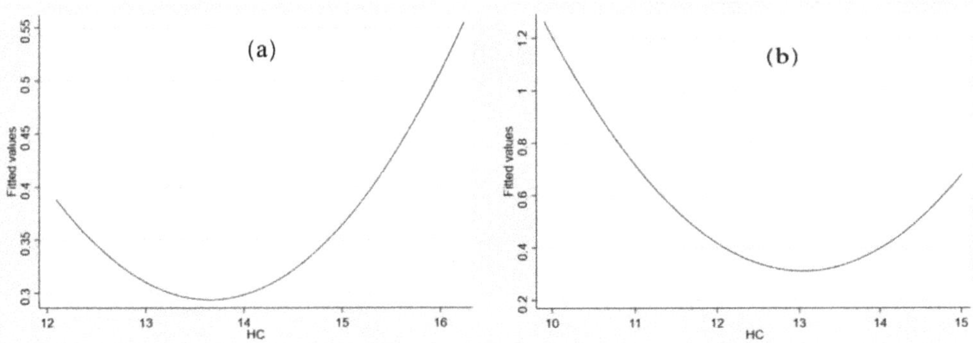

Figure 6. Fitting diagram of HC and GEE for different city size. (**a**) Fitting diagram of HC and GEE in large cities; (**b**) fitting diagram of HC and GEE in small and medium cities.

Table 8. Test of the relationship between HC and GEE for different city scale.

Variables	Big Cities	Small and Medium−Sized Cities
	GEE	*GEE*
HC	−1.166 ***	−1.020 ***
	(−6.54)	(−3.55)
HC^2	0.043 ***	0.044 ***
	(6.45)	(3.80)
FTZ	−0.015 **	0.021 *
	(−2.15)	(1.85)
LED	0.113 ***	−0.041
	(4.88)	(−1.63)
GI	−1.354 ***	−0.624 *
	(−4.43)	(−1.75)
CS	−0.197 ***	0.111 *
	(−2.63)	(1.84)
FDI	−0.722	2.376
	(−0.57)	(1.22)
FD	0.213 ***	0.065
	(3.70)	(0.88)
U test	13.665 ***	11.566 ***
	(5.11)	(2.37)
U test lower bound interval	9.906	9.906
U test upper bound interval	16.256	16.256
_cons	8.490 ***	5.914 ***
	(5.94)	(3.10)
N	1323.000	1159.000
R^2	0.732	0.789

Note: (1) *t* statistics in parentheses; (2) * $p < 0.1$, ** $p < 0.05$, *** $p < 0.01$.

4.6. Contribution of HC to GEE

Given other factors affecting the *GEE*, we further use LightGBM to measure the contribution of *HC* to *GEE*. Based on the six indices (*HC*, *FTZ*, *LED*, *GI*, *CS*, *FDI* and *FD*), this study uses the LightGBM machine learning method to predict urban *GEE* and fit it with the actual *GEE*. The fitting result is shown in Figure 7 and Table 9. Figure 7 shows that the general trend of the predicted value and actual values is the same. Further, the prediction performance results of LightGBM presented in Table 9 indicate that the R-squared value (R^2) of the training set is 0.886, and the R^2 value of the test set is 0.695, which implies that *HC* and the selected control variables are the main factors affecting the *GEE*.

Table 9. The performance measurement of GEE by LightGBM.

	MSE	RMSE	MAE	MAPE	R^2
training set	0.003	0.054	0.034	9.834	0.886
test set	0.007	0.082	0.059	16.618	0.695

An analysis of the relative importance of each variable to the *GEE* (Figure 8) reveals that the contribution of *CS* is the highest, reaching 21%. The contribution of *HC* and *LED* are 17%, respectively, second only to *CS*. The contributions of *GI*, *FDI* and *FD* are 15%, 14%, and 14%, respectively. The contribution of *FTZ* is the smallest, only 1%. Based on the contribution of various independent variable and control variables, we find that *CS*, *LED* and *HC* are the three main factors affecting *GEE*. *HC* is the second largest factor affecting *GEE*, second only to *CS*. *CS* and *HC* reflect the quantity and quality of urban population, and their total contribution is 38%. This is because people are the intrinsic factors that affect *GEE*; the other control variables include the extrinsic factors that affect *GEE*.

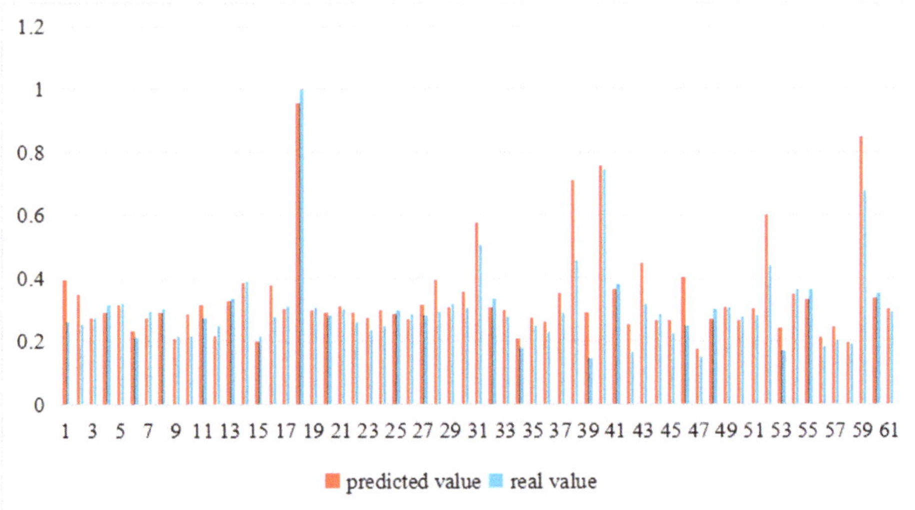

Figure 7. The comparison of GEE predicted value and real value.

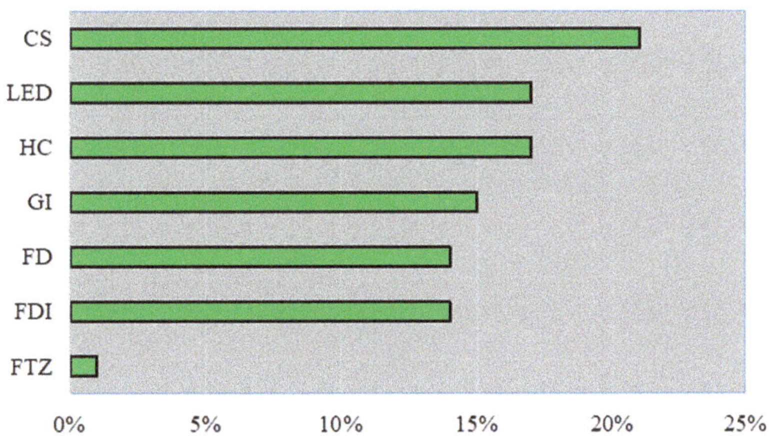

Figure 8. The contribution for determinants of GEE.

5. Conclusions and Policy Implications

5.1. Conclusions

With the rapid industrialization and urbanization, the increasing imbalance between economic system and ecosystem causes serious problems such as global warming, extreme climate, and frequent natural disasters, posing a great threat to human society. Therefore, the key path for countries around the world toward sustainable development is to transform to a "green growth" model that takes into account both economic growth and environmental protection. Meanwhile, China's human capital level has been continuously improving since the reform and opening up. Naturally, then, it raises a question for academics and policy authorities: what is the relationship between human capital and green growth? To answer this question, this paper selects the sample city data of China's 281 prefecture-level cities (including municipalities directly under the Central Government) and analyses the question in great detail from a theoretical perspective and at an empirical level. First, by reviewing classical literature, we put forward the hypothesis of U-shaped relationship between human capital and green growth. Then, we introduce the NDDF-DEA model to measure China's sample cities' green growth level during the statistical period. On this basis, we empirically test the previous research hypotheses by using econometric model, and measure the contribution of human capital to green growth by ML algorithm.

This paper has the following main findings: (1) China's human capital and green growth have a U-shaped relationship. Before reaching a certain threshold, human capital will inhibit green growth. After exceeding a certain threshold, human capital will promote green growth. Green innovation and industrial upgrading are transmission channels when human capital impacts green growth. (2) When considering other factors influencing green growth, human capital is very important. HC and economic growth have the same contribution weight to GG (17%), ranking two, second only to city size (21%). (3) The influence of human capital on green growth in China is characterized by regional imbalance and urban scale imbalance. It is good to hear that in the southern regions, human capital has surpassed the "U-shaped" threshold, promoting green growth. In contrast, however, human capital in northern China negatively impacts green growth. There is a significant U-shaped relationship between human capital and green growth in the eastern and western regions while the U-shaped relationship between the two in the central region is not significant. In the eastern regions, the current level of human capital is closer to the U-shaped inflection point. That is, when human capital level continues to improve, It will soon have a positive impact on green growth. But the human capital level in the western region still cannot reach the U-shaped threshold, which currently inhibits green growth. From the perspective of urban scale, the human capital of large cities has exceeded the U-shaped inflection point and played a role in promoting green growth; the human capital of small and medium-sized cities is still far from the U-shaped inflection point, which has a restraining effect on green growth. The level of urban scale development will accelerate the threshold crossing between HC and GG, and promote the positive correlation effect of HC on GG.

5.2. Policy Implications

Based on these findings, we provide the following relevant policy implications:

(1) Developing economies should pay full attention to the important value of education investment and talent cultivation in green transformation. Decision makers should regularly and dynamically assess human capital stock, accurately estimate human capital, and classify different talent development levels in various regions. They then can formulate matching talent development strategies and industrial policies to help improve human capital development levels and promote green growth.

(2) Companies (especially environmentally sensitive companies) should work hard to shape a corporate culture centred on knowledge management, green innovation, and people-orientation. They should spare no effort to build a talent echelon, greatly enhance the training of employees' skills, give full play to talents' subjective initiative, motivate employees' innovative practices, and realize the marginal incremental effect of human capital on companies' GTFP, promoting the transformation of green innovation achievements and industrial upgrading.

(3) Urban governance authorities in northern, central and western regions and small and medium-sized cities should rationally recognize the current shortcomings of HC development. On the one hand, they should increase the ratio of education expenditure in public expenditure, and gradually improve local population's education and skill level, so as to promote human capital development to a high level. On the other hand, given the location characteristics and resource endowments, they should actively explore the talent introduction policy for sustainable development and improve the supporting software and hardware infrastructure to attract top talents and value conversion. By adopting these measures, they can gradually use top talents

to promote local green growth so as to narrow the regional gap of green growth with the eastern, southern regions and large cities.

5.3. Limitations of This Paper

The deficiency of the paper is that it fails to further distinguish human capital into academic education and skill education. In the future, we can do more detailed research on the impact of different types of education and different levels of HC on GG.

Author Contributions: Conceptualization, data curation, methodology, project administration, writing-original draft and review, and editing, visualization, funding acquisition, X.L.; data analysis, model construction, writing—original draft, F.C.; software, model construction, modification, writing—original draft, proofread, S.F. All authors have read and agreed to the published version of the manuscript.

Funding: This research received no external funding.

Data Availability Statement: The data used to support the findings of this study are available from the corresponding author upon request.

Acknowledgments: Thanks to all the reviewers and editors for the publication of this paper.

Conflicts of Interest: The authors declare no conflict of interest.

Appendix A

Table A1. Full term and the abbreviations.

Number	Full Term	Abbreviation	Number	Full Term	Abbreviation
1	Green economic efficiency	GEE	12	Years of education	HC_1
2	green growth	GG	13	carbon dioxide emissions	CO_2
3	Human capital	HC	14	green total factor productivity	GTFP
4	Green innovation	GIN	15	sulfur dioxide	SO_2
5	Industrial upgrading	IU	16	total factor productivity	TFP
6	Free trade zone	FTA	17	data envelopment analysis	DEA
7	Level of economic development	LED	18	non-radial direction distance function	NDDF
8	Government intervention	GI	19	machine learning	ML
9	City scale	CS	20	Shephard distance function	SDF
10	Foreign direct investment	FDI	21	directional distance function	DDF
11	Fiscal decentralization	FD			

Appendix B

Table A2. The parameter values based on LightGBM machine learning algorithm.

Parameter	Parameter Value
Training time	0.219 s
Data segmentation	0.8
Data shuffle	Yes
Base learner	GBDT
Base learner number	130
Learning rate	0.1
L1 regular term	0
L2 regular term	1
Sample sign sampling rate	1
Tree feature sampling rate	1
Maximum depth of tree	10
Leaf node minimum sample	15

References

1. Huang, J.; Chen, X.; Yu, K.; Cai, X. Effect of technological progress on carbon emissions: New evidence from a decomposition and spatiotemporal perspective in China. *J. Environ. Manag.* **2020**, *274*, 110953. [CrossRef] [PubMed]
2. Kwakwa, P.A.; Alhassan, H.; Aboagye, S. Environmental Kuznets curve hypothesis in a financial development and natural resource extraction context: Evidence from Tunisia. *Quant. Financ. Econ.* **2018**, *2*, 981–1000. [CrossRef]
3. Lorek, S.; Spangenberg, J.H. Sustainable consumption within a sustainable economy beyond green growth and green economies. *J. Clean. Prod.* **2014**, *63*, 33–44. [CrossRef]
4. Xu, G.; Dong, H.; Xu, Z.; Bhattarai, N. China can reach carbon neutrality before 2050 by improving economic development quality. *Energy* **2022**, *243*, 123087. [CrossRef]
5. Wang, Y.; Xie, M.; Wu, Y.; Zhang, X.; Wang, M.; Zhang, Y.; Xie, Y. Ozone-related co-benefits of China's climate mitigation Policy. *Resour. Conserv. Recycl.* **2022**, *182*, 106288. [CrossRef]
6. Li, J.; Dong, X.; Jiang, Q.; Dong, K.; Liu, G. Natural gas trade network of countries and regions along the belt and road: Where to go in the future? *Resour. Policy* **2021**, *71*, 101981. [CrossRef]
7. Kasztelan, A. Green growth, green economy and sustainable development: Terminological and relational discourse. *Prague Econ. Pap.* **2017**, *26*, 487–499. [CrossRef]
8. Zhang, Z. Evolution paths of green economy modes and their trend of hypercycle economy. *Chin. J. Popul. Resour. Environ.* **2022**, *20*, 1–11. [CrossRef]
9. Pegels, A.; Altenburg, T. Latecomer development in a "greening" world: Introduction to the special issue. *World Dev.* **2020**, *135*, 105084. [CrossRef]
10. Shuai, S.; Fan, Z. Modeling the role of environmental regulations in regional green economy efficiency of China: Empirical evidence from super efficiency DEA-Tobit model. *J. Environ. Manag.* **2020**, *261*, 110227. [CrossRef]
11. Shen, F.; Liu, B.; Luo, F.; Wu, C.; Chen, H.; Wei, W. The effect of economic growth target constraints on green technology innovation. *J. Environ. Manag.* **2021**, *292*, 112765. [CrossRef] [PubMed]
12. Wang, K.; Zhao, B.; Fan, T.; Zhang, J. Economic growth targets and carbon emissions: Evidence from China. *Int. J. Environ. Res. Public Health* **2022**, *19*, 8053. [CrossRef] [PubMed]
13. Li, G.; Gao, D.; Li, Y. Dynamic environmental regulation threshold effect of technical progress on green total factor energy efficiency: Evidence from China. *Environ. Sci. Pollut. Res.* **2022**, *29*, 8804–8815. [CrossRef]
14. Lee, C.; Wang, C.; Ho, S. The dimension of green economy: Culture viewpoint. *Econ. Anal. Policy* **2022**, *74*, 122–138. [CrossRef]
15. Yuan, Y.; Zhang, Y.; Wang, L.; Wang, L. Coping Decisions of production enterprises under low-carbon economy. *Sustainability* **2022**, *14*, 9593. [CrossRef]
16. Zhou, X.; Jia, M.; Wang, L.; Sharma, G.D.; Zhao, X.; Ma, X. Modelling and simulation of a four-group evolutionary game model for green innovation stakeholders: Contextual evidence in lens of sustainable development. *Renew. Energy* **2022**, *197*, 500–517. [CrossRef]
17. Nehrebecka, N. Climate risk with particular emphasis on the relationship with credit-risk Assessment: What we learn from Poland. *Energies* **2021**, *14*, 8070. [CrossRef]
18. Becker, G.S. *Human Capital: A Theoretical and Empirical Analysis, with Special Reference to Education*; University of Chicago Press: Chicago, IL, USA, 2009.
19. Sulaiman, C.; Bala, U.; Tijani, B.A.; Waziri, S.I.; Maji, I.K. Human capital, technology, and economic growth: Evidence from Nigeria. *Sage Open* **2015**, *5*, 1474538686. [CrossRef]
20. Sultana, T.; Dey, S.R.; Tareque, M. Exploring the linkage between human capital and economic growth: A look at 141 developing and developed countries. *Econ. Syst.* **2022**, *2022*, 101017. [CrossRef]
21. Sun, X.; Li, H.; Ghosal, V. Firm-level human capital and innovation: Evidence from China. *China Econ. Rev.* **2020**, *59*, 101388. [CrossRef]
22. Wang, M.; Xu, M.; Ma, S. The effect of the spatial heterogeneity of human capital structure on regional green total factor productivity. *Struct. Chang. Econ. D* **2021**, *59*, 427–441. [CrossRef]
23. Fleisher, B.; Li, H.; Zhao, M.Q. Human capital, economic growth, and regional inequality in China. *J. Dev. Econ.* **2010**, *92*, 215–231. [CrossRef]
24. Ahmed, Z.; Zafar, M.W.; Ali, S. Linking urbanization, human capital, and the ecological footprint in G7 countries: An empirical analysis. *Sustain. Cities Soc.* **2020**, *55*, 102064. [CrossRef]
25. Akram, V.; Jangam, B.P.; Rath, B.N. Examining the linkage between human capital and energy consumption: Cross-country evidence. *OPEC Energy Rev.* **2020**, *44*, 3–26. [CrossRef]
26. Huang, C.; Zhang, X.; Liu, K. Effects of human capital structural evolution on carbon emissions intensity in China: A dual perspective of spatial heterogeneity and nonlinear linkages. *Renew. Sustain. Energy Rev.* **2021**, *135*, 110258. [CrossRef]
27. Mahmood, N.; Wang, Z.; Hassan, S.T. Renewable energy, economic growth, human capital, and CO_2 emission: An empirical analysis. *Environ. Sci. Pollut. Res.* **2019**, *26*, 20619–20630. [CrossRef]
28. Nathaniel, S.P.; Yalçiner, K.; Bekun, F.V. Assessing the environmental sustainability corridor: Linking natural resources, renewable energy, human capital, and ecological footprint in BRICS. *Resour. Policy* **2021**, *70*, 101924. [CrossRef]

29. Sheraz, M.; Deyi, X.; Ahmed, J.; Ullah, S.; Ullah, A. Moderating the effect of globalization on financial development, energy consumption, human capital, and carbon emissions: Evidence from G20 countries. *Environ. Sci. Pollut. Res.* **2021**, *28*, 35126–35144. [CrossRef]
30. Bano, S.; Zhao, Y.; Ahmad, A.; Wang, S.; Liu, Y. Identifying the impacts of human capital on carbon emissions in Pakistan. *J. Clean. Prod.* **2018**, *183*, 1082–1092. [CrossRef]
31. Bayar, Y.; Smirnov, V.; Danilina, M.; Kabanova, N. Impact of institutions and human capital on CO2 emissions in EU Transition Economies. *Sustainability* **2021**, *14*, 353. [CrossRef]
32. Haini, H. Examining the impact of ICT, human capital and carbon emissions: Evidence from the ASEAN economies. *Int. Econ.* **2021**, *166*, 116–125. [CrossRef]
33. Li, P.; Ouyang, Y. The dynamic impacts of financial development and human capital on CO2 emission intensity in China: An ARDL approach. *J. Bus. Econ. Manag.* **2019**, *20*, 939–957. [CrossRef]
34. Cakar, N.D.; Gedikli, A.; Erdogan, S.; Yildirim, D.C. Exploring the nexus between human capital and environmental degradation: The case of EU countries. *J. Environ. Manag.* **2021**, *295*, 113057. [CrossRef]
35. Chen, S.; Ma, J.; Ding, D.; Yu, S.; Tang, Y. The impact of green R&D cctivities on SO2 emissions: Evidence from China. *Math. Probl. Eng.* **2021**, *2021*, 6680560. [CrossRef]
36. Liu, F.; Lv, N. The threshold effect test of human capital on the growth of agricultural green total factor productivity: Evidence from China. *Int. J. Electr. Eng. Educ.* **2021**, 1486192314. [CrossRef]
37. Maranzano, P.; Cerdeira Bento, J.P.; Manera, M. The role of education and income inequality on environmental quality: A panel data analysis of the EKC hypothesis on OECD countries. *Sustainability* **2022**, *14*, 1622. [CrossRef]
38. Melton, R.B. Schultz's theory of "human capital". *Southwest. Soc. Sci. Q.* **1965**, *46*, 264–272.
39. Zafar, M.W.; Zaidi, S.A.H.; Khan, N.R.; Mirza, F.M.; Hou, F.; Kirmani, S.A.A. The impact of natural resources, human capital, and foreign direct investment on the ecological footprint: The case of the United States. *Resour. Policy* **2019**, *63*, 101428. [CrossRef]
40. Fleisher, B.M. Human capital and labor markets. *China Econ. Rev.* **2016**, *38*, 250. [CrossRef]
41. Timothy, V.L. The effect of top managers' human capital on SME productivity: The mediating role of innovation. *Heliyon* **2022**, *8*, e9330. [CrossRef]
42. Benos, N.; Karagiannis, S. Do education quality and spillovers matter? Evidence on human capital and productivity in Greece. *Econ. Model.* **2016**, *54*, 563–573. [CrossRef]
43. Wallenius, J. Human capital accumulation and the intertemporal elasticity of substitution of labor: How large is the bias? *Rev. Econ. Dynam.* **2011**, *14*, 577–591. [CrossRef]
44. Danquah, M.; Amankwah-Amoah, J. Assessing the relationships between human capital, innovation and technology adoption: Evidence from sub-Saharan Africa. *Technol. Forecast. Soc. Chang.* **2017**, *122*, 24–33. [CrossRef]
45. Chang, C.; Wang, P.; Liu, J. Knowledge spillovers, human capital and productivity. *J. Macroecon.* **2016**, *47*, 214–232. [CrossRef]
46. Aina, C.; Sonedda, D. Sooner or later? The impact of child education on household consumption. *Empir. Econ.* **2022**, 1–29. [CrossRef]
47. Broadstock, D.C.; Li, J.; Zhang, D. Efficiency snakes and energy ladders: A (meta-) frontier demand analysis of electricity consumption efficiency in Chinese households. *Energy Policy* **2016**, *91*, 383–396. [CrossRef]
48. Campbell, S.G.; Ungor, M. Revisiting human capital and aggregate income differences. *Econ. Model.* **2020**, *91*, 43–64. [CrossRef]
49. Schanes, K.; Giljum, S.; Hertwich, E. Low carbon lifestyles: A framework to structure consumption strategies and options to reduce carbon footprints. *J. Clean. Prod.* **2016**, *139*, 1033–1043. [CrossRef]
50. Shahbaz, M.; Gozgor, G.; Hammoudeh, S. Human capital and export diversification as new determinants of energy demand in the United States. *Energy Econ.* **2019**, *78*, 335–349. [CrossRef]
51. Fernandes, C.I.; Veiga, P.M.; Ferreira, J.J.; Hughes, M. Green growth versus economic growth: Do sustainable technology transfer and innovations lead to an imperfect choice? *Bus. Strat. Environ.* **2021**, *30*, 2021–2037. [CrossRef]
52. Dong, J.; Dou, Y.; Jiang, Q.; Zhao, J. How Does Industrial structure upgrading affect the global greenhouse effect? Evidence from RCEP and non-RCEP Countries. *Front. Energy Res.* **2021**, *9*, 683166. [CrossRef]
53. Capozza, C.; Divella, M. Human capital and firms' innovation: Evidence from emerging economies. *Econ. Innov. New Technol.* **2019**, *28*, 741–757. [CrossRef]
54. Chang, C.; Nguyen-Van, D. Human capital and firm innovation: New Evidence from Asean Countries. *J. Econ. Forecast.* **2021**, *3*, 52–71.
55. Buenechea-Elberdin, M.; Sáenz, J.; Kianto, A. Exploring the role of human capital, renewal capital and entrepreneurial capital in innovation performance in high-tech and low-tech firms. *Knowl. Manag. Res. Pract.* **2017**, *15*, 369–379. [CrossRef]
56. Ibarra, E.R.B.; Herrera, L. Innovation capability and organizational resources configuration. *Intang. Cap.* **2009**, *5*, 301–320.
57. Aalbers, R.; Dolfsma, W.; Koppius, O. Individual connectedness in innovation networks: On the role of individual motivation. *Res. Policy* **2013**, *42*, 624–634. [CrossRef]
58. Suseno, Y.; Standing, C.; Kiani-Mavi, R.; Jackson, P. National innovation performance: The role of human capital and social capital. *Innov. Eur. J. Soc. Sci. Res.* **2020**, *33*, 296–310. [CrossRef]
59. Hu, Y.; Fisher-Vanden, K.; Su, B. Technological spillover through industrial and regional linkages: Firm-level evidence from China. *Econ. Model.* **2020**, *89*, 523–545. [CrossRef]
60. Lin, J.; Liao, W.; Tseng, C. Effects of technology flows on industrial performance: The contingent role of diversity of R&D human capital. *J. Bus. Bus. Mark.* **2016**, *23*, 111–133.

61. Chen, L.; Huo, C. Impact of green innovation efficiency on carbon emission reduction in the Guangdong-Hong Kong-Macao GBA. *Sustainability* **2021**, *13*, 13450. [CrossRef]
62. Shi, T.; Si, S.; Chan, J.; Zhou, L. The carbon emission reduction effect of technological innovation on the transportation industry and its spatial heterogeneity: Evidence from China. *Atmosphere* **2021**, *12*, 1169. [CrossRef]
63. Hu, S.; Zeng, G.; Cao, X.; Yuan, H.; Chen, B. Does technological innovation promote green development? A case study of the Yangtze River Economic Belt in China. *Int. J. Environ. Res. Public Health* **2021**, *18*, 6111. [CrossRef] [PubMed]
64. Gao, Y.; Hafsi, T. R & D spending among Chinese SMEs: The role of business owner's characteristics. *Manag. Decis* **2015**, *53*, 1714–1735.
65. Liu, Y.; Li, Z.; Yin, X. Environmental regulation, technological innovation and energy consumption—A cross-region analysis in China. *J. Clean. Prod.* **2018**, *203*, 885–897. [CrossRef]
66. Zhang, C.; Chen, P. Industrialization, urbanization, and carbon emission efficiency of Yangtze River Economic—Empirical analysis based on stochastic frontier model. *Environ. Sci. Pollut. Res.* **2021**, *28*, 66914–66929. [CrossRef] [PubMed]
67. Fang, W.; Liu, Z.; Putra, A.R.S. Role of research and development in green economic growth through renewable energy development: Empirical evidence from South Asia. *Renew Energy* **2022**, *194*, 1142–1152. [CrossRef]
68. Liu, J.; Hull, V.; Godfray, H.C.J.; Tilman, D.; Gleick, P.; Hoff, H.; Pahl-Wostl, C.; Xu, Z.; Chung, M.G.; Sun, J. Nexus approaches to global sustainable development. *Nat. Sustain.* **2018**, *1*, 466–476. [CrossRef]
69. Qin, X.; Wang, J.; Liu, Y. Efficiency measurement and inefficiency environmental factors of China's green economy. *Prague Econ. Pap.* **2022**, *31*, 25–57. [CrossRef]
70. Shi, L. Labor industry allocation, industrial structure optimization, and economic growth. *Discret. Dyn. Nat. Soc.* **2021**, *2021*, 5167422. [CrossRef]
71. Lo, K.L.; Zhang, J.; Xia, F. Does digital technology innovation work better for industrial upgrading? An empirical analysis of listed Chinese manufacturing firms. *Appl. Econ. Lett.* **2022**, 1–6. [CrossRef]
72. Zhou, Y. Human capital, institutional quality and industrial upgrading: Global insights from industrial data. *Econ. Chang. Restruct.* **2018**, *51*, 1–27. [CrossRef]
73. Wu, N.; Liu, Z. Higher education development, technological innovation and industrial structure upgrade. *Technol. Forecast. Soc. Chang.* **2021**, *162*, 120400. [CrossRef]
74. Zhao, Y.; He, F.; Feng, Y. Research on the industrial structure upgrading effect of the employment mobility of graduates from China "double first-class" colleges and universities. *Sustainability* **2022**, *14*, 2353. [CrossRef]
75. He, D.; Zheng, M.; Cheng, W.; Lau, Y.Y.; Yin, Q. Interaction between higher education outputs and industrial structure evolution: Evidence from Hubei province, China. *Sustainability* **2019**, *11*, 2923. [CrossRef]
76. Sun, H.; Sun, W.; Geng, Y.; Kong, Y. Natural resource dependence, public education investment, and human capital accumulation. *Pet. Sci.* **2018**, *15*, 657–665. [CrossRef]
77. Kusi-Sarpong, S.; Mubarik, M.S.; Khan, S.A.; Brown, S.; Mubarak, M.F. Intellectual capital, blockchain-driven supply chain and sustainable production: Role of supply chain mapping. *Technol. Forecast. Soc. Chang.* **2022**, *175*, 121331. [CrossRef]
78. Muafi, M.; Sulistio, J. A nexus between green intellectual capital, supply chain integration, digital supply chain, supply chain agility, and business performance. *J. Ind. Eng. Manag.* **2022**, *15*, 275–295.
79. Zhu, W.; Zhu, Y.; Lin, H.; Yu, Y. Technology progress bias, industrial structure adjustment, and regional industrial economic growth motivation—Research on regional industrial transformation and upgrading based on the effect of learning by doing. *Technol. Forecast. Soc. Chang.* **2021**, *170*, 120928. [CrossRef]
80. Wei, W.; Zhang, M. The non-linear impact of industrial structure on CO_2 emissions in China. *Appl. Econ. Lett.* **2020**, *27*, 576–579. [CrossRef]
81. Liang, G.; Yu, D.; Ke, L. An empirical study on dynamic evolution of industrial structure and green economic growth—Based on data from china's underdeveloped areas. *Sustainability* **2021**, *13*, 8154. [CrossRef]
82. Yang, Y.; Wei, X.; Wei, J.; Gao, X. Industrial structure upgrading, green total factor productivity and carbon emissions. *Sustainability* **2022**, *14*, 1009. [CrossRef]
83. Zhang, H.; Zhang, J.; Song, J. Analysis of the threshold effect of agricultural industrial agglomeration and industrial structure upgrading on sustainable agricultural development in China. *J. Clean. Prod.* **2022**, *341*, 130818. [CrossRef]
84. Lu, S. Research on innovation-driven development strategy, comparative advantage and industrial transformation and upgrading. *Risus-J. Innov. Sustain.* **2019**, *10*, 146–152.
85. Xi, B.; Zhai, P. Economic growth, industrial structure upgrading and environmental pollution: Evidence from China. *Kybernetes* **2022**. [CrossRef]
86. Song, T.; Wang, E.; Lu, X.; Chen, H.; Zhang, J. Research on the calculation and influencing factors of the green development of regional industry in China. *Reg. Econ. Dev. Res.* **2020**, *1*, 33–43. [CrossRef]
87. Cheong, T.S.; Wu, Y. The impacts of structural transformation and industrial upgrading on regional inequality in China. *China Econ. Rev.* **2014**, *31*, 339–350. [CrossRef]
88. Xiao, H.; You, J. The heterogeneous impacts of human capital on green total factor productivity: Regional diversity perspective. *Front. Environ. Sci.* **2021**, *9*, 3562. [CrossRef]
89. Han, Y.; Zhang, F.; Huang, L.; Peng, K.; Wang, X. Does industrial upgrading promote eco-efficiency? –A panel space estimation based on Chinese evidence. *Energy Policy* **2021**, *154*, 112286. [CrossRef]

90. Qi, G.; Wang, Z.; Wang, Z.; Wei, L. Has industrial upgrading improved air pollution?—Evidence from China's digital economy. *Sustainability* **2022**, *14*, 8967. [CrossRef]
91. Sun, J.; Tang, D.; Kong, H.; Boamah, V. Impact of industrial structure upgrading on green total factor productivity in the Yangtze River Economic Belt. *Int. J. Environ. Res. Public Health* **2022**, *19*, 3718. [CrossRef]
92. Cheng, Z.; Li, X.; Wang, M. Resource curse and green economic growth. *Resour. Policy* **2021**, *74*, 102325. [CrossRef]
93. Wang, Y.; Chen, X. Natural resource endowment and ecological efficiency in China: Revisiting resource curse in the context of ecological efficiency. *Resour. Policy* **2020**, *66*, 101610. [CrossRef]
94. Xie, X.; Li, K.; Liu, Z.; Ai, H. Curse or blessing: How does natural resource dependence affect city level economic development in China? *Aust. J. Agric. Resour. Econ.* **2021**, *65*, 413–448. [CrossRef]
95. Hf, A.; Wlw, B. A directional slacks-based measure of technical inefficiency. *Socio-Econ. Plan. Sci.* **2009**, *43*, 274–287.
96. Zhou, P.; Ang, B.W.; Wang, H. Energy and CO_2 emission performance in electricity generation: A non-radial directional distance function approach. *Eur. J. Oper. Res.* **2012**, *221*, 625–635. [CrossRef]
97. Zhang, M.; Li, B. How to design regional characteristics to improve green economic efficiency: A fuzzy-set qualitative comparative analysis approach. *Environ. Sci. Pollut. Res.* **2022**, *29*, 6125–6139. [CrossRef]
98. Li, C.; Ji, J. Spatial-temporal characteristics and driving factors of green economic efficiency in China. *Ann. Oper. Res.* **2021**. [CrossRef]
99. Xiang, J. The Estimation of the Chinese Cities' Fixed Capital Stock. Master's Thesis, Hunan University, Changsha, China, 2011. (In Chinese)
100. Cheng, D.; Xue, Q.; Hubacek, K.; Fan, J.; Shan, Y.; Zhou, Y.; Coffman, D.M.; Managi, S.; Zhang, X. Inclusive wealth index measuring sustainable development potentials for Chinese cities. *Glob. Environ. Chang.* **2022**, *72*, 102417. [CrossRef]
101. Annabi, N.; Harvey, S.; Lan, Y. Public expenditures on education, human capital and growth in Canada: An OLG model analysis. *J. Policy Model.* **2011**, *33*, 852–865. [CrossRef]
102. Atems, B.; Blankenau, W. The 'time-release', crime-reducing effects of education spending. *Econ. Lett.* **2021**, *209*, 110143. [CrossRef]
103. Li, K.; Lin, B. The nonlinear impacts of industrial structure on China's energy intensity. *Energy* **2014**, *69*, 258–265. [CrossRef]
104. Liao, H.; Fan, Y.; Wei, Y. What induced China's energy intensity to fluctuate: 1997–2006? *Energy Policy* **2007**, *35*, 4640–4649. [CrossRef]
105. Yao, Y.; Ivanovski, K.; Inekwe, J.; Smyth, R. Human capital and energy consumption: Evidence from OECD countries. *Energy Econ.* **2019**, *84*, 104534. [CrossRef]
106. Liu, Z.K.; Cao, F.F.; Dennis, B. Does the China-Korea Free Trade Area Promote the Green Total Factor Productivity of China's Manufacturing Industry? *J. Korea Trade* **2019**, *23*, 27–44. [CrossRef]
107. Yuan, H.; Feng, Y.; Lee, C.; Cen, Y. How does manufacturing agglomeration affect green economic efficiency? *Energy Econ.* **2020**, *92*, 104944. [CrossRef]
108. Dhahri, S.; Omri, A. Entrepreneurship contribution to the three pillars of sustainable development: What does the evidence really say? *World Dev.* **2018**, *106*, 64–77. [CrossRef]
109. Droste, N.; Hansjurgens, B.; Kuikman, P.; Otter, N.; Antikainen, R.; Leskinen, P.; Pitkanen, K.; Saikku, L.; Loiseau, E.; Thomsen, M. Steering innovations towards a green economy: Understanding government intervention. *J. Clean. Prod.* **2016**, *135*, 426–434. [CrossRef]
110. Park, S.; Yoo, S.H. Government intervention in environmental performance considering working capital issues: Control the supplier or the buyer? *Comput. Ind. Eng.* **2022**, *163*, 107790. [CrossRef]
111. Fang, W.; Wu, T.; Chang, T.; Hung, C. What could entrepreneurial vision do for sustainable development? Explore the cross-level impact of organizational members' green shared vision on green creativity. *Sustainability* **2021**, *13*, 5364. [CrossRef]
112. Islam, R.; Ghani, A. Link among energy consumption, carbon dioxide emission, economic growth, population, poverty, and forest area Evidence from ASEAN country. *Int. J. Soc. Econ.* **2018**, *45*, 275–285. [CrossRef]
113. Song, Y.; Hao, F.; Hao, X.; Gozgor, G. Economic policy uncertainty, outward foreign direct investments, and green total factor productivity: Evidence from firm-level data in China. *Sustainability* **2021**, *13*, 2339. [CrossRef]
114. Cong, T.P.; Jain, V.; Purnomo, E.P.; Islam, M.M.; Mughal, N.; Guerrero, J.W.G.; Ullah, S. Controlling environmental pollution: Dynamic role of fiscal decentralization in CO_2 emission in Asian economies. *Environ. Sci. Pollut. Res.* **2021**, *28*, 65150–65159.
115. You, D.; Zhang, Y.; Yuan, B. Environmental regulation and firm eco-innovation: Evidence of moderating effects of fiscal decentralization and political competition from listed Chinese industrial companies. *J. Clean. Prod.* **2019**, *207*, 1072–1083. [CrossRef]
116. Fan, S.; Liu, X.; Sun, G. Evaluating the performance of inclusive growth based on the BP neural network and machine learning approach. *Comput. Intell. Neurosci.* **2022**, *2022*, 9491748. [CrossRef]
117. Liu, S.; Hou, P.; Gao, Y.; Tan, Y. Innovation and green total factor productivity in China: A linear and nonlinear investigation. *Environ. Sci. Pollut. Res.* **2022**, *29*, 12810–12831. [CrossRef]
118. Barro, R.J.; Lee, J.W. International data on educational attainment: Updates and implications. *Oxf. Econ. Pap.* **2001**, *53*, 541–563. [CrossRef]
119. Barro, R.J.; Lee, J.W. A new data set of educational attainment in the world, 1950–2010. *J. Dev. Econ.* **2013**, *104*, 184–198. [CrossRef]

International Journal of
Environmental Research and Public Health

MDPI

Article

Is Green Spread? The Spillover Effect of Community Green Interaction on Related Green Purchase Behavior

Jianming Wang [1], Xincheng Yang [1], Yini Xi [1] and Zhengxia He [2,*]

1 School of Business Administration, Zhejiang University of Finance & Economics, Hangzhou 310018, China; sjwjm@zufe.edu.cn (J.W.); yangxincheng1226@163.com (X.Y.); zcxyn@zufe.edu.cn (Y.X.)
2 School of Economics, Hangzhou Normal University, Hangzhou 311121, China
* Correspondence: hezhengxia79@163.com

Abstract: In the era of digital economy and mobile internet, many platforms or brands have built various online or offline green communities to guide customers or fans to engage in green interactions. Obviously, community green interaction can enhance brand emotional value and enhance customer stickiness, but whether community green interaction can further have a spillover effect on related or other green purchase behaviors has become an important topic for the theoretical and practical departments. This paper selects the "Little Bear Fuel Consumption Community" as the research object. Based on the theoretical framework of "Green Interaction—Environmental Emotion—Related Green Purchasing Behavior", this paper examines the spillover effect and impact mechanism of community green interaction on consumers' related green purchasing behavior. This paper uses a structural equation model and bootstrapping method to test the causal relationship between variables. This study lasted for 6 months, and a total of 348 valid questionnaires were collected in this study. We used SPSS 25 and AMOS 24 for data analysis. The results showed that the two dimensions of community green interaction (community green information interaction and community green interpersonal interaction) have a positive spillover effect on consumers' related green purchase behavior; community green interaction can positively spill over to consumers' related green purchase behavior through the psychological path of environmental emotion; community green information interaction and community green interpersonal interaction have positive effects on consumers' positive and negative environmental emotions; positive and negative environmental emotions positively affect consumers' related green purchase behavior; and in the two paths of community green information interaction—related green purchase behavior and community green interpersonal interaction—related green purchase behavior, both positive environmental emotion and negative environmental emotion play a role of partial mediation; product involvement has a negative moderating effect on the path of "community green interaction—environmental emotion". This paper opens the "black box" of the diffusion mechanism of community green interaction and provides a new explanatory framework for the spillover effect of community green interaction on related green purchase behavior.

Keywords: community green interaction; environmental emotion; related green purchase behavior; spillover effect; social diffusion mechanism

Citation: Wang, J.; Yang, X.; Xi, Y.; He, Z. Is Green Spread? The Spillover Effect of Community Green Interaction on Related Green Purchase Behavior. *Int. J. Environ. Res. Public Health* **2022**, *19*, 6571. https://doi.org/10.3390/ijerph19116571

Academic Editors: Taoyuan Wei and Qin Zhu

Received: 16 April 2022
Accepted: 26 May 2022
Published: 27 May 2022

1. Introduction

In the era of digital economy and mobile Internet, many platforms or brands have built diverse communities. In the traditional brand community, consumers or users often interact on products, services, and industries. Studies have confirmed that community interaction can improve consumers' loyalty, satisfaction [1–3], product purchase intention [4], value co-creation [5], and product innovation behavior [6]. With the development of the industry and the increasing attention of enterprises to green environmental protection, more and more enterprises begin to build a brand green community. For example, Patagonia has launched

Patagonia Action Works, a digital platform de-signed to connect environmental volunteers with environmental activists. The platform is similar to an "online social networking site". After entering the platform, users can enter the name of their area, and then search for corresponding environmental protection organizations based on different information, such as land, water, climate, community, biodiversity, and species protection. In the brand green community, consumers often interact on the topics of resource conservation, green products, green industry exhibitions, and so on. With the development of mobile Internet technology, there are more and more cases in which consumers pay attention to and share green consumption information on brand green community or other social media platforms, so as to drive more people to participate in the follow-up related green purchase behavior. However, at present, such community green interaction and its influence have not been paid enough attention in the academic circles. Obviously, community green interaction can enhance brand emotional value and enhance customer stickiness, but will this kind of community green interaction make consumers' subsequent or other related purchase behavior "greener"? This is a topic worthy of attention and discussion.

The relationship between community green interaction and associated green purchase behavior belongs to the research category of spillover effect. Research on behavior spillover effect focuses on the causal relationship between consumers' previous and subsequent behaviors, that is, the impact of initial behavior on subsequent behavior. For example, residents tend to buy more green products in subsequent consumption after garbage classification. Studies have proved that water saving, electricity-saving behavior, residents' recycling, reducing the use of hotel towels, reducing fuel consumption, and recycling packaging bags have affected consumers' subsequent environmental behavior [7–10], and even have affected consumers' support for some environmental policies [11–14]. However, most research is limited to the study of some behaviors that consumers can complete independently, such as water saving, waste classification, etc.; behaviors that need to interact with others, such as community interaction, have not received extensive attention at present. In addition, most studies related to spillover effects focus on two directly related behaviors of consumers (such as waste recycling behavior and energy-saving behavior) and lack of exploration on the spillover effects and their mechanism between green interaction and related green behaviors. Based on the theoretical perspectives of spillover effect and social diffusion, this study aims to explore the following four issues: (1) the direction of the spillover effect of community green interaction on related green purchase behavior. That is, after participating in community green interaction, do consumers prefer to make subsequent green purchases or are they more reluctant to make subsequent green purchases? (2) The strength of the spillover effect of community green interaction on related green purchase behavior. That is, the impact of consumers' participation in community green interaction on consumers' related green purchase behavior in their daily life. (3) The diffusion mechanism of the spillover effect of community green interaction on related green purchase behavior. That is, what is the mechanism of community green interaction affecting related green purchase behavior? (4) The boundary condition of the spillover effect of community green interaction on related green purchase behavior. That is, under what conditions does the spillover effect of community green interaction on related green purchase behavior hold?

The structure of the rest of this paper is as following: Section 2 is theoretical basis and research model; Section 3 is about research methods and sample analysis, which contains participants, instrument, data collection and description, reliability and validity test, and common method deviation test; Section 4 is data analysis and empirical results (including model fitness and path coefficient test, test of mediating effect of environmental emotion, and test of moderating effect of product involvement); Section 5 is discussion; and the final section, Section 6, is conclusions, which includes theoretical contributions, management implications, research limitations, and future prospects.

2. Theoretical Basis and Research Model

2.1. Theoretical Basis

2.1.1. Spillover Effect Theory

Scholars have defined the spillover effect in environmental problems as the impact of participating in a behavior on the probability of subsequent behavior. Existing studies mainly divide spillover effects into behavior spillover, time spillover, and environment spillover. Behavior spillover refers to behavior A leading to behavior B, which is the most common type of spillover in previous studies [15,16]. Time spillover focuses on how to formulate a behavior conducive to the environment to affect the frequency of the same behavior in the future, that is, how a behavior changes with the change of time or environment [17]. Environmental spillovers focus on how specific behaviors spread across the environment [18,19]. From the perspective of results, spillover effects can be divided into positive spillover and negative spillover. Positive spillover means that participating in the first behavior will increase the possibility of participating in the second behavior. For example, Xu et al. confirmed that recycling has a positive spillover effect on consumers' green consumption [20]. Negative spillover effect means that participating in the first behavior will reduce the possibility of participating in the second behavior [16]. For example, the laboratory experimental results of Chatelain et al. (2018) support negative spillovers between residents' private environmental protection behaviors [21]. Ma et al. found that after consumers are forced to make pro environmental behavior, they will later make behavior that destroys the environment [22].

2.1.2. The Theory of Cognitive Dissonance

The theory of cognitive dissonance was put forward by Fistinger in 1957. This theory holds that the individual's cognitive structure is composed of many cognitive elements, such as thought, concept, attitude, own behavior, and so on. If there is disharmony or conflict between cognitive elements, cognitive imbalance will occur, which will make individuals feel uncomfortable. In order to eliminate such negative psychological state, individuals have three means: choose to change one of the elements to maintain cognitive consistency; add new cognition; or emphasize the importance of one of them [23]. The theory of cognitive dissonance can be used to explain the generation of positive spillover effect. From the perspective of cognitive dissonance, the spillover effect of positive green behavior occurs because people want to avoid the unpleasant feeling of inconsistent performance between different pro environmental behaviors.

2.1.3. Self-Perception Theory

Self-perception can mainly be used to explain the impact of behavior on self-cognition. When people form evaluation cognition (such as attitude, norms, and values), they will take their own behavior as a clue [24], that is, people will understand their attitude, emotion and psychological state according to their own behavior and the situation in which the behavior occurs. A key point of self-perception theory is that behavior comes before attitude, that is, there is behavior first, then emotion, and then further cognition [25]. This theory can be used to explain the spillover effect of pro-environmental behavior. After making behaviors related to environmental protection, consumers will further judge their attitude towards environmental protection, so they are more likely to engage in pro environmental behaviors consistent with their self-perception in subsequent behaviors.

2.1.4. Social Diffusion Theory

Social diffusion theory originates from innovation diffusion theory, which originally refers to the process in which new technologies and products diffuse from innovation providers to social systems over time and are gradually applied or accepted by potential adopters [26]. With the passage of time, the innovation diffusion theory is not only limited to the fields of new technologies and new products, but also gradually applied to the fields of policy innovation diffusion and consumption behavior innovation diffusion [27]. In the era of digital economy and mobile Internet, with the help of mobile Internet technology, innovative consumption concepts and behavior patterns quickly penetrate through brand communities or other social media platforms, reflecting the "diffusion effect" and "herding effect" in the field of consumption, which shows that consumers influence their purchase decisions by observing others' consumption behavior and learning online interactive information [28].

2.1.5. Research Framework

At present, there are few studies on community interaction in the field of green consumption. However, it is found that "interaction-psychological change-response" is a research framework suitable for community interaction. Based on the research framework of "customer interaction—customer emotion—post purchase satisfaction", Jing et al. explored the influence mechanism of customer interaction on consumers' post purchase satisfaction [29]. Based on the path of "customer interaction-self-determination-community satisfaction", Wang et al. analyzed the role of self-determination in customer interaction and community satisfaction [30]. Therefore, this study adopts the framework of "interaction-psychological change- response" for the construction of theoretical model.

2.2. *Literature Review and Research Hypothesis*

2.2.1. Community Green Interaction

Community interaction mainly refers to the communication among community individuals [31]. Interaction is essentially the exchange of information between communicating individuals [32], and the community is the platform for consumers to communicate and interact. Community interaction enables consumers to establish contact with other members of the community, and makes consumers' understanding of products more comprehensive and three-dimensional through continuous communication and exchange [33]. Nowadays, with the upgrading of green consumption, more and more communities take green environmental protection as their interactive content. To sum up, this study proposes that community green interaction mainly refers to the interaction of community members around resources saving and environmental protection. Previous studies confirmed that community interaction can affect consumers' purchase intention, purchase decision, purchase behavior, and repeated purchase intention [4,33–35]. Compared with the traditional community interaction, community green interaction pays more attention to environmental protection, showing the characteristics of pro-environment and pro-society, so it may be closely related to consumers' related green purchase behavior. Combined with previous studies on the impact of community interaction on consumers, this study believes that community green interaction may spill over to consumers' related green purchase behavior.

2.2.2. Dimension of Community Green Interaction

Scholars have proposed a variety of ways to divide the dimensions of community interaction from different perspectives. Based on the starting point of community green interaction, this study believes that members participate in green interaction mainly for two purposes: one is to obtain professional information related to green products, the use of green products, and the recent development of green industry; the second is to establish emotional contact with other members of the community through interaction. Therefore, referring to the division method of Jing et al. (2013), this study divides the community green interaction into two dimensions: green information interaction and green interpersonal interaction [29]. Green information interaction is an interaction based on the topic of enterprise green products and industries; green interpersonal interaction is an interaction based on the topics of resource conservation, environmental protection, and daily life, and it mainly focuses on interpersonal communication and exchange among members, rather than professional green information sharing and discussion.

2.2.3. Spillover Effect of Community Green Interaction on Consumers' Related Green Purchase Behavior

First of all, after participating in the community green information interaction, on the one hand, consumers can obtain professional information about green products, industries, energy conservation, and environmental protection; on the other hand, they can share their own green information with other members of the community. In the process of constantly exchanging information, consumers' green cognition will be continuously improved and strengthened. After participating in the green interpersonal interaction of the community, consumers have established a close relationship with other members, and their attitudes and concepts on environmental protection will be more vulnerable to the influence of other members. To sum up, participating in community green interaction (information interaction and interpersonal interaction) will improve consumers' cognition and attitude. Further, according to the theory of cognitive dissonance, perceived inconsistencies between cognitive or behavioral elements will lead to uncomfortable feelings [36]. This discomfort, in turn, stimulates dysregulation reduction strategies, such as behavior change, or the balance of the two behaviors [37]. Therefore, in order to maintain cognitive consistency and avoid the unhappiness caused by cognitive imbalance, consumers participating in community green interaction will be more inclined to related green purchase behavior, that is, community green interaction has a positive spillover on consumers' related green purchase behavior.

Secondly, according to self-perception theory, people will know themselves according to their behavior and the situation in which the behavior occurs. As the theme of community green interaction is to protect the environment and reduce resource waste, it can be regarded as a pro environmental behavior to a certain extent. After consumers communicate and discuss in the green community, they will have environmental self-identity. Environmental self-identity refers to the degree to which individuals regard themselves as environmentalists. Individuals with strong environmental self-identity are more likely to save resources and reduce waste generation [38]. In other words, after participating in the community green interaction, consumers will have the self-cognition of "I am an environmental protection person" and "I am a green consumer", and think that they have the responsibility to protect the environment and save resources, so as to match their own behavior with their own environmental protection identity, and they are more likely to carry out related green purchase behavior in the future.

Finally, consumers' participation in community green interaction is often regarded as spontaneous behavior. According to attribution theory, if individuals attribute the initial behavior to internal causes, it is more likely to produce positive spillover. That is, consumers will regard green interactive behavior as something they take the initiative to do, rather than due to the promotion of the external environment. Therefore, compared with negative spillover, community green interaction is more likely to have positive spillover on subsequent related green purchase behavior. To sum up, this study puts forward the following hypotheses:

H1a. *Community green information interaction has a positive spillover effect on consumers' related green purchase behavior.*

H1b. *Community green interpersonal interaction has a positive spillover effect on consumers' related green purchase behavior.*

2.2.4. Community Green Interaction and Environmental Emotion

Environmental emotion refers to people's sensitivity to the significance of saving resources and protecting the environment, the waste of resources and the pollution of the environment, or the emotion expressed by people when participating in environmental protection actions and the subsequent attitude experience [39]. Wang (2015) explored the structural dimension of environmental emotion through qualitative research and found that environmental emotion is divided into positive dimension and negative dimension [40].

Specifically, positive environmental emotion refers to the feelings of pleasure, pride, approval, and love for the improvement of environmental problems or the implementation of better environmental behaviors, while negative environmental emotion refers to the feelings of guilt, worry, anger, and hatred for the deterioration of environmental problems or the implementation of worse environmental behaviors.

This study argues that participating in community green interaction can improve consumers' environmental emotion. First, community green information interaction can effectively improve consumers' green cognition. Community green information interaction can enable consumers to have more professional environmental knowledge, such as which products are more beneficial to the environment and which behaviors can lead to better resource saving. Meanwhile, community green information interaction can also make consumers have strong environmental awareness and have high sensitivity to environmental issues. According to the "cognition-emotion-behavior" model in psychology, cognition is the antecedent variable of emotion, and individual cognition of external things and stimuli produces related emotions [39,41,42]. The level of green cognition can be improved through community green information interaction, and then the environmental emotion will be further enhanced. Therefore, consumers will have a stronger sense of pleasure (approval) for their (others') good environmental behavior, that is, they have a strong positive environmental emotion; conversely, consumers will have a strong sense of guilt (worrying) about their (others') bad environmental behavior, that is, they have a strong negative environmental emotion. Accordingly, this study puts forward the following hypotheses:

H2a. *Community green information interaction positively affects the positive environmental emotion of community members.*

H2b. *Community green information interaction positively affects the negative environmental emotion of community members.*

Secondly, compared with the cognitive improvement brought by community green information interaction, participating in community green interpersonal interaction will affect consumers' attitudes and concepts to a greater extent, and then improve environmental emotion. Social identity theory holds that when an individual realizes that he belongs to a specific social group, he will also realize the emotional and value significance brought to him as a group member. The awareness of belonging to a certain group strongly affects our perception, attitude, and behavior, and we endow ourselves with the characteristics in line with the group [43]. Community green interpersonal interaction can enhance the mutual trust and intimacy of community members [30,44], and make community members gain a sense of identity [45] and belonging. After having a sense of identity with the community, consumers' attitudes and emotions are more likely to be affected by the green values of the community, and then have stronger environmental emotions. Therefore, participating in community green interpersonal interaction makes consumers have higher positive environmental emotion and negative environmental emotion.

H3a. *Community green interpersonal interaction positively affects the positive environmental emotion of community members.*

H3b. *Community green interpersonal interaction positively affects the negative environmental emotion of community members.*

2.2.5. The Mediating Role of Consumers' Environmental Emotion

Different from emotion, environmental emotion is a lasting and stable emotion, so the same consumer can have positive and negative environmental emotion at the same time, which can change consumers' purchase behavior to a certain extent [46,47]. Wang (2015) verified that the two dimensions of environmental emotion (positive emotion and negative emotion) have a positive impact on consumers' low-carbon purchase behavior [40]. Koenig-Lewis et al. (2014) showed that positive environmental emotion and negative environmental emotion play a complete mediating role in the impact of cognitive benefits on purchase behavior [48]. He et al. (2013) found that green emotion plays a mediating role in the path of green cognition affecting consumer behavior [39]. To sum up, this study infers that the two dimensions of environmental emotion (positive emotion and negative emotion) can positively affect consumers' related green purchase behavior. Combined with the impact of community green interaction on environmental emotion, this study believes that environmental emotion plays a mediating role between community green interaction and related green purchase behavior. Therefore, this study puts forward the following hypotheses:

H4a. *Positive environmental emotion plays a mediating role between community green information interaction and consumers' related green purchase behavior.*

H4b. *Positive environmental emotion plays a mediating role between community green interpersonal interaction and consumers' related green purchase behavior.*

H4c. *Negative environmental emotion plays a mediating role between community green information interaction and consumers' related green purchase behavior.*

H4d. *Negative environmental emotion plays a mediating role between community green interpersonal interaction and consumers' related green purchase behavior.*

2.2.6. Moderating Effect of Product Involvement

Product involvement mainly explores the subjective psychological state of consumers according to their understanding of products. Product involvement will have an impact on consumers' information collection and processing behavior. According to the possibility model of fine processing, using different information processing paths (edge path vs. central path) will affect consumers' decision-making. Consumers with higher product involvement tend to choose the central path to process information. At this time, they will pay more cognitive efforts and pay more attention to the gains and losses brought by green consumption and its impact on society. Consumers with low product involvement tend to choose the edge path to process information, and consumers pay more attention to the feelings related to green consumption, such as pride, appreciation, guilt, contempt, and so on. Therefore, this study argues that community green interaction can better cause environmental emotional changes of consumers with low product involvement. In comparison, since consumers with high product involvement focus on what knowledge information interaction can bring, community information interaction has a low promotion effect on environmental emotion. In addition, community interpersonal interaction cannot bring consumers with high product involvement the information they want to obtain, so its impact on environmental emotion is also small. Therefore, this study puts forward the following hypotheses:

H5a. *Product involvement plays a negative moderating role between community green information interaction and positive environmental emotion.*

H5b. *Product involvement plays a negative moderating role between community green interpersonal interaction and positive environmental emotion.*

H5c. *Product involvement plays a negative moderating role between community green information interaction and negative environmental emotion.*

H5d. *Product involvement plays a negative moderating role between community green interpersonal interaction and negative environmental emotion.*

2.3. The Research Model

To sum up, based on the theoretical perspectives of spillover effect, self-perception, behavioral attribution, and social diffusion, this study takes community green interaction as the starting point and pays attention to whether it will have behavioral spillover effect on consumers' related green purchase behavior. According to the characteristics of community green interaction, community green interaction is divided into green information interaction and green interpersonal interaction as antecedent variables. Consumers' positive environmental emotion and negative environmental emotion are taken as mediating variables to explain consumers' psychological changes. Consumers' related green purchase behavior is regarded as the outcome variable reflecting consumers' response. This study constructs a theoretical framework of community green interaction (information interaction and interpersonal interaction)—environmental emotion (positive environmental emotion and negative environmental emotion)—related green purchase behavior. The hypothetical model examined in this study is shown in Figure 1.

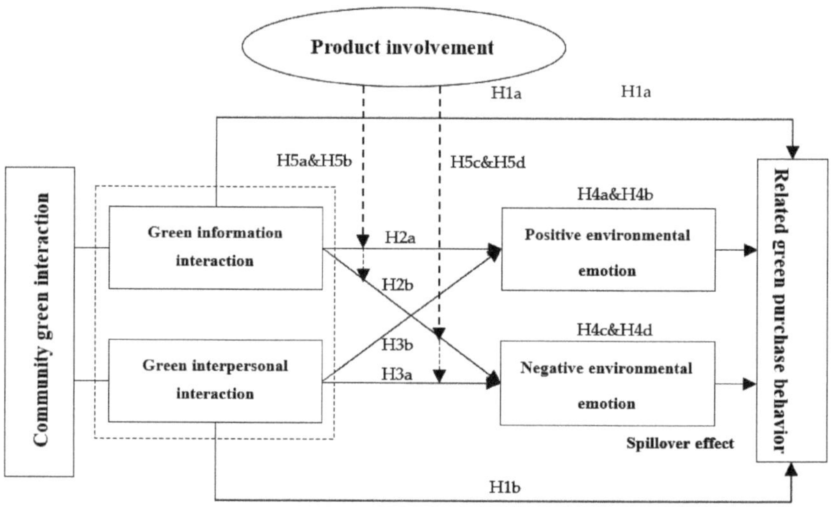

Figure 1. Hypothetical model of spillover effect of community green interaction.

3. Research Methods and Sample Analysis

3.1. Sample Selection

Considering the typicality of community green interaction, this study selects all consumers who have joined "little bear fuel consumption community" as the research sample. Little bear fuel consumption APP is a kind of vehicle fuel consumption calculation tool, which can accurately assist users in calculating single fuel consumption and average fuel consumption, and make statistics on single month fuel cost and average fuel cost, so as to help users save fuel and money and reduce emission. Users can join the little bear fuel consumption community and participate in interaction, share fuel saving information and daily environmental protection behaviors in the community. The samples we selected are all consumers who have joined the bear fuel consumption community for a period of time and have had community interaction experience.

3.2. Measurement

This study used a questionnaire to measure the construct, and the questionnaire consisted of three parts. In the first part, we inform the respondents that this survey is completely anonymous, the information will not be leaked, and the results are only used for academic research. At the same time, we emphasize that this questionnaire is only for members who participate in the green interaction of the bear fuel consumption community. The second part is the scale that measures the construct. The scales used in this study were all seven-point Likert scales, and the measurement items were adapted from the scales of previous studies and adjusted according to Chinese semantics. The measurement of community green information interaction and community green interpersonal interaction refers to the relevant scale of Nambisan and Baron [49], and five measurement items are selected for measuring these constructs respectively; the measurement of positive environmental emotion and negative environmental emotion refers to the research of Steenkamp [50], and six items are selected, respectively. The related green purchase behavior in this study is not limited to the automobile industry, but extended to other consumption fields based on the spillover effect theory. And combined with the scale of existing literature, a total of six items are designed. Product involvement was modified by referring to the scale of Laurent and Kapferer [51]. After the expert group discussion and the data from pre investigation, a formal research scale was finally formed. The items were all on Likert seven scale, "1" means totally disagree, "7" means totally agree. The specific measurement items are shown in Table 1. The third part is the collection of basic information on participants, including demographic variables such as gender, age, education, etc.

Table 1. Measurement items of the scale and results of confirmatory factor analysis.

Latent Variable	Item	Standardized Factor
1 Community green information interaction	I will discuss fuel consumption records in the community	0.722
	I will discuss how to save fuel consumption in the community	0.807
	I will exchange information about cars with low consumption in the community	0.776
	I will exchange information about new energy vehicles in the community	0.878
	I will exchange the trend of the automobile industry in the community	0.756
2 Community green interpersonal interaction	I will discuss environmental behavior in the community	0.669
	I will discuss oil price information in the community	0.731
	I will discuss daily traffic conditions in the community	0.705
	I think members of the community and I trust each other	0.859
	I think I have established a friendship with members of the community	0.789
3 Positive environmental emotion	I feel happy to contribute to environmental protection by getting information on fuel conservation	0.735
	I feel happy to get information about environmental protection and contribute to environmental protection.	0.656
	I feel happy that I can contribute to environmental protection by practicing environmental behavior every day.	0.693
	I commend my members for saving fuel consumption and contributing to environmental protection	0.797
	I commend members for their contribution to environmental protection by using environmentally friendly car accessories	0.664
	I applaud the members for their daily practice of environmental protection	0.762
4 Negative environmental emotion	I feel guilty for not saving fuel consumption and causing harm to the environment	0.735
	I feel guilty for using resource consuming auto parts to destroy the environment	0.854
	I feel guilty for the harm caused by my daily environmental damage	0.780
	I am worried that non community members do not save fuel consumption and do harm to the environment	0.805
	I'm worried that non community members use consumable accessories to damage the environment	0.728
	I am worried about the harm caused by environmental damage by non-members of the community	0.714
5 Related green purchase behavior	I'm willing to buy a car with less fuel consumption	0.711
	I am willing to buy new energy vehicles	0.835
	I am willing to buy environmental protection products in my daily life	0.879
	I would like to recommend my friends to buy cars with less fuel consumption consumption	0.687
	I would like to recommend my relatives and friends to buy new energy vehicles	0.653
	I would like to recommend relatives and friends to buy environmental protection products in daily life	0.806
6 Product involvement	I will spend time learning about the cars I buy	0.880
	Information about cars in community interaction is what I need	0.823
	The information about cars in community interaction is valuable to me	0.736
	The information about saving fuel consumption in community interaction is what I need	0.805

3.3. Data Collection and Description

Before the formal experiment, an online pre investigation was conducted on the members of the little bear fuel consumption community. A total of 100 questionnaires were distributed, 13 invalid questionnaires were deleted, and 87 valid questionnaires were obtained. The pre investigation results showed that the scale used to measure each construct has good reliability and validity. In the formal survey, the questionnaire was distributed to the members of the community. We have distributed online questionnaires on the Questionnaire Star online platform (https://www.wjx.cn/, accessed on 15 April 2022), which can restrict IP access and prevent respondents from repeating the questionnaire. The questionnaire was shared to the community through WeChat, QQ and other software or filled in by individuals. Participants can open the online questionnaire by clicking a link via mobile devices and terminate or quit at any time when filling in the questionnaire. The samples we selected are all members of the bear fuel consumption community and have had experience in participating in community interactions.

The different communities have been established in the little bear fuel consumption community according to different models, and car owners can interact with others for their models in the community. The small number of people in the community of some unpopular models leads to a poor interactive atmosphere. Therefore, after market research, we selected the 20 most popular models on the market and distributed questionnaires in the corresponding groups. From June to July 2020, we observed members in 20 communities and recorded members who had a record of interacting within the group (a total of 1328 people participated in the interaction). We excluded samples with fewer than two interactions or low participation (e.g., members with very few words spoken), resulting in a final sample of 954. We contacted the above-mentioned members and distributed questionnaires, which they filled out voluntarily. We ensured that a minimum of 20 samples (About 40–50% of the total community) were collected for each community. Therefore, our sample can better represent consumers who participate in the interaction of the little bear fuel consumption community. A total of 400 questionnaires were recovered, we eliminated the questionnaires with too short or too long answering time, and eliminated the questionnaires with too many choices of the same option. In the end, 348 valid questionnaires were obtained, with an effective rate of 87%. Our sample size accounts for 36.4% of the active users of the bear fuel consumption active community. From a data analysis point of view, the sample size required to use the structural equation model is 10–15 times the measurement indicators of the questionnaire. In this study, our measurement index is 31, and the sample size of 348 is sufficient. The descriptive statistics of samples are shown in Table 2. The proportion of men in the total sample was 63.8%, which was attributed to the fact that more men used cars than women. The number of community members aged 25–34 is the largest, mainly because these people are relatively young and willing to interact with others on the community platform. Our sampling mainly covered the central and eastern regions of China (this is the region with relatively developed economy and green development in China), and consumers in these regions are also most likely to engage in community green interactions. Therefore, our samples can better represent consumers who engage in community green interaction in China.

3.4. Reliability and Validity Test

3.4.1. Reliability Test

As shown in Table 3, the coefficients Cronbach's α of each scale ranged from 0.866 to 0.896, all higher than 0.7, indicating that the scale has good reliability.

3.4.2. Content Validity

This study mainly draws on the more mature scales in the literature, which are modified according to the results of expert and group discussion, combined with the research object, background, and purpose, so the scale of this study has good content validity.

3.4.3. Convergent Validity

Using AMOS 24.0 for confirmatory factor analysis, the following test results are obtained: absolute fitness index $\frac{\chi^2}{df} = 1.178 < 3$, RMSEA = 0.023, GFI = 0.915, AGFI = 0.900. value-added fitness index NFI = 0.921, TLI = 0.986, CFI = 0.987, indicating that the fitness of the model is good. As shown in Table 2, the standardization factor loadings of each measurement item on its corresponding latent variable are between 0.65–0.882. As shown in Table 3, the average variance extracted (AVE) of each latent variable is greater than 0.5, and the combined reliability (CR) is greater than 0.8. The above indicators show that the scale of this study has good convergent validity.

Table 2. Descriptive statistics of samples.

Demographics	Category	Number	Percentage
Gender	Male	222	63.8%
	Female	126	36.2%
Age	19–24 years old	89	25.6%
	25–34 years old	144	41.4%
	35–44 years old	72	20.7%
	45–55 years old	24	6.9%
	Over 55 years old	19	5.5%
Education	Junior school or below	19	5.5%
	Senior school or technical secondary school	28	8.0%
	College or vocational school	105	30.2%
	Undergraduate	152	43.7%
	Postgraduate and above	44	12.6%
Monthly income	Below 3500 yuan	74	21.3%
	3501–5000 yuan	67	19.3%
	5001–6500 yuan	81	23.3%
	6501–8000 yuan	47	13.5%

Table 3. Analysis results of correlation coefficient, reliability and discriminant validity.

Latent Variable	1	2	3	4	5	6
1 Community green information interaction	0.789					
2 Community green interpersonal interaction	0.319	0.754				
3 Positive environmental emotion	0.501	0.471	0.74			
4 Negative environmental emotion	0.471	0.442	0.546	0.77		
5 Related green purchase behavior	0.484	0.453	0.593	0.578	0.766	
6 Product involvement	0.283	0.147	0.295	0.334	0.353	0.812
Cronbach'salpha	0.89	0.866	0.875	0.896	0.892	0.885
AVE	0.623	0.568	0.547	0.594	0.587	0.660
CR	0.892	0.867	0.878	0.897	0.894	0.886

Note: the diagonal value is the square root of AVE, and the lower left part is the Pearson correlation coefficient between latent variables.

3.4.4. Discriminant Validity

As shown in Table 3, the square root of AVE value of each latent variable is greater than the correlation coefficient between this latent variable and other latent variables, indicating that each construct has both certain correlation and their own independence, so the discriminant validity of the scale in this study is good.

3.5. Common Method Deviation Test

Since the scale is used to measure the construct in this study, there may be a problem of common method deviation. We adopted a series of control procedures to reduce the interference caused by common method deviation, such as emphasizing the anonymity of this study, improving the items and order of the scale, etc. Referring to previous studies, this paper uses two methods to test whether there is a common method deviation. Firstly,

Harman single factor test was carried out by using SPSS 25 0 to make exploratory factor analysis on each item. The results showed that the variance interpretation rate of the first non-rotating factor was 33.554%, less than 50%, so there was no significant common method deviation. Secondly, confirmatory factor analysis with common method factors was conducted [52]. The confirmatory factor analysis model M1 and the confirmatory factor analysis model M2 with common factors are constructed respectively, and the main fitness indexes of the two models are compared: $\Delta\frac{\chi^2}{df} = 0.039$, ΔRMSEA = 0.003, ΔTLI = 0.003, ΔCFI = 0.003, all less than 0.05, indicating that the model fitness degree has not been significantly improved after adding the common method factor, so there is no significant common method deviation problem [53].

4. Data Analysis and Empirical Results

SPSS 25.0 and AMOS 24.0 are used to analyze the data in this study. First, the structural equation model includes both the path test model and the construct measurement model, which can estimate the parameters in the model as a whole. This method is very suitable for testing the relationship between multiple variables. In addition, the maximum likelihood method is a commonly used parameter estimation method. Therefore, the structural equation model is used in this study to test the relationship between variables, and the parameter estimation method is the maximum likelihood method. Secondly, bootstrapping uses the method of self-sampling to fit the probability distribution, which is suitable for hypothesis testing with unknown probability distribution. Therefore, we use the bootstrapping method to test the significance of the mediation effect value. Among them, the mediating effect test is conducted using the percentile Bootstrap method with deviation correction. The 95% confidence interval of the mediating effect value does not contain 0, indicating that the mediating effect is significant; otherwise, it is not significant. Finally, for the moderation effect test, the interaction term needs to be added to the regression model, so we use SPSS 25.0 to test the moderation effect.

4.1. Model Fitness and Path Coefficient Test

In order to explore the relationship and mechanism among community green interaction, consumer environmental emotion and consumer related green purchase behavior, AMOS 24.0 is used to test the data. The path test results and model fitness indicators are shown in Table 4. It can be seen from the Table 4 that the absolute fitness index $\frac{\chi^2}{df} = 1.248 < 3$, RMSEA = 0.027 < 0.05, GFI = 0.921, AGFI = 0.906, which are all greater than 0.9. Value added fitness index NFI = 0.926, RFI = 0.918, TLI = 0.983, CFI = 0.984, which are all greater than 0.9. Therefore, the fitness degree of the model is good.

It can be seen that green information interaction has a significant positive impact on related green purchase behavior ($\beta = 0.139$, $p = 0.009 < 0.05$, C.R. = 2.631 > 1.96); green interpersonal interaction has a significant positive impact on related green purchase behavior ($\beta = 0.119$, $p = 0.019 < 0.05$, C.R. = 2.336 > 1.96), indicating that Hypothesis H1a and H1b passed the test. Green information interaction has a significant positive impact on positive environmental emotion ($\beta = 0.332$, $p = 0.000 < 0.05$, C.R. = 6.561 > 1.96), and also has a significant positive impact on negative environmental emotion ($\beta = 0.349$, $p = 0.000 < 0.05$, C.R. = 6.253 > 1.96), indicating that H2a and H2b passed the test. Green interpersonal interaction has a significant positive impact on positive environmental emotion ($\beta = 0.292$, $p = 0.000 < 0.05$, C.R. = 5.945 > 1.96), and also has a significant positive impact on negative environmental emotion ($\beta = 0.306$, $p = 0.000 < 0.05$, C.R. = 5.608 > 1.96), showing that H3a and H3b are confirmed. Positive environmental emotion has a significant positive impact on related green purchase behavior ($\beta = 0.32$, $p = 0.000 < 0.05$, C.R. = 4.466 > 1.96). Similarly, negative environmental emotion has a significant positive impact on related green purchase behavior ($\beta = 0.278$, $p = 0.000 < 0.05$, C.R. = 4.557 > 1.96), indicating that H4a, H4b, H4c, H4d are partially supported.

Table 4. Path test results and model fitness indicators.

Path			Non-Standardized Coefficient	Standardized Coefficient	S.E.	C.R.	P
Community green information interaction→ Positive environmental emotion			0.332	0.398	0.051	6.561	***
Community green information interaction→ Negative environmental emotion			0.349	0.377	0.056	6.253	***
Community green interpersonal interaction→ Positive environmental emotion			0.292	0.356	0.049	5.945	***
Community green interpersonal interaction→ Negative environmental emotion			0.306	0.335	0.054	5.608	***
Community green information interaction→ Related green purchase behavior			0.139	0.157	0.053	2.631	0.009
Community green interpersonal interaction→ Related green purchase behavior			0.119	0.136	0.051	2.336	0.019
Positive environmental emotion→ Related green purchase behavior			0.32	0.302	0.072	4.466	***
Negative environmental emotion→ Related green purchase behavior			0.278	0.292	0.061	4.557	***
χ^2/df	RMSEA	GFI	AGFI	NFI	IFI	TLI	CFI
1.248	0.027	0.921	0.906	0.926	0.984	0.983	0.984

Note: *** $p < 0.001$.

4.2. Test of Mediating Effect of Environmental Emotion

In order to further test the mediating effect of positive environmental emotion and negative environmental emotion, this study adopted the Bootstrap mediating effect test method proposed by Preacher and Hayes [54] and adopted the Bias-corrected percentile Bootstrap method according to Fang et al. [55], using AMOS 24.0 for repeated sampling 5000 times to calculate the 95% confidence interval of the mediating effect.

As Table 5 shows, positive environmental emotion and negative environmental emotion play a mediating role in the impact of community green information interaction on related green purchase behavior. Among them, the mediation effect value of positive environmental emotion is 0.106, with the confidence interval being [0.057,0.175], excluding 0, indicating that the mediating effect is significant, and H4a is confirmed. The mediating effect value of negative environmental emotion is 0.097, with the confidence interval being [0.054,0.159], excluding 0, indicating that the mediating effect is significant, and H4b is confirmed. It can be seen from Table 4 that the direct effect of community green information interaction on related green purchase behavior is significant, accounting for 40.64%, while the mediating effect of positive environmental emotion and negative environmental emotion accounts for 30.99% and 28.36%, respectively. Therefore, both positive environmental emotion and negative environmental emotion play a partial mediating role in the path of community green information interaction-related green purchase behavior.

In the impact of community green interpersonal interaction on related green purchase behavior, positive environmental emotion, and negative environmental emotion also play a mediating role. Among them, the mediating effect value of positive environmental emotion is 0.094, with the confidence interval being [0.047,0.16], excluding 0, indicating that the mediating effect is significant, and H4c is confirmed. The mediating effect value of negative environmental emotion is 0.085, with the confidence interval being [0.043,0.144], excluding 0, indicating that the mediating effect is significant, and H4d is confirmed. Similarly, the direct effect of community green interpersonal interaction on related green purchase behavior is significant, accounting for 39.93%, while the mediating effect of positive environmental emotion accounted for 31.54%, and the mediating effect of negative environmental emotion accounted for 28.52%. Therefore, both positive environmental emotion and negative environmental emotion play a partial mediating role in the path of community green interpersonal interaction-related green purchase behavior.

Table 5. Mediating effect test by Bootstrap method.

Path	Mediator	Effect Value	SE	Bias-Corrected 95%CI		
				Lower	Upper	p
Community green information interaction—related green purchase behavior	Positive environmental emotion	0.106	0.029	0.057	0.175	***
	Negative environmental emotion	0.097	0.026	0.054	0.159	***
Community green interpersonal interaction —related green purchase behavior	Positive environmental emotion	0.094	0.028	0.047	0.16	***
	Negative environmental emotion	0.085	0.025	0.043	0.144	***

Note: *** $p < 0.001$.

4.3. Test of Moderating Effect of Product Involvement

This study uses SPSS 25.0 to verify the moderating effect of product involvement by using multi-layer regression analysis method, and constructs Models 1 to 6. The results are shown in Table 6.

Firstly, the moderating effect of product involvement on the path of "community green interaction—positive environmental emotion" was examined. The results of Model 3 showed that community green information interaction and community green interpersonal interaction have a significant positive impact on positive environmental emotion, and the explanatory power of the model is improved (R2 is improved) after the interaction item was added. The interaction coefficient of product involvement and community green information interaction is −0.130 ($p < 0.05$), and the interaction coefficient of product involvement and community green interpersonal interaction is −0.112 ($p < 0.05$), that is, product involvement has a significant negative moderating effect between community green interaction and positive environmental emotion (as shown in Figure 2), indicating that H5a and H5b are supported.

Table 6. Moderating effect of product involvement.

Variable	Positive Environmental Emotion			Negative Environmental Emotion		
	Model 1	Model 2	Model 3	Model 4	Model 5	Model 6
(Constant)						
Gender	0.131 **	0.131 **	0.115 **	0.192 ***	0.190 ***	0.171 ***
Age	−0.047	−0.044	−0.064	0.056	0.062	0.036
Education	0.182 ***	0.173 ***	0.184 ***	0.183 ***	0.166 ***	0.174 ***
Community green Information interaction	0.368 ***	0.346 ***	0.337 ***	0.351 ***	0.311 ***	0.295 ***
Community green interpersonal interaction	0.327 ***	0.320 ***	0.312 ***	0.327 ***	0.315 ***	0.309 ***
Product involvement		0.096 ***	0.096 *		0.175 ***	0.176 ***
Product involvement × Community green information interaction			−0.130 **			−0.200 ***
Product involvement × Community green interpersonal interaction			−0.112 *			−0.097 *
R^2	0.363	0.372	0.406	0.348	0.376	0.433
Adjusted R^2	0.354	0.361	0.392	0.339	0.365	0.419
F value	39.011	33.624	28.997	36.534	34.307	32.321

Note: * $p < 0.05$; ** $p < 0.01$; *** $p < 0.001$.

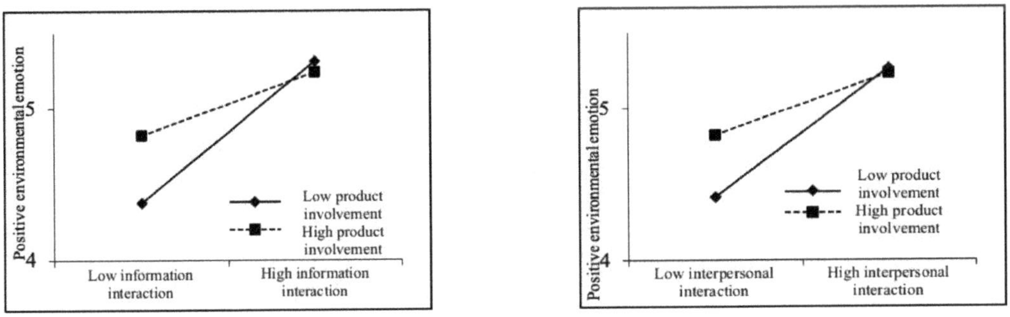

(**a**) (**b**)

Figure 2. Moderating effect of product involvement. (**a**) Moderating effect of product involvement on the path "green information interaction—positive environmental emotion". (**b**) Moderating effect of product involvement on the path "green interpersonal interaction—positive environmental emotion".

Secondly, the moderating effect of product involvement on the path of "community green interaction—negative environmental emotion" was examined. As shown in Model 6, the explanatory power of the model is improved (R^2 is improved) after the interaction item was added. The interaction item coefficient of product involvement and community green information interaction is −0.200 ($p < 0.05$), and the interaction item coefficient of product involvement and community green interpersonal interaction is −0.097 ($p < 0.05$), which shows that product involvement has a significant negative moderating effect on the path of "community green interaction-negative environmental emotion" (as shown in Figure 3). Thus, H5c and H5d are verified.

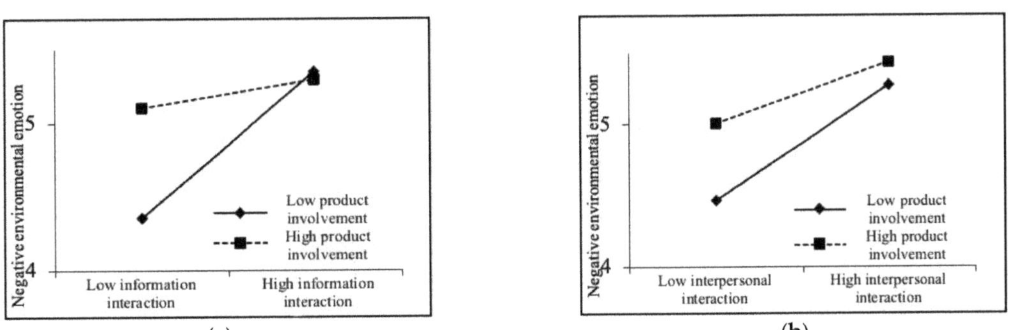

(**a**) (**b**)

Figure 3. Moderating effect of product involvement. (**a**) Moderating effect of product involvement on the path "green information interaction—negative environmental emotion". (**b**) Moderating effect of product involvement on the path "green interpersonal interaction—negative environmental emotion".

This study further used the Johnson-Neyman method to make Floodlight Analysis on the moderating effect. The results showed that in the path of "community green information interaction-positive environmental emotion", community green information interaction has a significant impact on positive environmental emotion when the level of product involvement is lower than 6.729. In the path of "community green information interaction- negative environmental emotion", community green information interaction has a significant impact on negative environmental emotion when the level of product involvement is lower than 6.189. In the path of "community interpersonal information interaction-positive environmental emotion", community green information interaction has a significant impact on positive environmental emotion when the level of product involvement is lower than 6.912. In the path of "community green information interaction-negative environmental emotion", green information interaction has a significant impact on negative environmental emotion when the level of product involvement is lower than 6.546.

5. Discussion

Based on cognitive dissonance theory, self-perception theory, and social diffusion theory, combined with the research paradigm of "interaction-psychological change-response", this study puts forward the hypothesis model of "interaction-emotion-behavior" spillover effect and empirically tests the applicability, explanatory power, and boundary conditions of the spillover effect model of community green interaction on related green purchase behavior. When consumers interact on the brand community or network platform, this kind of community green interaction will drive more people to participate in green consumption, that is, the influence of community green interaction in promoting the diffusion of green consumption in the whole society is becoming increasingly prominent with the dissemination and diffusion of green consumption information. This study analyzes the phenomenon that community green interaction spills over to related green purchase behavior and shows that consumers' "former green" in one field can effectively drive the "latter green" in other fields and even the "common green" in the whole field. This study also reveals the generation mechanism and underlying causes of the spillover effect of community green interaction. The core idea of this study is that community green interaction (consumers generate "first green " in a certain field, which is the breakthrough) can bring about the improvement of green cognition and community belonging, which will further affect consumers' environmental emotion. After consumers have higher environmental emotion, they will be more inclined to carry out related green purchase behavior in their daily life (that is, generate "latter green" in other fields and even "common green" in the whole field). Community green interaction helps to improve consumers' environmental emotion and finally spills over to the follow-up or other green purchase behaviors in daily life, which is the social diffusion mechanism of community green interaction. Our results reveal several interesting phenomena as follows.

(1) Our research proves that community green interaction (including two dimensions of community green information interaction and community green interpersonal interaction) has a positive spillover effect on consumers' related green purchase behavior. This is in line with previous studies which argued that social interaction indeed influences consumer buying behavior. Adjei et al. (2010) pointed out that online C2C communication has positive influence on immediate purchase intentions, and the depth and breadth of future purchase [34]. Jing and Yu (2020) found that online brand community interaction significantly improves consumers' purchase intention [35]. However, the purchase behavior referred to in past research is a broad concept.

Our research further refines consumers' purchasing behavior, focusing on consumers' related green purchasing behavior. Further, we divide community green interaction into two dimensions: community green information interaction and community green interpersonal interaction. Our study demonstrates that both have positive spillover effects on consumer-related green buying behavior. We believe that the reason for this spillover effect is the improvement of cognition and identity based on community interaction. Community green information interaction can effectively enhance consumers' green awareness. While community green interpersonal interaction can effectively enhance consumers' sense of identity and belonging.

Li (2011) believes that group identity will be generated after experiencing social activities with common cultural atmosphere with members, interaction between members and website, and then further contact activities [56]. To some extent, this also reflects the idea of Bandura's Social learning theory, that is, human behavior is the product of the interaction of internal processes and external influences. Khare et al. (2021) found that online communities and celebrities significantly predicted green clothing purchase behavior [57].

Liu and Liu (2020) showed that information interaction and interpersonal interaction in virtual brand community have significant positive effects on impulse buying [58]. It shows that community interaction does have an impact on consumer purchasing behavior. We believe that if the theme of consumer interaction is related to environmental protection, then it will help promote their green buying behavior. Therefore, how to effectively guide consumers to engage in green community interaction is a key issue.

(2) Secondly, this study proves that community green interaction (two dimensions of green information interaction and green interpersonal interaction) has a positive effect on consumers' environmental emotion (including positive environmental emotion and negative environmental emotion). This is because green information interaction can improve consumers' green cognitive level, and the improvement of cognition can enhance their emotion. Previous research has proved that increased awareness increases consumer's environmental emotion [39]. Ye (2019) showed that the improvement of green awareness can lead to the improvement of consumers' green emotion [57]. The reason why community interpersonal interaction can improve environmental emotion is that consumers have identity [45] and emotional belonging to their community after participating in interpersonal interaction, which is more vulnerable to the influence of other members' concept of green environmental protection. Trust, identification and interaction factors in online communities can trigger consumers' cognitive and emotional responses. Frequent communication and interaction among members can mobilize consumers' positive emotions and product cognition [56]. The behavior of others can also become the reference of their own behavior, and the environmental emotion has been enhanced imperceptibly. This shows that the antecedent of environmental emotion is not only cognition, but also behavior, which is an important supplement to the previous studies.

Secondly, both positive and negative environmental emotion positively affect consumers' related green purchase behavior, and positive environmental emotion has a stronger impact on related green purchase behavior ($\beta_{\text{Path coefficient of positive environmental emotion}} = 0.32 > \beta_{\text{Path coefficient of negative environmental emotion}} = 0.278$), which is consistent with the conclusions of previous studies [40,59]. Our study proves once again the applicability and extensibility of the broaden-and-build theory of positive emotions proposed by Fredrickson [60] in the field of green consumption, that is, compared with negative emotions, positive emotions are more helpful for individuals to make appropriate behavior choices. This study argues that the reason may be that consumers will avoid negative environmental emotion in order to maintain a positive self. Therefore, compared with positive environmental emotion, negative environmental emotion has less effect. Some recent studies, such as Khan and Mohsin (2017) and Joshi et al. (2021), showed that positive emotions act as a powerful driving factor for green purchase behavior [61,62].

In summary, Community green interaction can positively spillover to consumers' related green purchase behavior through the psychological path of environmental emotion. In the two paths of community, green information interaction-related green purchase behavior and community green interpersonal interaction-related green purchase behavior, both positive environmental emotion and negative environmental emotion are partial mediating variables, and there is no significant difference in the mediating effect value. Positive environmental affect and negative environmental affect can play a mediating role at the same time, which is consistent with previous research conclusions [46–48]. It is very important that environmental emotion is different from emotion, which is a stable psychological variable. Therefore, our conclusions help shape consumers' green buying behavior and green buying habits. Combined with the above discussion, the theoretical model of this study can be summarized as "interaction-emotion-behavior", that is, interaction can improve environmental emotion and further spillover to subsequent behavior. On the one hand, it reflects the dynamic relationship and influence mechanism between behavior and behavior; on the other hand, it is also the development of previous theoretical models.

(3) Product involvement has a significant negative moderating effect on the path of "community green interaction—environmental emotion". That is to say, both community green information interaction and community green interpersonal interaction have a great effect on the environmental emotion of consumers with low product involvement, while the effect on the environmental emotion of consumers with high product involvement is weak. This shows that "community green interaction—environmental emotion" is not applicable to all consumers. Consumers with high product involvement tend to use the central path to process information [63], so they pay more attention to the professional information and knowledge they can get in the community interaction. Therefore, community green interaction has a weak effect on their environmental emotion. However, consumers with low product involvement use edge path to process information [64,65], they are more likely to have emotional changes after participating in community interaction. This is consistent with Areni (2003) and Liu et al. (2015), that consumers with low product involvement pay more attention to emotional and transformational information, while consumers with high product involvement pay more attention to rational and functional information [53,66].

6. Conclusions

This study focuses on the direction, mechanism and boundary conditions of the spillover effect of community green interaction on consumer's related green purchase behavior. The main conclusions of this paper are as follows:

(1) Community green information interaction and community green interpersonal interaction have a significant impact on consumers' related green purchase behavior;
(2) Community green interaction (two dimensions of green information interaction and green interpersonal interaction) has a positive effect on consumers' environmental emotion (two dimensions of positive environmental emotion and negative environmental emotion);
(3) Both positive and negative environmental emotion positively affect consumers' related green purchase behavior;
(4) Community green interaction can positively spillover to consumers' related green purchase behavior through the psychological path of environmental emotion;
(5) Product involvement has a significant negative moderating effect on the path of "community green interaction—environmental emotion".

6.1. Theoretical Contributions

The main theoretical contributions of this study are as follows:

First, it fills the lack of research on the application of community interaction in the field of green consumption, enriches the research on spillover effect, and finds the social diffusion mechanism of community green interaction. Existing studies generally examined the role of community interaction in products, industries, after-sales services, and other fields, and few studies pay attention to community green interaction and its impact on follow-up and other related behaviors. For the first time, this study selects "little bear fuel consumption community" as the research object, and empirically tests the spillover effect and social diffusion mechanism of community green interaction on related green purchase behavior in the Chinese context.

Second, this study proposes that community green interaction is divided into two dimensions: community green information interaction and community green interpersonal interaction, which provides more space and possibilities for follow-up research. From the perspective of outcome variables, past studies often regarded psychological variables such as consumer loyalty, satisfaction and purchase intention as the outcome variables of community interaction. This study explored the dynamic relationship between the two coherent behaviors of community green interaction and green purchase, which is a supplement to the research paradigm of community interaction.

Third, this study constructs and tests the theoretical model of "community green interaction—environmental emotion—related green purchase behavior", and further confirms that community green interaction can spillover to related green purchase behavior through the path of environmental emotion from the perspective of environmental emotion, which opens the "black box" of the diffusion mechanism of community green interaction and provides a new perspective for the explanation of spillover effect. In addition, this study reveals the negative moderating effect of consumer product involvement on the path of "community green interaction environmental emotion", and clarifies the boundary conditions of the theoretical model of this study. This study enriches the research on the antecedents of product involvement and consumer environmental emotion and expands the application of relevant theories.

6.2. Management Implications

First, it is proposed that community members should be guided to actively participate in community green information interaction. This study shows that professional green information interaction in the community can have a positive spillover effect on related green purchase behavior. Therefore, the government should encourage various enterprises and network platforms to actively build various forms of online or offline green communities, such as online WeChat groups, discussion groups, green communities, and offline green teams, to provide consumers with a platform for information interaction. In the community, professional green information can be published regularly, such as knowledge related to enterprise green products, popular science videos, usage tips, etc. On the one hand, it can stimulate members' intentions to participate in green information interaction; on the other hand, this is also a way for enterprises to export environmental protection values, which can not only enhance the image of enterprises, but also publicize their green products, make the content of green information interaction more closely related to the purchase behavior of green products, and give full play to its positive spillover effect on the related green purchase behavior. In addition, various answer competitions on green, green product experience exchange and green knowledge sharing meetings can be regularly organized within the community, and certain rewards can be given to enhance the enthusiasm of community members to participate in green information interaction.

Second, guide community members should be guided to actively participate in community green interpersonal interaction. This study shows that green interpersonal interaction in the community can have a positive spillover effect on related green purchase behaviors, such as discussing oil prices, exchanging daily environmental behaviors, etc. Therefore, enterprises should strengthen the guidance of community green interpersonal interaction. Governments and enterprises can regularly push some current hot environmental topics, such as "is plastic a great invention or a bad invention?" "Do you choose to order takeout without tableware?" and guide members to discuss within the community, so as to strengthen the interpersonal interaction among community members. In addition, enterprises should also maintain a good community interaction atmosphere, and can select members with strong environmental awareness as the main managers of the community. On the one hand, they have the ability to call on other members of the group to join the interaction and can play a "catalyst" role in the process of green interpersonal interaction in the community. On the other hand, group management by community members can create a more relaxed and pleasant interactive atmosphere, help to improve members' sense of participation and emotional belonging, and give better play to the spillover effect of community green interpersonal interaction on related green purchase behavior.

Third, the connection of environmental emotion should be established through community interaction. Environmental emotion is the intermediate mechanism of the spillover effect of community green interaction on related green purchase behavior, which should be paid attention to by enterprises. On the one hand, enterprises can convey the benefits of green environmental protection to individuals and society through video, pictures, text, and other forms; trigger discussions among community members; and then improve the positive environmental emotion of community members. On the other hand, enterprises can appropriately show community members the bad environmental behaviors existing in the current society and the harm caused by these behaviors to the environment and individuals, and trigger members' discussion, so as to improve the negative environmental emotion of community members. Through the above methods, consumers' emotional experience in the process of community green interaction should be strengthened, and consumers' subsequent green purchase behavior should be further stimulated. In addition, in view of the stronger impact of positive environmental emotion, enterprises can appropriately increase the transmission of positive emotion.

Fourth, consumers with different product involvement should be distinguished and the spillover effect of community green interaction should also be given better play to. Community green interaction has different effects on consumers with high product involvement and low product involvement. Therefore, enterprises should distinguish consumers with different product involvement and guide them accordingly. Specifically, enterprises can divide consumers with different product involvement into several groups. For consumers with low product involvement, enterprises can regularly push emotion oriented green information and environmental protection videos, so as to better play the role of community interaction in improving environmental emotion, and impose the spillover effect on related green purchase behavior. For consumers with high product involvement, community green interaction has little impact on environmental emotion, so professional green information related to green products, industries and energy conservation, and emission reduction can be pushed to stimulate their interest in participating in community interaction and strengthen the direct spillover of community green interaction on related green purchase behavior.

6.3. Research Limitations and Future Prospects

There are still some limitations in this paper, which is worthy of further research in the future. This study concludes that consumers' positive environmental emotion and negative environmental emotion have partial mediating effects, and some mechanisms have not been explored, which is worthy of further exploration. Is the community green interaction environmental emotion related green purchase behavior model proposed in this paper applicable to a wider range of samples? Are there other boundary conditions? Can the "interaction-emotion-behavior" model proposed in this paper be applied to explain a wider range of phenomena? Can it be integrated with the traditional "knowledge-emotion-action" model? The above problems need to be further solved and improved in the future.

Author Contributions: Conceptualization, J.W.; Data curation, J.W; Investigation, X.Y.; Methodology, Y.X.; Software, Y.X. and Z.H.; Validation, X.Y.; Visualization, X.Y.; Writing—original draft, J.W.; Writing—review & editing, Z.H. All authors have read and agreed to the published version of the manuscript.

Funding: This study is supported by Major Project of The National Social Science Fund of China (20ZDA087).

Institutional Review Board Statement: Not Applicable.

Informed Consent Statement: Not applicable.

Data Availability Statement: Some or all data and models that support the findings of this study are available from the corresponding author upon reasonable request.

Conflicts of Interest: The authors declare no conflict of interest.

References

1. Mcalexander, J.H.; Schouten, J.W.; Koenig, H.F. Building brand community. *J. Mark.* **2002**, *66*, 89–92. [CrossRef]
2. Sen, S.; Bhattacharya, C.B.; Korschun, D. The Role of Corporate Social Responsibility in Strengthening Multiple Stakeholder Relationships: A Field Experiment. *J. Acad. Mark. Sci.* **2006**, *34*, 198–201. [CrossRef]
3. Wang, X.H.; Feng, W.Q. Research on the impact of social interpersonal interaction and quasi social interaction on the quality of brand relationship in virtual brand community. *Collect. Essays Financ. Econ.* **2017**, *220*, 78–88. (In Chinese) [CrossRef]
4. Zhao, J.B. A Study on the Influence of Online Brand Community Interaction Information on Lurker's Purchase Intention. *China Bus. Mark.* **2018**, *32*, 71–82. (In Chinese) [CrossRef]
5. Kwangyong, S.; Peng, X.; Qin, P. Research on the Effect of Customer-to-Customer Interaction of Virtual Brand Community on Customer Participation in Value Co-creation by Taking Experiential Value as Mediating Variable. *Chin. J. Manag.* **2016**, *13*, 1808–1816. (In Chinese) [CrossRef]
6. Dong, J.; Xu, Z.; Fang, Q.; Han, X. Influence of Customer-to-customer Interactions on Product Innovation Behavior in Virtual Brand Community. *Technol. Econ.* **2019**, *38*, 48–54, 126. (In Chinese) [CrossRef]
7. Lanzini, P.; Thogersen, J. Behavioural spillover in the environmental domain: An intervention study. *J. Environ. Psychol.* **2014**, *40*, 381–390. [CrossRef]
8. Tiefenbeck, V.; Staake, T.; Roth, K.; Sachs, O. For better or for worse? Empirical evidence of moral licensing in a behavioral energy conservation campaign. *Energy Policy* **2013**, *57*, 160–171. [CrossRef]
9. Baca-Motes, K.; Brown, A.; Gneezy, A.; Keenan, E.A.; Nelson, L.D. Commitment and behavior change: Evidence from the field. *J. Consum. Res.* **2013**, *39*, 1070–1084. [CrossRef]
10. Geng, L.; Chen, Y.; Ye, L.; Zhou, K.; Chen, Y. How to predict future pro-environmental intention? The spillover effect of electricity-saving behavior under environmental and monetary framing. *J. Clean. Prod.* **2019**, *233*, 1029–1037. [CrossRef]
11. Lacasse, K. The importance of being green: The influence of green behaviors on Americans' political attitudes toward climate change. *Environ. Behav.* **2015**, *47*, 754–781. [CrossRef]
12. Lacasse, K. Don't be satisfied, identify! Strengthening positive spillover by connecting pro-environmental behaviors to an "environmentalist" label. *J. Environ. Psychol.* **2016**, *48*, 149–158. [CrossRef]
13. Truelove, H.B.; Yeung, K.L.; Carrico, A.R.; Gillis, A.J.; Raimi, K.T. From plastic bottle recycling to policy support: An experimental test of pro-environmental spillover. *J. Environ. Psychol.* **2016**, *46*, 55–66. [CrossRef]
14. Ling, M.L. Study on the Internal Mechanism and Influencing Factors of the Spillover Effect of Residents' Environmental Behavior. Ph.D. Dissertation, Zhejiang University, Hangzhou, China, 2020. (In Chinese).
15. Dolan, P.; Galizzi, M.M. Like ripples on a pond: Behavioral spillovers and their implications for research and policy. *J. Econ. Psychol.* **2015**, *47*, 1–16. [CrossRef]
16. Truelove, H.B.; Carrico, A.R.; Weber, E.U.; Raimi, K.T.; Vandenbergh, M.P. Positive and negative spillover of pro-environmental behavior: An integrative review and theoretical framework. *Glob. Environ. Chang.* **2014**, *29*, 127–138. [CrossRef]
17. Littleford, C.; Ryley, T.J.; Firth, S.K. Context, control and the spillover of energy use behaviours between office and home settings. *J. Environ. Psychol.* **2014**, *40*, 157–166. [CrossRef]
18. Frezza, M.; Whitmarsh, L.; Schfer, M.; Schrader, U. Spillover effects of sustainable consumption: Combining identity process theory and theories of practice. *Sustain. Sci. Pract. Policy* **2019**, *15*, 15–30. [CrossRef]
19. Verfuerth, C.; Jones, C.R.; Gregory-Smith, D.; Oates, C. Understanding Contextual Spillover: Using Identity Process Theory as a Lens for Analyzing Behavioral Responses to a Workplace Dietary Choice Intervention. *Front. Psychol.* **2019**, *10*, 345. [CrossRef]
20. Xu, J.Q.; She, S.X.; Tian, Y.Z.; Chen, J. The spillover effect of recycling efforts on green consumption and its impact mechanism. *J. Nanjing Univ. Technol. (Soc. Sci. Ed.)* **2020**, *19*, 94–103, 112. (In Chinese) [CrossRef]
21. Chatelain, G.; Hille, S.L.; Sander, D.; Patel, M.; Hahnel, J.J.; Brosch, T. Feel good, stay green: Positive affect promotes pro-environmental behaviors and mitigates compensatory "mental bookkeeping" effects. *J. Environ. Psychol.* **2018**, *56*, 3–11. [CrossRef]
22. Ma, B.; Li, X.; Jiang, Z.; Jiang, J. Recycle more, waste more? When recycling efforts increase resource consumption. *J. Clean. Prod.* **2019**, *206*, 870–877. [CrossRef]
23. Jiang, Z.X. Experimental Study on Negative Spillover Effect of Green Consumption. Master's Dissertation, Zhejiang University of Finance and Economics, Hangzhou, China, 2020. (In Chinese).
24. Bem, D.J.; McConnell, H.K. Testing the self-perception explanation of dissonance phenomena: On the salience of premanipulation attitudes. *J. Pers. Soc. Psychol.* **1970**, *14*, 100–104. [CrossRef] [PubMed]
25. Yao, G.A.; Qing, Z. Research on the influence of consumers' experience in the process of maintaining trust in e-commerce enterprises. *Nankai Bus. Rev.* **2010**, *13*, 99–107. (In Chinese) [CrossRef]
26. Ren, B.; Shao, L.N.; You, J.X. Generalized Bass model of Chinese electric vehicles based on innovation diffusion theory. *Soft Sci.* **2013**, *27*, 17–22. (In Chinese) [CrossRef]
27. Guo, L.; Su, T.Y.; Qing, Y. Whether and how to innovate: A dynamic comparative study on the diffusion mechanism of policy innovation. *China Soft Sci.* **2019**, *5*, 160–167. (In Chinese) [CrossRef]
28. Feng, J.; Yao, Z. Research on the influence of online comment information on purchase decision based on social learning theory. *Chin. J. Manag. Sci.* **2016**, *24*, 106–114. (In Chinese) [CrossRef]
29. Jing, F.J.; Zhao, J.B.; Yu, Y. Analysis on the relationship between customer interaction, emotion and post purchase satisfaction—from the perspective of online brand community. *China Bus. Mark.* **2013**, *27*, 86–93. (In Chinese) [CrossRef]

30. Wang, Y.G.; Ma, S.; Sun, B. The mediating role of self-determination between customer interaction and community satisfaction—An Empirical Study Based on SOR theory and Self-determination theory. *J. Shanxi Univ. Financ. Econ.* **2012**, *34*, 99–107. (In Chinese) [CrossRef]
31. Sicilia, M.; Palazón, M. Brand communities on the internet. *Corp. Commun.* **2008**, *13*, 255–270. [CrossRef]
32. Gruen, T.W.; Osmonbekov, T.; Czaplewski, A.J. eWOM: The impact of customer-to-customer online know-how exchange on customer value and loyalty. *J. Bus. Res.* **2005**, *59*, 449–456. [CrossRef]
33. Teo, H.H.; Chan, H.C.; Wei, K.K.; Zhang, Z. Evaluating information accessibility and community adaptivity features for sustaining virtual learning communities. *Int. J. Hum.-Comput. Stud.* **2003**, *59*, 671–697. [CrossRef]
34. Adjei, M.T.; Noble, S.M.; Noble, C.H. The Influence of C2C communications in online brand communities on customer purchase behavior. *J. Acad. Mark. Sci.* **2010**, *5*, 634–653. [CrossRef]
35. Jing, L.; Yu, H.Y. The influence of customer interaction on customer purchase intention in online brand community. *Tax. Econ.* **2020**, *2*, 36–45. (In Chinese)
36. Toro-Alvarez, F. Coherence and dissonance: A new understanding in management and organizations. *Psychology* **2020**, *11*, 748–762. [CrossRef]
37. Gawronski, B. Meaning, violation of meaning, and meaninglessness in meaning maintenance. *Psychol. Inq.* **2012**, *23*, 100–106. [CrossRef]
38. Carrico, A.R.; Raimi, K.T.; Truelove, H.B.; Eby, B. Putting Your Money Where Your Mouth Is: An Experimental Test of Pro-Environmental Spillover From Reducing Meat Consumption to Monetary Donations. *Environ. Behav.* **2018**, *50*, 723–748. [CrossRef]
39. He, A.Z.; Du, J.; Chen, M.L. The influence mechanism of retail enterprises' green cognition and green emotion on green behavior. *China Soft Sci.* **2013**, *4*, 117–127. (In Chinese) [CrossRef]
40. Wang, J.M. The dimensional structure of environmental emotion and its impact on consumers' carbon emission reduction behavior – the two factor theory hypothesis of emotion behavior and its verification. *J. Manag. World* **2015**, *12*, 82–95. (In Chinese) [CrossRef]
41. Westbrook, R.A.; Oliver, R.L. The dimensionality of consumption emotion patterns and consumer satisfaction. *J. Consum. Res.* **1991**, *18*, 84–91. [CrossRef]
42. Frijda, N.H. The place of appraisal in emotion. *Cognit. Emot.* **1993**, *7*, 357–387. [CrossRef]
43. Zhang, Y.R.; Zuo, B. Social identity theory and its development. *Adv. Psychol. Sci.* **2006**, *3*, 475–480. (In Chinese) [CrossRef]
44. Köhler, C.F.; Rohm, A.J.; De Ruyter, K.; Wetzels, M. Return on interactivity: The impact of online agents on newcomer adjustment. *J. Mark.* **2011**, *75*, 93–108. [CrossRef]
45. Xu, J.; Xing, Y.F.; Yan, J.H. The impact of multiple interactions on Farmers' willingness to participate in regional brand Co-construction of agricultural products: The mediating effect of psychological contract. *J. Agro-For. Econ. Manag.* **2021**, *20*, 42–50. (In Chinese) [CrossRef]
46. Newton, J.D.; Tsarenko, Y.; Ferraro, C.; Sands, S. Environmental concern and environmental purchase intentions: The mediating role of learning strategy. *J. Bus. Res.* **2015**, *68*, 1974–1981. [CrossRef]
47. Ritter, A.M.; Borchardt, M.; Vaccaro, G.; Pereira, G.M.; Almeida, F. Motivations for promoting the consumption of green products in an emerging country: Exploring attitudes of Brazilian consumers. *J. Clean. Prod.* **2015**, *106*, 507–520. [CrossRef]
48. Koenig-Lewis, N.; Palmer, A.; Dermody, J.; Urbye, A. Consumers' evaluations of ecological packaging—Rational and emotional approaches. *J. Environ. Psychol.* **2014**, *37*, 94–105. [CrossRef]
49. Nambisan, S.; Baron, R.A. Virtual customer environments: Testing a model of voluntary participation in value co-creation activities. *J. Prod. Innov. Manag.* **2009**, *26*, 388–406. [CrossRef]
50. Laros, F.; Steenkamp, J. Emotions in consumer behavior: A hierarchical approach. *J. Bus. Res.* **2005**, *58*, 1437–1445. [CrossRef]
51. Laurent, G.; Kapferer, J.-N. Measuring consumer involvement profiles. *J. Mark. Res.* **1985**, *22*, 41–53. [CrossRef]
52. Xiong, H.X.; Zhang, J.; Ye, B.J.; Zheng, X.; Sun, P.Z. Common method variance effects and the models of statistical approaches for controlling it. *Adv. Psychol. Sci.* **2012**, *20*, 757–769. (In Chinese) [CrossRef]
53. Liu, S.M.; Liu, K.T.; Li, T.T.; Lu, L. The influence of college students' mindfulness on subjective well-being: The mediating effect of emotional regulation and psychological elasticity. *Psychol. Sci.* **2015**, *38*, 889–895. (In Chinese) [CrossRef]
54. Preacher, K.J.; Hayes, A.F. Asymptotic and resampling strategies for assessing and comparing indirect effects in multiple mediator models. *Behav. Res. Methods* **2008**, *40*, 879–891. [CrossRef] [PubMed]
55. Fang, J.; Zhang, M.Q.; Qiu, H.Z. Test methods and effect measurement of mediating effect: Review and prospect. *Psychol. Dev. Educ.* **2012**, *28*, 105–111. (In Chinese) [CrossRef]
56. Li, Z. Research on the Interactive Relationship and Related Activities of Online Brand Community—From the Perspective of Web Content Analysis. *Lanzhou Acad. J.* **2011**, *7*, 61–64.
57. Khare, A.; Sadachar, A.; Chakraborty, S. Influence of celebrities and online communities on Indian consumers' green clothing involvement and purchase behavior. *J. Fash. Mark. Manag.* **2021**. [CrossRef]
58. Liu, C.; Liu, F. The Study of the Effect of Virtual Brand Community Interaction on Impulse Buying: The Moderating Role of Self-Construal. *Open J. Bus. Manag.* **2020**, *08*, 414–427. [CrossRef]
59. Meneses, G.D. Refuting fear in heuristics and in recycling promotion. *J. Bus. Res.* **2010**, *63*, 104–110. [CrossRef]
60. Fredrickson, B.L. What good are positive emotions. *Rev. Gen. Psychol.* **1998**, *2*, 300–319. [CrossRef]

61. Khan, S.N.; Mohsin, M. The power of emotional value: Exploring the effects of values on green product consumer choice behavior. *J. Clean. Prod.* **2017**, *150*, 65–74. [CrossRef]
62. Joshi, Y.; Uniyal, D.P.; Sangroya, D. Investigating consumers' green purchase intention: Examining the role of economic value, emotional value and perceived marketplace influence. *J. Clean. Prod.* **2021**, *328*, 129638. [CrossRef]
63. Ba Umeister, R.F.; Finkel, E.J. *Advanced Social Psychology: The State of the Science*; Oxford University Press: New York, NY, USA, 2010.
64. Hou, J.D.; Chang, X.P.; Zhang, X.L. How does Negative Online Word-of-Mouth Destination affect Image: Two Scenario Experiments based on Samples of University Students. *J. China Univ. Geosci. (Soc. Sci. Ed.)* **2020**, *20*, 121–137. [CrossRef]
65. Ye, S.H.; Wu, G.B.; Hao, S. Effects of Congruence between Employee communication Behavior and Brand Personality on Consumers' Brand Attitudes. *Foreign Econ. Manag.* **2017**, *39*, 91–104. [CrossRef]
66. Areni, C. The effects of structural and grammatical variables on persuasion: An elaboration likelihood model perspective. *Psychol. Mark.* **2003**, *20*, 349–375. [CrossRef]

International Journal of
Environmental Research and Public Health

MDPI

Article

Exploring the Mechanism of the Impact of Green Finance and Digital Economy on China's Green Total Factor Productivity

Jianfeng Guo [1,2], Kai Zhang [2,*] and Kecheng Liu [1]

1 Henley Business School, University of Reading, Berkshire RG9 3AU, UK
2 Economics and Management School, Xi'an University of Posts and Telecommunications, Xi'an 710061, China
* Correspondence: ken@stu.xupt.edu.cn

Abstract: In the context of the "double cycle," promoting the development of a green economy is an important goal for China's high-quality economic development in the digital age. This paper uses data from 30 provinces (municipalities and autonomous regions) in China during the 2006–2019 period using the Compiled Green Finance Index (GF) and Digital Economy Index (DE). The interrelationship between green finance, digital economy and green total factor productivity ($GTFP$) is empirically tested by conducting multiple regressions on panel data from 2006–2019 to perform an empirical analysis. Based on this, further analysis was performed with the threshold model. This study found that green finance and digital economy can contribute well to green total factor productivity, but the combination of the two does not have a good effect on green total factor productivity. Further study found that the green finance and digital economy's contribution to green total factor productivity is mainly derived from technological progress. The regression results based on the panel threshold model show that the more underdeveloped the digital economy is in certain regions, the stronger the role of green finance in promoting efficiency improvement. Therefore, policymakers should formulate differentiated green financial policies according to the level of development of the digital economy and give play to the role of green finance and the digital economy in promoting green total factor productivity.

Keywords: green finance; digital economy; green total factor productivity

Citation: Guo, J.; Zhang, K.; Liu, K. Exploring the Mechanism of the Impact of Green Finance and Digital Economy on China's Green Total Factor Productivity. *Int. J. Environ. Res. Public Health* **2022**, *19*, 16303. https://doi.org/10.3390/ijerph192316303

Academic Editors: Taoyuan Wei and Qin Zhu

Received: 30 October 2022
Accepted: 2 December 2022
Published: 5 December 2022

1. Introduction

Since the reform and opening up, China's economy has been maintaining high growth, but the high economic growth stage has been accompanied by a dependence on resources and the pollution of the environment [1,2]. The report of the 19th National Congress of the Communist Party of China points out that "China's economy has shifted from a stage of high-speed growth to a stage of high-quality development, and it is necessary to promote quality change, efficiency change, power change in economic development and improve total factor productivity." In today's recurring epidemic, the need to energize productivity growth is even more pronounced. As the concept of green development continues to penetrate the concept of national governance, how to improve the quality of the ecological environment and the development of the green economy has gradually become a greater concern for people [3–5]. Promoting $GTFP$ and improving the efficiency of the green economy play important parts in promoting the development of China's green economy [6,7]. The improvement of total factor productivity is mainly reflected in two aspects of technological progress and efficiency progress, and each general technological innovation in human history has been able to significantly promote the leapfrog development of these two aspects [8–10]. Green finance incorporates environment and pollution into endogenous factors, and through the development of related green credit and green investment business, it guides the flow of funds into green environmental protection projects, optimizes resource allocation, and achieves win-win interaction between green finance and green economy [11,12]. Therefore, green finance plays an important role in promoting high-quality economic development [13–15].

In recent years, thanks to the continuous breakthrough of information technology in China, the rapid development and wide application of digital technology have given rise to the digital economy. The digital economy is different from the traditional agricultural and industrial economies. As a new economy, it is deeply integrated with all industries in China, triggering great social and economic changes while providing a new path for China to achieve environmentally friendly and sustainable development because of its improved efficiency and reduced dependence on resources and the environment. The 14th Five-Year Plan of the National Economic and Social Development of the People's Republic of China and the Outline of the Vision 2035 clearly proposed to "give full play to the advantages of massive data and rich application scenarios, promote the deep integration of digital technology and the real economy, empower the transformation and upgrading of traditional industries, give birth to new industries, new business models and grow new engines of economic development." It is easy to see that the digital economy has become an important driving force for China's economic development [16–18].

In the new era, China has the will and motivation to promote the two-pronged approach of green finance and digital economy to drive high-quality economic development. Green finance refers to all financial innovation and management activities that help achieve environmental improvement, enhance eco-efficiency and promote sustainable development functions. It includes not only environmental finance, low-carbon finance, and sustainable finance activities but also financial policies, financial services, risk management, and other related financial resource allocation activities adopted by governments, enterprises, and other economic agents that are conducive to promoting green investment and financing. Green finance is a financial innovation based on ecological and environmental protection, strengthening the link between the green industry and the financial industry, focusing on issues such as environmental pollution and ecological and environmental protection [19,20]. A digital economy is a new form of economic and social development after the agricultural and industrial economies. The G20 Initiative on Digital Economy Development and Cooperation, released at the 2016 Group of Twenty (G20) Summit, defines the digital economy as a series of economic activities in which the use of digitized knowledge and information is a key factor in production. Additionally, modern information networks are an important carrier of information, and the effective use of information and communication technologies is an important driving force for efficiency improvement and economic structure optimization. Thus, the digital economy has become an important engine for China's high-quality economic development due to its efficient use of resources. $GTFP$ is defined as the integration of input variables such as capital, energy and labor, economic benefits representing desired outputs and environmental pollution representing undesired outputs into the productivity measurement framework taking into account both increases in desired outputs and decreases in undesired outputs. We usually use $GTFP$ as an indicator to measure and evaluate the quality of growth of an economy.

Based on the above background, in order to better propose countermeasures to promote high-quality economic development, this article estimates the level of green finance and digital economy by constructing a multidimensional indicator system and verifies the effects of green finance and digital economy on $GTFP$ and its decomposition term using multiple regressions in a unified framework. We also use a threshold model to analyze the intensity of the impact of green finance and the digital economy on $GTFP$. This article first compares the existing relevant studies and then introduces the selection of variables and the setting of the model. Second, this article discusses the main sources driving $GTFP$ by analyzing its decomposition term. On this basis, this article conducts an empirical analysis of green finance and digital economy acting on $GTFP$, its decomposition term separately, and green finance and digital economy acting on $GTFP$, and its decomposition term together. We discuss in depth the influence mechanism in the process of green finance and the digital economy affecting *GTFP*. After that, the threshold effects on the roles of green finance and digital economy in the decomposition term efficiency progress of *GTFP* are further investigated and discussed. Finally, the conclusions of this study are drawn, thus providing a theoretical basis for relevant policy formulation. The main conclusions provide not only new ideas for developing green finance, promoting the construction of the digital economy and enhancing *GTFP*, but also provide important references in the implementation of green development concepts for local governments.

2. Literature Review

Green finance, as a link between the financial and green industries, completes the measures to upgrade the industrial structure by means of financial support for the green industry to continuously improve technological innovation, in line with the law of energy development, transforming energy use from fossil to clean energy, optimizing fossil energy, strengthening the global energy transition, and promoting green development [21]. Green finance is a new financial innovation that combines the concept of finance with the green industry, which introduces financial market volatility and geopolitical uncertainty, but is generally beneficial to the development of green finance and the green industry [22]. At the same time, the dual strategic transformation underscores the undisputed complementary relationship between green finance and digital transformation [23]. Additionally, the digital economy has promoted the development of the green economy well due to its resource allocation optimization and technological innovation-driven industrial structure upgrading. Therefore, it is necessary to study the impact of green finance and the digital economy on the green economy in depth. A review of the available research results shows the following main aspects:

(1) Research related to green finance and green economy.

Green finance is mainly through the guidance of financial institutions to make them invest in green projects that can bring energy saving and environmental protection to improve *GTFP*, as well as through social supervision to restrict the financial channels of high-polluting enterprises to either promote their transformation or green technology research and development, thus promoting *GTFP*. On the one hand, as the original energy-intensive production method is transformed into a green and environment-friendly production method, which has a very high cost, this requires green finance to provide financial support for green industries to optimize capital allocation [24–26]. In order to obtain more support from green loans, enterprises are more willing to take the initiative of carrying out the research and development of green technology and improve their own productivity. The incentivizing effect of green finance on enterprise technological innovation has well-promoted the development of *GTFP* [27–29]. By supporting green projects, green finance has greatly promoted environmental protection and played an important role in promoting China's high-quality economic development [30]. On the other hand, green finance is a special fund used to promote green development projects. After assessing green finance projects and approving them for financial support from green financial services, enterprises that want to develop green projects have the obligation and responsibility to fulfill corresponding social and environmental protections. At this time, their production and operation behaviors need to be supervised by relevant supervision departments, and the funds they obtain through green finance channels need to be used in green-related industries, thus improving *GTFP* [31]. The impact of green finance on *GTFP* tends to show different effects in stages [32]. In the short term, the transformation of highly polluting industries and the establishment of new green industries often require large amounts of financial support. The long transformation cycle of highly polluting industries and the establishment of new green industries leads to high input and low output of green financial inputs, which will reduce *GTFP*, while in the later stage, the transformation of highly polluting industries and the establishment of new green industries bring output returns, which will increase *GTFP*. Therefore, green finance and *GTFP* often show a U-shaped fitting curve [33]. Secondly, due to its unique loan conditions, green finance will have a corresponding loan threshold when providing services to enterprises. After receiving the loan, enterprises will also be supervised by regulatory authorities to monitor whether the enterprise loan is used for green projects, so there is often a threshold effect in the process of promoting *GTFP* in green finance [34–36]. In order to obtain capital loans, enterprises need to purchase equipment and upgrade corresponding green technologies to meet the requirements of green development, which will increase the cost burden on enterprises in the short term [37]. As enterprises continue to expand their business, improve their production efficiency level, and meet the requirements of green development, they are able to obtain more financial support, be regulated by the corresponding regulatory authorities, pay more and more attention to green development, actively carry out green production, improve production efficiency, and thus increase the total green factor [38].

(2) Research related to digital and green economies.

As an emerging economy, the digital economy has a significant impact on the new information industry revolution, so the development of the digital economy puts forward new requirements for the policy system in the industrial economy era. On the one hand, the digital economy improves the efficiency of resource allocation through digital technologies, and this more efficient way of production contributes to *GTFP* [39,40]. It has been shown in the literature that the digital economy contributes to *GTFP* mainly through green technological change [41]. At the same time, the digital economy itself relies on network infrastructure and information tools, such as smart machines, which break the limitations of time and space through information technology and internet mode, giving human beings the ability to process big data and continuously disseminate a large amount of information. The development of this ability relies on continuous technological innovation and research and development, so there are financial thresholds as well as technical thresholds in the digital economy for *GTFP* development [42–44]. On the other hand, the digital economy can be deeply integrated into all walks of life by upgrading and positively impacting the transformation of the industrial structure [45,46]. In addition, the upgrading of the industrial structure has a significant impact on the improvement of *GTFP* [47,48]. As the development of China's digital economy continues to promote China's economy in a more equitable and efficient direction, the combination of traditional production industries and the digital economy tends to promote the flow of production factors from the primary industry to the secondary and tertiary industries, and the continuous optimization of resource allocation to more efficient sectors, effectively improving the degree of dependence of economic development on energy resources and promoting the transformation and upgrading of industrial structure to digitalization, rationalization, and greening [49,50]. Other literature has empirically tested the impact of the digital economy on *GTFP* at the provincial or city level, affirming a positive and significant impact of the digital economy on *GTFP*, but often with regional heterogeneity [51–54].

In summary, it can be seen that, although there have been rich discussions in the academic community about the impact of the digital economy and green finance on total factor productivity, the discussion on how to promote *GTFP* and the impact of the digital economy and green finance on *GTFP* in the context of new development concepts and digital economy is still insufficient, mainly in the following aspects. First, very little of the literature analyzes the impact of the digital economy and green finance on *GTFP* within the same framework and also ignores the specific sources of total factor productivity gains. Second, few studies have examined the mechanisms at play in the process of green finance and the digital economy affecting *GTFP*. Finally, the established literature is more concerned with analyzing the direct effects of green finance and the digital economy on the impact of *GTFP*, and is less concerned with the impacts of both technological progress and efficiency improvement.

3. Methods

3.1. Data Sources

The starting point of the study chosen for this paper is 2006, and the endpoint is 2019. The initial data of each indicator are mainly obtained from the statistical yearbooks of all Chinese provinces, China Statistical Yearbook, China Science and Technology Statistical Yearbook, China Energy Statistical Yearbook, CSMAR and CCER databases, China Foreign Direct Investment Statistical Bulletin, China Insurance Yearbook and China Industrial Statistical Yearbook, etc. Some of the missing data are filled in by linear interpolation. In addition, in the selection of inter-provincial samples, due to the problem of more missing data and inconsistent data caliber, this paper selected 30 provinces and cities, except for Tibet, Hong Kong, Macao and Taiwan, as the research subjects.

3.2. Indicator Setting

1. The explanatory variable, Green Total Factor Productivity (*GTFP*). Since Data Envelopment Analysis (DEA) has the advantage of not requiring functional assumptions, and the non-angle and non-radial distance of the Malmquist index (*ML*) can treat pollution emissions as a non-desired output and solve the problem of radial distance function, it can achieve a decrease in non-desired output and an increase in desired output at the same time. This paper draws on the measure of Chung [55] to measure the *GTFP* index using the DEA–SBM non-angle, non-radial distance Malmquist index. The input indicators in this paper include labor, capital, and energy, using the total number of employees at the end of the year to measure labor and capital input: $K_{i,t} = K_{i,t-1}(1 - \delta_{i,t}) + I_{i,t}$, where, K denotes physical capital stock, δ denotes depreciation rate (the value of δ was taken as δ = 9.6% by referring to Zhang [56]) and I denotes real fixed-asset investment in each province. Energy inputs are measured using society-wide electricity consumption, and desired output indicators are measured using the gross product. The entropy value method is applied to collate industrial wastewater emissions, industrial SO_2 emissions, and industrial smoke (dust) emissions into a comprehensive index of environmental pollution to measure non-desired output indicators. The *ML* index can be further decomposed into technical efficiency change (*EC*) and technical progress change (*TC*). The specific expressions are as follows,

$$ML_t^{t+1} = EC \times TC \tag{1}$$

$$EC_t^{t+1} = \frac{1 + \vec{D}_0^t(\chi^t, y^t, b^t; y^t, -b^t)}{1 + \vec{D}_0^{t+1}(\chi^{t+1}, y^{t+1}, b^{t+1}; y^{t+1}, -b^{t+1})} \tag{2}$$

$$TC_t^{t+1} = \{\frac{[1 + \vec{D}_0^t(\chi^{t+1}, y^{t+1}, b^{t+1}; y^{t+1}, -b^{t+1})]}{[1 + \vec{D}_0^t(\chi^t, y^t, b^t; y^t, -b^t)]} \times \frac{[1 + \vec{D}_0^{t+1}(\chi^{t+1}, y^{t+1}, b^{t+1}; y^{t+1}, -b^{t+1})]}{[1 + \vec{D}_0^{t+1}(\chi^t, y^t, b^t; y^t, -b^t)]}\}^{\frac{1}{2}} \tag{3}$$

where *EC* means that a change in pure technical efficiency and a change in the efficiency of the scale of production causes a change in the internal efficiency of the producer, and the increase in industrial output resulting from this change is called a change in technical efficiency. *TC* means a change in industrial output caused by pure technological progress. The *ML* index is multiplied cumulatively to obtain the final *GTFP*.

2. Green Finance Index (*GF*). Some scholars have used green corporate bank loans, green credit share, green investment level, and green credit policy dummy variables as proxy variables for green finance. For the sake of comprehensiveness, this paper calculates provincial green finance development indicators using the composite index method based on data from 30 Chinese provinces and cities from 2006–2019. According to the definition of green finance, it mainly integrates four aspects: green credit, green investment, green insurance, and government support, among which green credit is the most important part of green finance, while other green financial products have gradually diversified in recent years, so green credit cannot be taken as the only indicator to measure the level of

green finance. Referring to LY He's [57] research ideas, while considering the validity and availability of data, the entropy value method is used to calculate the level of green finance in each province. The index system was constructed as shown in Table 1.

Table 1. Green Finance Index.

Tier 1 Indicators	Characterization Indicators	Indicator Description
Green credit	Percentage of interest expenses in high-energy-consuming industries	Six high-energy-consuming industrial industries' interest expenses/total industrial interest expenses
Green investment	Investment in environmental pollution control as a percentage of GDP	Environmental pollution control investment/GDP
Green insurance	Agricultural insurance depth	Agricultural insurance income/total agricultural output
Government support	Percentage of financial expenditure on environmental protection	Financial expenditure on environmental protection/financial general budget expenditure

3. The Digital Economy (*DE*). The concept of the digital economy in economics refers to the use of big data, the rapid integration, optimization and regeneration of resources to achieve the optimal allocation of resources to achieve high-quality economic development from all digital integration of resources can be considered the digital economy, generally speaking. The digital economy is a major economic form after the development of agricultural and industrial economies. With modern information networks as the main carrier and data resources as the key element, it promotes the integration and application of modern information technology, facilitates modern digital transformation, changes people's current life, production and governance, and is a new economic form that is more equitable and efficient. At present, the academic community continues to dig deeper into the digital economy as well as supplement and improve the evaluation system of digital economy indicators, mainly combined with infrastructure construction, internet level, and a series of elements to measure, compared with the previous single way of measurement methods, its measurement methods and levels continue to expand and deepen on the original basis but has not yet reached a unified standard. Nowadays, it can be determined that the core of the digital economy is digital resources, through modern information technology applications to provide consumers with convenient and fast services and products so that digital transactions become an emerging economic form of producers and consumers trading ties. Based on the existing literature and considering the availability and completeness of data, this paper constructs a measurement system containing four primary indicators and 27 secondary indicators, covering various elements such as digital infrastructure, digital penetration rate, digital technology talent benefits, and digital research. The data were mainly obtained from the China Statistical Yearbook, the Electronic Information Industry Statistical Bulletin and the provincial statistical yearbooks. Based on the construction of the index system, the KMO and the Bartlett test were conducted, and it was found that the KMO = 0.863 and the Bartlett test results proved that there were significant differences among the indicators. Principal component analysis can be used for dimensionality reduction. The construction of the index system is shown in Table 2.

Table 2. Digital Economy Development Index.

Tier 1 Indicators	Secondary Indicators
Digital infrastructure	Long-distance cable density; Mobile phone switch capacity per capita; Number of internet ports per capita
Digital penetration rate	Internet penetration; Cell phone penetration; Number of websites per capita Number of websites owned by unit companies; Express business volume per capita
Digital technology talent benefits	Information transmission; Software and electronic technology service industry employees The proportion of employees in information transmission, software, and electronic technology services The number of legal entities in the information transmission, software, and electronic technology services industry Software business revenue; The number of resident populations at the end of the year Software business revenue per capita; Software business revenue as a percentage of GDP Total telecommunication services; Total telecom services per capita; Total telecom business as a proportion of GDP Electronic information manufacturing industry's main business income actual value Actual value of main business income of electronic information manufacturing industry per capita Electronic information manufacturing industry's main business income as a proportion of GDP
Digital research	Human capital; Education level; Education Funding; Education level Number of patent applications; Number of patent applications per capita

4. Other variables. With reference to existing studies, the following control variables are selected in this paper: The level of openness to foreign investment ($OPEN$) is expressed as the share of total foreign direct investment in real terms in local GDP; The level of industrial structure (OIS) is expressed as the share of secondary industry output in local GDP; The level of urbanization (URB) is expressed as the share of the urban resident population within the resident population; Research, development, and investment (RD) is expressed as the number of local patents; Finally, the level of government spending (GOV) is expressed as a share of government fiscal spending in regional GDP. The above data are from the "China Statistical Yearbook," "China Environmental Statistical Yearbook," and the provincial statistical yearbooks.

3.3. Model Construction

In order to test the relationship between green finance, digital economy and $GTFP$, as well as the relationship between green finance, digital economy and the decomposition term of $GTFP$, and to deeply investigate the path of action on $GTFP$, Equations (4)–(6) are constructed in this paper.

$$GTFP_{it} = \alpha_0 + \alpha_1 GF_{it} + \alpha_2 DE_{it} + \alpha_i \, Controls_{it} + \lambda_i + \mu_{it} + \varepsilon_{it} \quad (4)$$

$$TC_{it} = \beta_0 + \beta_1 GF_{it} + \beta_2 DE_{it} + \beta_i \, Controls_{it} + \lambda_i + \mu_{it} + \varepsilon_{it} \quad (5)$$

$$EC_{it} = \gamma_0 + \gamma_1 GF_{it} + \gamma_2 DE_{it} + \gamma_i \, Controls_{it} + \lambda_i + \mu_{it} + \varepsilon_{it} \quad (6)$$

where $GTFP_{it}$ denotes $GTFP$ in province, i, in year, t, GF_{it} is a measure of green finance, DE_{it} is the digital economy, and $Controls_{it}$ represent control variables. TC_{it} represents technological progress, EC_{it} represents efficiency progress, λ_i denotes time-fixed effects, μ_{it} denotes individual-fixed effects, and ε_{it} is a random error term.

In order to clarify the mechanism of the interaction term between green finance and digital economy on $GTFP$ and the effect of the interaction term between green finance and digital economy on the decomposition term of $GTFP$, the cross-product term $GF_{it} \times DE_{it}$ is introduced in the model to test the role played by the interaction term between green finance and digital economy in $GTFP$. The models constructed in this paper are Equations (7)–(9).

$$GTFP_{it} = \alpha_0 + \alpha_1 GF_{it} + \alpha_2 DE_{it} + \alpha_3 GF_{it} \times DE_{it} + \alpha_i\ Controls_{it} + \lambda_i + \mu_{it} + \varepsilon_{it} \quad (7)$$

$$TC_{it} = \beta_0 + \beta_1 GF_{it} + \beta_2 DE_{it} + \beta_3 GF_{it} \times DE_{it} + \beta_i\ Controls_{it} + \lambda_i + \mu_{it} + \varepsilon_{it} \quad (8)$$

$$EC_{it} = \gamma_0 + \gamma_1 GF_{it} + \gamma_2 DE_{it} + +\gamma_3 GF_{it} \times DE_{it} + \gamma_i\ Controls_{it} + \lambda_i + \mu_{it} + \varepsilon_{it} \quad (9)$$

where $GF_{it} \times DE_{it}$ represents the cross-product term of green finance and digital economy. Finally, we verified whether the coefficient α_3 in Equation (7), the coefficient β_3 in Equation (8), and the coefficient γ_3 in Equation (9) are significant.

4. Results

4.1. Descriptive Statistics and Correlation Analyses

The descriptive statistics of all analyses are listed in Table 3. The mean value of $GTFP$ is 1.514, the maximum value is 4.979, and the minimum value is 0.608, which indicates a large difference in $GTFP$ between regions. The minimum value of green finance development level is 0.050 and the maximum value is 0.0793. The minimum value of digital economy development level is 0.11 and the maximum value is 0.77. This indicates that the level of green finance development and the level of digital economy development in that there are also large differences between regions.

Table 3. Descriptive statistics.

VarName	Obs	Mean	SD	Min	Max
GTFP	420	1.514	0.573	0.6080	4.9789
GF	420	0.160	0.099	0.0500	0.7930
DE	420	0.260	0.108	0.1104	0.7695
OIS	420	45.494	8.537	16.2000	61.5000
OPEN	420	0.022	0.020	0.0001	0.1210
URB	420	0.546	0.136	0.2746	0.8960
RD	420	0.010	0.006	0.0000	0.0324
GOV	420	3721.738	2686.780	174.54	17,297.85

The development of $GTFP$ with EC and TC in China from 2006–2019 is shown in Figure 1. As we can see in Figure 1a, the $GTFP$ level increased significantly from 0.98 in 2006 to 2.45 in 2019. Comparing Figure 1b,c, it can be found that both technological progress and efficiency improvement have a significant increase from 2006–2019, but the increase in technological progress is closer to the $GTFP$ improvement curve, which indicates that the green total factor improvement mainly comes from technological progress.

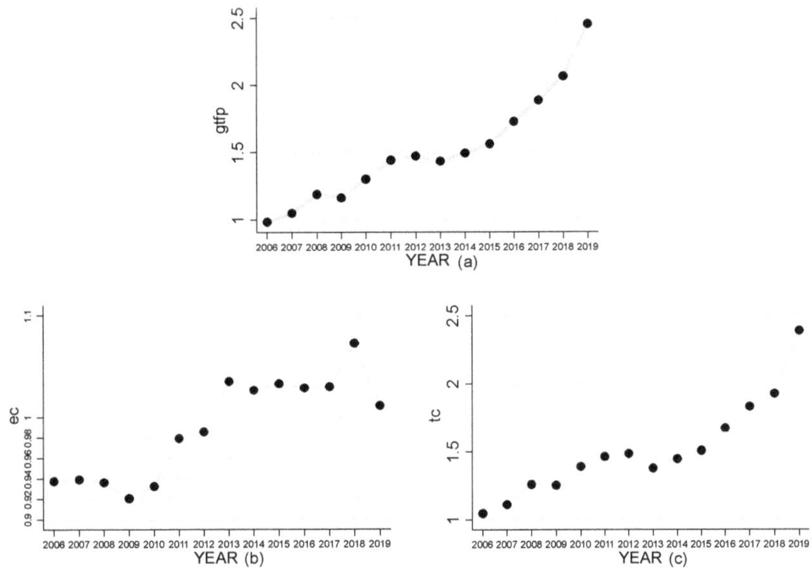

Figure 1. Technical efficiency change, Technical progress change and *GTFP* trend graphs. (**a**) indicates the trend of *GTFP*; (**b**) indicates the trend of technical efficiency change; (**c**) indicates the trend of technical progress change.

Green finance enables financial institutions to invest in green projects that can bring energy savings and environmental protection through the guidance of financial institutions. China's economic transformation also promotes the concept of low carbon, energy saving, and environmental protection. For this reason, high-pollution enterprises face policy constraints as well as loan restrictions; Thus, they are more willing to comply with the concept of energy conservation and environmental protection through technological innovation and industrial structure innovation, which is conducive to raising the level of *GTFP*. The fitted graph of green finance and *GTFP* is shown in Figure 2a, which shows that there is a positive correlation between green finance and *GTFP*.

On the one hand, the development of the digital economy has freed the traditional economy from its heavy dependence on energy and the environment, as well as significantly reduced the excessive consumption of energy in the industrial economy model, improved the efficiency of factor utilization, and promoted green and high-quality development through energy conservation and emission reduction. On the other hand, the development of the digital economy has broken the geographical barriers and realized the cross-regional flow of talents, information, and technology, which is conducive to stimulating green technological innovation and improving *GTFP*. The fitted graph of the digital economy and *GTFP* is shown in Figure 2b, which shows that there is a positive correlation between the digital economy and *GTFP*.

The relationship between green finance and the digital economy is complementary. By combining with the high-energy digital economy industry, green finance can promote the related industries to continuously develop technology to reduce the energy consumption of the digital economy and help achieves low-carbon sustainable development. The digital economy, with its unique data and information technology, provides numerous benefits, such as a linked upstream and downstream platform for the green finance system, established information sharing and security mechanisms, greatly improved matching efficiency of investment and financing, the increased scale of green finance, and allows green finance to be well-integrated into the industry [10]. The fitted graph of green finance and the digital economy is shown in Figure 2c, which shows that there is a positive correlation between green finance and the digital economy.

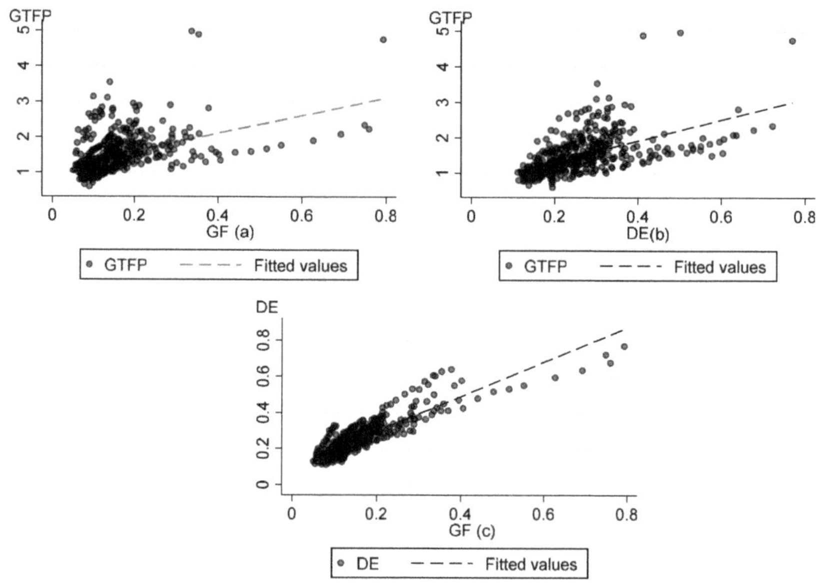

Figure 2. The relationship between *GF*, *DE* and *GTFP*. (**a**) shows the relationship between *GF* and *GTFP*; (**b**) shows the relationship between *DE* and *GTFP*; (**c**) shows the relationship between *GF* and *DE*.

Overall, green finance, the digital economy, and *GTFP* are generally on an upward trend. In terms of relationships, green finance–*GTFP*, digital economy–*GTFP*, and green finance–digital economy all have linear relationships.

This paper performs correlation test indicators, as shown in Table 4. The significant results between the independent and the dependent variables are extremely significant, which indicates that the independent variables selected in this paper are strongly correlated with the dependent variable. Secondly, the significance levels of the remaining variables have been tested, except for the level of foreign openness and the level of government expenditure, which are not significant.

Table 4. Correlation analyses.

	GTFP	*GF*	*DE*	*OIS*	*OPEN*	*URB*	*RD*	*GOV*
GTFP	1							
GF	0.430 ***	1						
DE	0.554 ***	0.877 ***	1					
OIS	0.224 ***	0.444 ***	0.415 ***	1				
OPEN	0.002	0.159 ***	0.012	0.080	1			
URB	0.375 ***	0.591 ***	0.574 ***	0.336 ***	0.320 ***	1		
RD	0.312 ***	0.432 ***	0.452 ***	0.150 ***	0.371 ***	0.649 ***	1	
GOV	−0.078	0.243 ***	0.244 ***	−0.037	0.304 ***	0.390 ***	0.320 ***	1

Note: *** means $p < 0.01$.

In this paper, LLC tests were conducted on the independent and dependent variables, the significance levels were passed, and the test results are shown in Table 5.

Table 5. LLC test results.

VarName	Statistic	p-Value
$GTFP$	−11.2467	0.0000
GF	−9.3897	0.0000
DE	−12.2336	0.0000

4.2. Regression Analysis Results of Green Finance, Digital Economy, and GTFP

Column (1) of Table 6 shows the test results of Equation (1). It can be seen that the coefficient 2.115 of GF is significantly positive at the 1% level, which indicates that the development of GF contributes to the growth of $GTFP$, as it increases by 2.115% for every 1% increase in the level of GF. The coefficient of 1.991 for DE is significantly positive at the 1% level, which indicates that the development of the digital economy also contributes to the growth of $GTFP$, as it increases by 1.991% for every 1% increase in the level of development of the digital economy. The rest of the variables are economic variables essential to the operation of the economy. The coefficient of OIS −0.008 is significant at the 5% level, which indicates that the higher the share of secondary industry structure, the higher the inhibitory effect on $GTFP$. The coefficient of $OPEN$ −4.426 is significant at the 1% level, which indicates that foreign investment has a suppressive effect on $GTFP$ to some extent. The coefficient of URB 3.125 is significantly positive at the 1% level, which indicates that as urbanization continues, it can also contribute to the growth of $GTFP$. The coefficient of RD −40.160 is significant at the 1% level, which indicates that R&D and input inhibit the development of $GTFP$ to some extent, which is due to the fact that enterprises are more inclined to deepen R&D on existing technologies in the process as they expect to achieve higher energy use efficiency in order to achieve the purpose of energy saving and environmental protection, rather than to develop new projects. This kind of R&D and input to improve energy efficiency rather than update to the green energy-saving industry as the purpose of R&D and investment, to a certain extent, caused the phenomenon of energy rebound, thus, it has a certain inhibitory effect on the growth of $GTFP$. Finally, GOV and $GTFP$ are not significant at the level of significance.

Table 6. Multiple regression results.

	(1)	(2)	(3)	(4)	(5)	(6)
	$GTFP$	TC	EC	$GTFP$	TC	EC
GF	2.115 ***	1.011 **	0.434 *	1.885	−2.719 ***	2.687 ***
	(3.25)	(2.29)	(1.87)	(1.32)	(−2.88)	(5.48)
DE	1.991 ***	3.000 ***	−0.416 *	1.930 ***	2.005 ***	0.184
	(3.14)	(6.97)	(−1.85)	(2.68)	(4.21)	(0.75)
OIS	−0.008 **	−0.008 ***	0.001	−0.008 **	−0.011 ***	0.003 *
	(−2.07)	(−3.20)	(0.93)	(−2.07)	(−4.07)	(1.92)
$OPEN$	−4.462 ***	−1.911 *	−0.256	−4.536 ***	−3.099 ***	0.461
	(−3.05)	(−1.93)	(−0.49)	(−2.98)	(−3.08)	(0.88)
URB	3.125 ***	1.398 ***	1.156 ***	3.214 ***	2.841 ***	0.284
	(5.63)	(3.72)	(5.86)	(4.33)	(5.80)	(1.12)
RD	−40.160 ***	−36.677 ***	−0.966	−39.665 ***	−28.658 ***	−5.808 **
	(−4.96)	(−6.68)	(−0.34)	(−4.64)	(−5.07)	(−1.98)
GOV	0.000	−0.000	0.000	0.000	−0.000	0.000
	(0.11)	(−0.24)	(0.42)	(0.11)	(−0.21)	(0.40)
$GF \times DE$				0.305	4.940 ***	−2.982 ***
				(0.18)	(4.45)	(−5.16)
_cons	−0.192	0.596 ***	0.351 ***	−0.200	0.460 **	0.433 ***
	(−0.62)	(2.86)	(3.21)	(−0.64)	(2.24)	(4.05)
N	420	420	420	420	420	420
r2	0.672	0.762	0.186	0.672	0.774	0.239

Note 1: The t statistics are in parentheses; * means $p < 0.1$, ** means $p < 0.05$, and *** means $p < 0.01$. Note 2: Due to the length of this article, only the double regression results are shown here.

Columns (2) and (3) are the Equations (2) and (3) constructed from the decomposition terms TC and EC of $GTFP$. It can be found that GF is significantly positive in both Equations (2) and (3). While DE is significantly positive in Equation (2), the coefficient is −0.416 and also significant in Equation (3). This suggests that GF increased both TC and EC, and contributed to the growth of $GTFP$ by the coupling of the two. Finally, DE only increased TC and did not contribute well to EC, but in general, it also contributed to the growth of $GTFP$.

When combining Columns (1), (2) and (3), GF can increase the level of $GTFP$, and DE is able to improve $GTFP$. This is consistent with the findings of existing studies. However, the boost to $GTFP$ comes mainly from TC.

Columns (4), (5), and (6) are the results after adding $GF \times DE$.

The results of $GF \times DE$ in column (4) are not significant, and the results of GF are not significant. Only the coefficient of DE 1.930 is significantly positive at the 1% level. This suggests that the combined effect of GF and DE does not contribute well to the improvement of $GTFP$.

Columns (5) and (6) are the Equations (5) and (6) constructed for the decomposition terms TC and EC of $GTFP$. The coefficients of GF are −2.719, which is significant in Equation (5), and 2.867, which is significantly positive in Equation (6). The coefficient of DE is 2.005, which is significantly positive in Equation (5), but insignificant in Equation (6). The interaction of $GF \times DE$ have coefficients of 4.94, which is significantly positive in Equation (5), and −2.982, which is significant in Equation (6). This indicates that GF is able to produce significant positive effects when acting on $GTFP$ as well as when EC acts on $GTFP$, but when GF is combined with DE, it has an inhibitory effect on EC during production. Since $GTFP = EC \times TC$, the joint effects of GF and DE shows a significant increase in TC but also a significant inhibition in EC; Thus, the joint effects of GF and DE do not significantly increase $GTFP$.

The combined Columns (4), (5) and (6) show that GF and DE together have not been very good at significantly enhancing $GTFP$. Both of them were significantly positive when acting together on TC, but significantly negative when acting on EC. It is possible that GF and DE together did not achieve the expected results when acting on EC due to some limitation, which requires further analysis.

4.3. Threshold Effect

There is a complex relationship between GF, DE, and $GTFP$. On the one hand, GF helps $GTFP$ by supporting green industries. On the other hand, DE is also able to improve $GTFP$ by optimizing the allocation of resources. Although both of them can contribute to the improvement of $GTFP$ in our country, they do not improve $GTFP$ very well when there is a combined effect of the two. This "threshold effect" may exist in $GTFP$ or in the decomposition of $GTFP$, TC, or EC, and it is because of this threshold effect that GF and DE together do not contribute well to the growth of $GTFP$. Based on this, this paper conducted threshold effect tests on $GTFP$ and its decomposition terms TC and EC with GF and DE as threshold variables, respectively, and repeated the samples 300 times using the Bootstrap method. The results obtained are shown in Table 7.

Table 7. Threshold effect test.

Core Explanatory Variables	Threshold Variables	Explained Variables	Models	Fstat	Prob	Crit1	Crit5	Crit10
DE	GF	$GTFP$	Single threshold	7.83	0.6533	31.1242	22.7221	19.7758
		TC	Single threshold	11.03	0.6533	43.2799	27.9036	24.6994
		EC	Single threshold	15.22	0.3033	41.3621	30.4086	24.0252
GF	DE	$GTFP$	Single threshold	6.80	0.6467	37.3207	23.3565	19.7139
		TC	Single threshold	16.23	0.4300	40.7447	30.5658	25.9309
		EC	Single threshold	27.46	0.0367 **	31.8307	22.9264	19.6111
			Double threshold	16.21	0.1033	33.2990	21.249	18.7108

Note: ** means $p < 0.05$.

As seen in row (1) of Table 7, the *p*-value of 0.6533 for the single threshold of *GF* to *GTFP* did not pass the significance test, thus concluding that there is no single threshold of *GF* to *GTFP*.

As seen in row (2) of Table 7, the *p*-value of 0.6533 for the single threshold of *GF* to *TC* did not pass the significance test, thus concluding that there is no single threshold of *GF* to *TC*.

From row (3) of Table 6, it can be seen that the *p*-value of 0.3033 for the single threshold of *GF* to *EC* does not pass the significance test, thus concluding that there is no single threshold of *GF* to *EC*.

As can be seen from row (4) of Table 7, the *p*-value of 0.6467 for the single threshold of *DE* to green *GTFP* did not pass the significance test, thus concluding that there is no single threshold for *DE* to *GTFP*.

From row (5) of Table 7, it can be seen that the *p*-value of 0.43 for the single threshold of *DE* to *TC* does not pass the significance test, thus concluding that there is no single threshold of *DE* to *TC*.

From row (6) of Table 7, it can be seen that the *p*-value of the single threshold of the decomposition term *EC* for *DE* on *GTFP* is 0.0367, which indicates that there is a threshold effect of *DE* on *EC* at the 5% significance level, while the *p*-value of the double threshold is 0.1033, which fails the test of a double threshold. Thus, there is a single threshold effect of *DE* on *EC* with a threshold value of 0.2020, and a single threshold effect is chosen. To verify the accuracy of the threshold estimates, Figure 3 gives the relationship between the threshold estimates and the likelihood ratio statistic, from which it can be seen that the threshold estimate of *DE* and *EC* is 0.2020 with a confidence interval of [0.1981, 0.2025], where the value of the likelihood ratio statistic is less than the critical value at the 5% level. Thus, the threshold effect estimate is considered to be true.

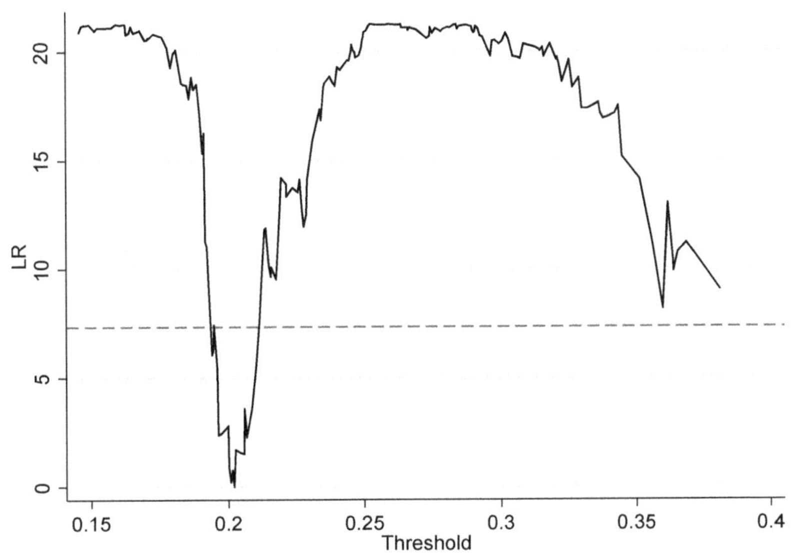

Figure 3. Threshold estimates and confidence intervals.

Based on the threshold test, we need to estimate the threshold model of *DE* on *EC* further, as shown in Table 8.

Table 8. Results of *DE* regression estimation on *EC* panel threshold model.

Variable Name		Coefficient	t	$p > \|t\|$	[95% Conf.	Interval]
GF	$EC < 0.2020$	0.6178002	1.65	0.100 *	−0.11998	1.35558
GF	$EC > 0.2020$	−0.6190305	−3.17	0.002 ***	−1.00325	−0.23481
OIS		0. 0034252	2.73	0.007 ***	0.00010	0.00590
OPEN		0.0611079	0.12	0.902	−0.91623	1.03845
URB		0.813084	4.52	0.000 ***	0.45942	1.16675
RD		−3.97947	−1.44	0.150	−9.40768	1.44874
GOV		1.46×10^{-7}	0.06	0.953	-4.68×10^{-6}	4.97×10^{-6}

Note: The t statistics are in parentheses; * means $p < 0.1$ and *** means $p < 0.01$.

As can be seen from Table 8, the coefficient of *GF* on *EC* is 0.6178 and significantly positive at the 10% level when the *DE* does not exceed 0.2020, which indicates that *GF* has a significant contribution to *EC* when the *DE* level does not exceed the threshold value of 0.2020. The coefficient of *GF* on *EC* is −0.6190 and significant at the 1% level when the *DE* level exceeds 0.2020, which indicates that *GF* has a significant inhibitory effect on *EC* when the *DE* level exceeds the threshold value of 0.2020. To sum up these results, *DE* has a threshold effect on *EC*. For less developed regions of digital economy, such as Guizhou, Yunnan, and Shanxi, we should improve the construction of digital economy infrastructure and promote the development of *DE*, while also promoting the development of *GF*, promoting the upgrading of *TC* and *EC*, and promoting the development of *GTFP*. For regions such as Beijing, Guangdong, Anhui, etc., the infrastructure construction of *DE* is nearly perfect, and there are already good results for enterprise efficiency improvement. The inflow of *GF* funds to enterprises does not achieve good results for enterprise efficiency improvement, at which time we should restrict the inflow of *GF* funds to the efficiency improvement aspect of enterprises and encourage the flow of *GF* funds to aspects of technological progress, such as the research and development of independent intellectual property rights and the introduction of advanced equipment, so as to promote the continuous progress of *GTFP* in China.

We show no significant change in the number of thresholds and threshold coefficients by replacing the control variables section, which indicates that the threshold regression results are robust.

5. Conclusions

This paper empirically examined the impact and mechanism of green finance and the digital economy on *GTFP*, based on relevant data from 30 provinces, municipalities, and autonomous regions in China from 2006–2019. It was found that green finance and the digital economy have a significantly positive effect on *GTFP*, but they did not have a good effect when they acted together with *GTFP*. Based on this, a panel threshold model was introduced to investigate why the joint effects of green finance and the digital economy did not have a good effect on *GTFP* in depth. This paper hopes to provide a research basis for studies in related fields and provide a reference for government policy formulation.

Our study finds that:

(1) *GF* has a significant impact on green *GTFP*, with a coefficient of 3.25. Combined with existing studies, the mechanism is that green finance provides financial support to green industries, promotes the upgrading of industrial structure, fosters technological innovation, optimizes mineral resources for clean energy [58], and thus promotes *GTFP*. *DE* has a significant impact on *GTFP*, with a coefficient of 3.14. Combined with existing research, the mechanism is to optimize resource allocation by means of digital technology and promote the transformation and upgrading of industrial structure to digitalization, rationalization and greening, thus promoting *GTFP*.

(2) The coefficient of the effect of *GF* acting on the *GTFP* decomposition term, *TC*, was 2.29, which was significant at the 5% level. The coefficient of the effect of *GF* acting on the *GTFP* decomposition term, *EC*, was 1.87, which was significant at the 10% level.

The coefficient of the effect of DE acting on the $GTFP$ decomposition term, TC, was 6.07, which was significant at the 1% level. The coefficient of the effect of DE acting on the $GTFP$ decomposition term, EC, was -1.85, which was significant at the 10% level. Combining the available studies with this trend chart, GF mainly drives $GTFP$ for the advancement of TC, and DE mainly drives $GTFP$ for the advancement of TC.

(3) When GF and DE acted together on $GTFP$, the effect was not significant. The "threshold effect" test reveals that there is a single threshold effect when GF and DE act together on EC. The threshold estimate is 0.2020, and the confidence interval is [0.1981, 0.2025], in which the likelihood ratio statistic is less than the critical value at the 5% level, and the threshold effect estimate is true. For regions with developed DE, the digital infrastructure is better built, and the productivity level of enterprises is relatively high when the GF funds acting on the EC side cannot produce good results. For the less developed areas of DE, the digital infrastructure construction has not yet reached a comprehensive level of perfection, and GF funds are able to flow into the enterprises to help improve the digital infrastructure construction, thus improving the level of enterprise productivity. At this time, GF funds can have a better impact on EC.

6. Policy Recommendations

Based on the above findings and analysis, this paper explores the following aspects of GF and DE to promote $GTFP$ and makes the following recommendations:

(1) Enhance the ability of green finance to drive green total factor productivity development.

In terms of green finance, financial institutions with green credit businesses should strengthen the support of green credit and establish a special green department to promote the development of green finance. Actively promote green investment business, accelerate the innovation of green financial products, expand green financial channels, encourage the participation of financial institutions and related enterprises, and disclose the development of green business. There should be the issuance of green credit guidelines and credit policies to provide credit support to green industries, such as photovoltaics, energy conservation, environmental protection, and new energy vehicles. The government should play an active role in forming an organic unity with green credit as the starting point and green investment and green insurance developing together to promote the progress of $GTFP$ effectively.

(2) Improve the construction of digital infrastructure.

In terms of the digital economy, first of all, the green value of the digital economy should be fully explored, cross-regional allocation of digital resources should be promoted, the rapid development of 5G projects should be accelerated, and the business environment in each region should be optimized. Secondly, digital economy empowerment relies on institutional innovation related to the development of the digital economy and the introduction and improvement of laws and regulations, such as the Personal Information Protection Law and the Data Security Law, should be completed as soon as possible to explore the protection of intellectual property rights and personal privacy data security in the digital economy, as well as broaden the space for the development and the promotion of the quality of the digital economy. Finally, it is necessary to strengthen the construction of digital talents, and relevant universities and research institutes should open digital economy majors as soon as possible to enhance the effectiveness of training-related talents, which will promote the high-quality development of the digital economy.

(3) Pay attention to the driving effect of technological progress on green total factor productivity.

Technological progress is the main driving force for green total factor productivity improvement. Therefore, policymakers need to improve laws and regulations on the protection of independent intellectual property rights research and development to ensure that new technologies developed to promote green development are protected by law and to encourage enterprises to actively apply technologies related to green development in their business operations. In addition, the funds financed by enterprises through green finance channels should be supervised accordingly to ensure that these funds are applied to green production and operation projects of enterprises, which effectively promote the technological progress of enterprises and thus promote the development of $GTFP$.

(4) Implement differentiated green financial policies.

Focus on the "threshold effect" in the development of green finance and digital economy, and implement differentiated green finance policies according to local conditions and scientific guidance. For the less developed areas of China's digital economy, policymakers should promote the development of green finance and the digital economy to promote $GTFP$ development levels by driving technological progress and enterprise efficiency improvements. For the developed regions of China's digital economy, policymakers should restrict the flow of green financial funds to the improvement of enterprise efficiency, avoid the blind flow of funds and disorderly development, and actively guide the flow of green financial funds to the research and development of independent intellectual property rights, the introduction of advanced equipment, industrial structure upgrading and other technological progress, so as to promote the green development of China's economy.

Author Contributions: Conceptualization, J.G.; methodology, K.Z.; software, J.G.; validation, K.Z.; formal analysis, J.G.; investigation, K.Z.; resources, K.L.; data curation, K.Z.; writing—original draft preparation, K.Z.; writing—review and editing, J.G.; visualization, K.L.; supervision, J.G.; project administration, J.G.; funding acquisition, J.G. All authors have read and agreed to the published version of the manuscript.

Funding: This research was funded by Shaanxi San Qin Scholarship Research Funding, Joint Program on Innovation Technology Research along Belt and Road, and the establishment of a research project on computational finance and risk management by the University of Reading (UK), and Xi'an University of Posts and Telecommunications (China).

Institutional Review Board Statement: Not applicable.

Informed Consent Statement: Not applicable.

Data Availability Statement: Not applicable.

Conflicts of Interest: The authors declare no conflict of interest.

References

1. Zhou, W.J.; Chen, L.; Feng, L.L. Effectiveness evaluation on third-party governance model for environmental pollution in China. *Environ. Sci. Pollut. Res.* **2019**, *26*, 17305–17320. [CrossRef]
2. Hao, Y.; Gao, S.; Guo, Y.X.; Gai, Z.Q.; Wu, H.T. Measuring the nexus between economic development and environmental quality based on environmental Kuznets curve: A comparative study between China and Germany for the period of 2000–2017. *Environ. Dev. Sustain.* **2021**, *23*, 16848–16873. [CrossRef]
3. Zhang, M.L.; Li, B.Z. How to design regional characteristics to improve green economic efficiency: A fuzzy-set qualitative comparative analysis approach. *Environ. Sci. Pollut. Res.* **2022**, *29*, 6125–6139. [CrossRef] [PubMed]
4. Kang, D.; Zhai, X.; Chen, F.; Wang, W.; Lu, J. How to promote the development of a green economy: Talent or technology?-Evidence from China's high-speed rail. *Front. Psychol.* **2022**, *13*, 953506. [CrossRef] [PubMed]
5. Rong, B.; Chu, C.J.; Zhang, Z.; Li, Y.T.; Yang, S.H.; Wang, Q. Assessing the Coordinate Development Between Economy and Ecological Environment in China's 30 Provinces from 2013 to 2019. *Environ. Model. Assess.* **2022**. [CrossRef]
6. Liu, T.; Li, Y. Green development of China's Pan-Pearl River Delta mega-urban agglomeration. *Sci. Rep.* **2021**, *11*, 15717. [CrossRef]
7. Shapiro, J.S.; Walker, R. Why Is Pollution from US Manufacturing Declining? The Roles of Environmental Regulation, Productivity, and Trade. *Am. Econ. Rev.* **2018**, *108*, 3814–3854. [CrossRef]
8. Ren, Y.J. Research on the green total factor productivity and its influencing factors based on system GMM model. *J. Ambient Intell. Humaniz. Comput.* **2020**, *11*, 3497–3508. [CrossRef]
9. Wang, N.; Xu, C.Y.; Kong, F.B. Value Realization and Optimization Path of Forest Ecological Products-Case Study from Zhejiang Province, China. *Int. J. Environ. Res. Public Health* **2022**, *19*, 7538. [CrossRef] [PubMed]
10. Tian, X.Y.; Zhang, Y.X.; Qu, G.H. The Impact of Digital Economy on the Efficiency of Green Financial Investment in China's Provinces. *Int. J. Environ. Res. Public Health* **2022**, *19*, 8884. [CrossRef]
11. He, F.; Wang, M.; Zhou, P. Evaluation of market risk and resource allocation ability of green credit business by deep learning under internet of things. *PLoS ONE* **2022**, *17*, e0266674. [CrossRef] [PubMed]

12. Gong, C.; Wan, Y.Z. A Study of Fiscal Expenditure Structure and Green Economic Growth Effects: A Sample from Asian Economies. *J. Environ. Public Health* **2022**, *2022*, 2180532. [CrossRef]
13. Wang, R.; Wang, F.Y. Exploring the Role of Green Finance and Energy Development towards High-Quality Economic Development: Application of Spatial Durbin Model and Intermediary Effect Model. *Int. J. Environ. Res. Public Health* **2022**, *19*, 8875. [CrossRef] [PubMed]
14. Li, C.G.; Fan, X.B.; Hu, Y.X.; Yan, Y.; Shang, G.F.; Chen, Y.Z. Spatial spillover effect of green finance on economic development, environmental pollution, and clean energy production across China. *Environ. Sci. Pollut. Res.* **2022**. [CrossRef] [PubMed]
15. Dong, H.Y.; Xu, G.Y.; Rehman, H.U. The spatial panel econometric diagnosis of the influence of green finance on Chinese economic ecologization. *Environ. Sci. Pollut. Res.* **2022**. [CrossRef] [PubMed]
16. Zhang, W.; Zhao, S.Q.; Wan, X.Y.; Yao, Y. Study on the effect of digital economy on high-quality economic development in China. *PLoS ONE* **2021**, *16*, e0257365. [CrossRef] [PubMed]
17. Zhao, M.; Liu, R.; Dai, D.B. Synergistic Effect between China's Digital Transformation and Economic Development: A Study Based on Sustainable Development. *Sustainability* **2021**, *13*, 13773. [CrossRef]
18. Jiao, S.T.; Sun, Q.B. Digital Economic Development and Its Impact on Econimic Growth in China: Research Based on the Prespective of Sustainability. *Sustainability* **2021**, *13*, 10245. [CrossRef]
19. Iqbal, S.; Taghizadeh-Hesary, F.; Mohsin, M.; Iqbal, W. Assessing the Role of the Green Finance Index in Environmental Pollution Reduction. *Estud. Econ. Appl.* **2021**, *39*. [CrossRef]
20. Liu, C.; Xiong, M.X. Green finance reform and corporate innovation: Evidence from China. *Financ. Res. Lett.* **2022**, *48*, 102993. [CrossRef]
21. Islam, M.M.; Sohag, K.; Hammoudeh, S.; Mariev, O.; Samargandi, N. Minerals import demands and clean energy transitions: A disaggregated analysis. *Energy Econ.* **2022**, *113*, 106205. [CrossRef]
22. Sohag, K.; Hammoudeh, S.; Elsayed, A.H.; Mariev, O.; Safonova, Y. Do geopolitical events transmit opportunity or threat to green markets? Decomposed measures of geopolitical risks. *Energy Econ.* **2022**, *111*, 106068. [CrossRef]
23. Husain, S.; Sohag, K.; Wu, Y. The response of green energy and technology investment to climate policy uncertainty: An application of twin transitions strategy. *Technol. Soc.* **2022**, *71*, 102132. [CrossRef]
24. Feng, Y.C.; Liang, Z. How does green credit policy affect total factor productivity of the manufacturing firms in China? The mediating role of debt financing and the moderating role of environmental regulation. *Environ. Sci. Pollut. Res.* **2022**, *29*, 31235–31251. [CrossRef] [PubMed]
25. Zhang, K.; Zhou, X.M. Is Promoting Green Finance in Line with the Long-Term Market Mechanism? The Perspective of Chinese Commercial Banks. *Mathematics* **2022**, *10*, 1374. [CrossRef]
26. Wang, Y.L.; Lei, X.D.; Long, R.Y.; Zhao, J.J. Green Credit, Financial Constraint, and Capital Investment: Evidence from China's Energy-intensive Enterprises. *Environ. Manag.* **2020**, *66*, 1059–1071. [CrossRef] [PubMed]
27. Li, W.Q.; Fan, J.J.; Zhao, J.W. Has green finance facilitated China's low-carbon economic transition? *Environ. Sci. Pollut. Res.* **2022**, *29*, 57502–57515. [CrossRef] [PubMed]
28. Chen, X.; Chen, Z.G. Can Green Finance Development Reduce Carbon Emissions? Empirical Evidence from 30 Chinese Provinces. *Sustainability* **2021**, *13*, 12137. [CrossRef]
29. Zhang, H.F.; Wang, Y.X.; Li, R.; Si, H.Y.; Liu, W. Can green finance promote urban green development? Evidence from green finance reform and innovation pilot zone in China. *Environ. Sci. Pollut. Res.* **2022**. [CrossRef]
30. Zhou, X.G.; Tang, X.M.; Zhang, R. Impact of green finance on economic development and environmental quality: A study based on provincial panel data from China. *Environ. Sci. Pollut. Res.* **2020**, *27*, 19915–19932. [CrossRef]
31. Liu, Y.H.; Lei, J.; Zhang, Y.H. A Study on the Sustainable Relationship among the Green Finance, Environment Regulation and Green-Total-Factor Productivity in China. *Sustainability* **2021**, *13*, 11926. [CrossRef]
32. Hu, Q.Q.; Li, X.; Feng, Y.H. Do Green Credit Affect Green Total Factor Productivity? Empirical Evidence from China. *Front. Energy Res.* **2022**, *9*, 821242. [CrossRef]
33. Lei, X.D.; Wang, Y.L.; Zhao, D.X.; Chen, Q. The local-neighborhood effect of green credit on green economy: A spatial econometric investigation. *Environ. Sci. Pollut. Res.* **2021**, *28*, 65776–65790. [CrossRef]
34. Chen, S.L.; Huang, Z.H.; Drakeford, B.M.; Failler, P. Lending Interest Rate, Loaning Scale, and Government Subsidy Scale in Green Innovation. *Energies* **2019**, *12*, 4431. [CrossRef]
35. Lai, Z.X.; Lou, G.X.; Zhang, T.T.; Fan, T.J. Financing and coordination strategies for a manufacturer with limited operating and green innovation capital: Bank credit financing versus supplier green investment. *Ann. Oper. Res.* **2021**. [CrossRef]
36. Zhang, J.J.; Li, F.Q.; Ding, X.H. Will green finance promote green development: Based on the threshold effect of R&D investment. *Environ. Sci. Pollut. Res.* **2022**, *29*, 60232–60243.
37. Zhang, Z.F.; Duan, H.Y.; Shan, S.S.; Liu, Q.Z.; Geng, W.H. The Impact of Green Credit on the Green Innovation Level of Heavy-Polluting Enterprises-Evidence from China. *Int. J. Environ. Res. Public Health* **2022**, *19*, 650. [CrossRef]
38. Hou, Y.J.; Fang, Z. Unleashing the mechanism between small and medium enterprises, and green financing in China: A pathway toward environmental sustainability and green economic recovery. *Environ. Sci. Pollut. Res.* **2022**. [CrossRef]
39. Chen, P.Y. Relationship between the digital economy, resource allocation and corporate carbon emission intensity: New evidence from listed Chinese companies. *Environ. Res. Commun.* **2022**, *4*, 075005. [CrossRef]

40. Ding, C.H.; Liu, C.; Zheng, C.Y.; Li, F. Digital Economy, Technological Innovation and High-Quality Economic Development: Based on Spatial Effect and Mediation Effect. *Sustainability* **2022**, *14*, 216. [CrossRef]
41. Hu, X.Y.; Guo, P.F. A spatial effect study on digital economy affecting the green total factor productivity in the Yangtze River Economic Belt. *Environ. Sci. Pollut. Res.* **2022**. [CrossRef] [PubMed]
42. Han, D.R.; Ding, Y.Y.; Shi, Z.Y.; He, Y. The impact of digital economy on total factor carbon productivity: The threshold effect of technology accumulation. *Environ. Sci. Pollut. Res.* **2022**, *29*, 55691–55706. [CrossRef] [PubMed]
43. Meng, F.S.; Zhao, Y. How does digital economy affect green total factor productivity at the industry level in China: From a perspective of global value chain. *Environ. Sci. Pollut. Res.* **2022**, *29*, 79497–79515. [CrossRef]
44. Jiang, Y.Y.; Deng, F. Multi-Dimensional Threshold Effects of the Digital Economy on Green Economic Growth?-New Evidence from China. *Sustainability* **2022**, *14*, 12888. [CrossRef]
45. Su, J.Q.; Su, K.; Wang, S.B. Does the Digital Economy Promote Industrial Structural Upgrading?-A Test of Mediating Effects Based on Heterogeneous Technological Innovation. *Sustainability* **2021**, *13*, 10105. [CrossRef]
46. Zhao, S.Q.; Peng, D.Y.; Wen, H.W.; Song, H.L. Does the Digital Economy Promote Upgrading the Industrial Structure of Chinese Cities? *Sustainability* **2022**, *14*, 10235. [CrossRef]
47. Liu, Y.; Yang, Y.L.; Li, H.H.; Zhong, K.Y. Digital Economy Development, Industrial Structure Upgrading and Green Total Factor Productivity: Empirical Evidence from China's Cities. *Int. J. Environ. Res. Public Health* **2022**, *19*, 2414. [CrossRef]
48. Lyu, Y.; Wang, W.; Wu, Y.; Zhang, J. How does digital economy affect green total factor productivity? Evidence from China. *Sci. Total Environ.* **2022**, *857 Pt 2*, 159428. [CrossRef]
49. Kohli, R.; Melville, N.P. Digital innovation: A review and synthesis. *Inf. Syst. J.* **2019**, *29*, 200–223. [CrossRef]
50. Zhang, W.; Zhou, H.; Chen, J.; Fan, Z. An Empirical Analysis of the Impact of Digital Economy on Manufacturing Green and Low-Carbon Transformation under the Dual-Carbon Background in China. *Int. J. Environ. Res. Public Health* **2022**, *19*, 13192. [CrossRef]
51. Gu, B.; Liu, J.; Ji, Q. The effect of social sphere digitalization on green total factor productivity in China: Evidence from a dynamic spatial Durbin model. *J. Environ. Manag.* **2022**, *320*, 115946. [CrossRef] [PubMed]
52. Wu, H.T.; Hao, Y.; Ren, S.Y.; Yang, X.D.; Xie, G. Does internet development improve green total factor energy efficiency? Evidence from China. *Energy Policy* **2021**, *153*, 112247. [CrossRef]
53. Zhao, S.Q.; Peng, D.Y.; Wen, H.W.; Wu, Y.Z. Nonlinear and spatial spillover effects of the digital economy on green total factor energy efficiency: Evidence from 281 cities in China. *Environ. Sci. Pollut. Res.* **2022**. [CrossRef]
54. Du, M.Y.; Hou, Y.F.; Zhou, Q.J.; Ren, S.Y. Going green in China: How does digital finance affect environmental pollution? Mechanism discussion and empirical test. *Environ. Sci. Pollut. Res.* **2022**. [CrossRef]
55. Chung, Y.H.; Färe, R.; Grosskopf, S. Productivity and Undesirable Outputs: A Directional Distance Function Approach. *J. Environ. Manag.* **1997**, *51*, 229–240. [CrossRef]
56. Jun, Z. Investment, investment efficiency, and economic growth in China. *J. Asian Econ.* **2003**, *14*, 713–734. [CrossRef]
57. He, L.; Liu, R.; Zhong, Z.; Wang, D.; Xia, Y. Can green financial development promote renewable energy investment efficiency? A consideration of bank credit. *Renew. Energy* **2019**, *143*, 974–984. [CrossRef]
58. Islam, M.M.; Sohag, K.; Alam, M.M. Mineral import demand and clean energy transitions in the top mineral-importing countries. *Resour. Policy* **2022**, *78*, 102893. [CrossRef]

International Journal of
Environmental Research and Public Health

MDPI

Article

Research on the Heterogeneity Threshold Effect of Foreign Direct Investment and Corporate Social Responsibility on Haze Pollution

Zhanjie Wang [1], Yongfeng Ma [1], Shasha Wang [2,*] and Yongjian Wang [3,*]

1 School of Business Administration, Guizhou University of Finance and Economics, Guiyang 550025, China
2 Graduate School, Guizhou University of Finance and Economics, Guiyang 550025, China
3 Business School, Jiangsu Normal University, Xuzhou 221116, China
* Correspondence: wangshasha0109@163.com (S.W.); wyj0104@jsnu.edu.cn (Y.W.)

Abstract: Carrying out environmental protection and governance in the process of using foreign capital to develop the economy is a realistic problem that China needs to solve urgently. In order to reduce environmental pollution, all enterprises are called upon by the local government to fulfil CSR and improve the quality of FDI use. However, previous studies have rarely explored the threshold effect of FDI and CSR on haze pollution. This paper employs the threshold effect model to explore the above problem based on panel data of 30 provinces in China from 2009 to 2018. The empirical study found the following: (1) FDI has a significantly positive double-threshold effect on haze pollution. Meanwhile, the promotion effect of FDI on haze pollution is the strongest in the two threshold ranges. (2) CSR has a significantly negative single-threshold effect on haze pollution; that is, the increase in CSR intensity inhibits haze pollution. Such a negative effect shows the characteristics of increasing marginal efficiency. (3) In addition, the provinces in different thresholds display obvious geographical distribution characteristics. Through the above analysis, it can be observed that FDI and CSR have distinct impacts on haze pollution. Thus, the country and the government can reduce haze pollution by improving the investment structure, using environmentally friendly technology, guiding enterprises to abide by business ethics and promoting social responsibilities fulfilment.

Keywords: foreign direct investment; corporate social responsibility; haze pollution; threshold effect; heterogeneity analysis; fixed-effect model

Citation: Wang, Z.; Ma, Y.; Wang, S.; Wang, Y. Research on the Heterogeneity Threshold Effect of Foreign Direct Investment and Corporate Social Responsibility on Haze Pollution. *Int. J. Environ. Res. Public Health* **2023**, *20*, 4802. https://doi.org/10.3390/ijerph20064802

Academic Editors: Taoyuan Wei and Qin Zhu

Received: 9 February 2023
Revised: 4 March 2023
Accepted: 6 March 2023
Published: 9 March 2023

1. Introduction

Since the reform and opening up, foreign direct investment (FDI) has played a vital role in promoting the economic development and optimizing the foreign trade structure of China. However, whilst boosting the economic strength of the country, FDI has also brought significant ecological damage and environmental pollution [1]. Using the Beijing–Tianjin–Hebei region as an example, according to China Statistical Yearbook data, the amount of FDI increased from USD 18.15 billion to USD 48.43 billion, with an increase of 166.8% during the period of 2009–2018. Meanwhile, the GDP in the same period increased from USD 484.43 billion to USD 11285.7 billion, an increase in more than 20 times. It can be seen that the GDP of the Beijing–Tianjin–Hebei region has also achieved continuous growth with the increase in the number of FDI. However, according to the Announcement of China's Environmental Status in 2018 issued by the Ministry of Environmental Protection, the number of days of light pollution, moderate pollution, heavy pollution and serious pollution in Beijing–Tianjin–Hebei region accounted for 27.1%, 10.50%, 6.0% and 3.20%, respectively, in a year. The announcement showed that the average concentration of PM 2.5 is 77 $\mu g/m^3$, 1.20 times higher than the national secondary standard [2].

Driven by industrialization, western countries have produced massive amounts of air pollution for over a century that have affected the economic development of China over the past 40 years. The frequent occurrence of haze pollution events, including PM2.5, PM10 and other significant sources of pollution, has gradually expanded the scope and degree of pollution. Therefore, many scholars have proposed the 'pollution paradise' effect of FDI. To circumvent environmental regulations in their own countries and transfer crude industries, some multinational enterprises take advantage of their capital to export highly polluting and energy-intensive industries to less developed countries that are in urgent need of economic development and have low environmental awareness, thereby becoming essential drivers of environmental degradation [3–5]. Liu and Gao highlighted a strong positive correlation between FDI and environmental pollution, arguing that the degree of regional pollution is aggravated along with an increasing FDI agglomeration [6]. Dong et al. confirmed the positive effect of FDI on haze pollution based on quantile regression and Shapley value decomposition [7].

However, some scholars also put forward the 'pollution halo' effect of FDI. Less-developed countries that introduced advanced clean production technologies and environmental governance from developed countries through FDI channels could improve their utilization rate of natural resources and the quality of their ecological environment [8,9]. He and Liu concluded that in China, FDI had a positive effect on pollution emissions (especially industrial sulfur dioxide emissions), and the impact of FDI on environmental pollution varied significantly across the eastern, central and western regions of China [10]. Nathaniel et al. pointed out that the hypothesis of FDI effect was not valid in Mediterranean coastal countries, but FDI could effectively promote the improvement of local environmental quality [11]. In response to these contrasting findings, some scholars have explored the green technology spillover effects of induced labor- and capital-based FDI [12]. Some scholars have taken another approach and analysed the problem from the perspective of the 'coordinated development of two-way FDI', concluding that the coordinated development of two-way FDI in China could significantly suppress haze pollution [13]. However, a unified conclusion on this topic is yet to be reached. This paper approves the 'pollution paradise' effect of FDI because the high incidence of haze pollution not only affects the health and well-being of the people, but also poses a huge threat to the ecological civilization construction and low-carbon green growth in China [14]. Therefore, how to encourage countries to commit themselves to increasing their levels of environmental protection whilst simultaneously boosting trade and investment via FDI has become a hot topic amongst scholars.

Very few scholars have explored the impact of CSR on haze pollution. As micro-entities of the regional economy, enterprises serve as creators of economic value and producers of environmental pollution. Their functional performance in their economic, social and environmental responsibilities contributes to improving regional environmental quality and overall social welfare [15,16]. In this sense, encouraging more companies to commit themselves to social and environmental issues can help societies increase their trust in business communities, enhance the social capital of companies and thereby lead to a synergistic win-win situation [17]. In terms of CSR transmissibility, if core firms in a region are active in haze reduction and management, then their activities will significantly affect the ecosystem of the industrial chain [18]. The core enterprises set an example of continuously making significant contributions to the overall environment, especially haze pollution, by promoting social responsibility awareness on the upper and lower levels of the industrial chain. Therefore, if more firms voluntarily commit to limiting their pollution emissions, even beyond the provisions of international protocols and treaties, then such environmentally responsible behavior can become a benchmark for competitors to follow, thus forming a virtuous catch-up cycle [19]. Some scholars have also pointed out that the CSR activities vigorously carried out by enterprises can stimulate active cognitive responsibility feedback from the society. Enterprises can attract those who care about social and environmental issues to act together by engaging in green and low-carbon production and designing environment friendly and energy efficient products to inform consumers that the production process of their final products minimizes harm to the environment [20].

Such behavior can also help consumers be generally aware that companies, even sometimes seen as purely economic actors, have made environmentally responsible commitments in this area [21].

Furthermore, in the context of the influx of FDI into China's economic construction, FDI may indirectly influence the level of haze pollution in China through corporate social responsibility (CSR). Foreign companies with high CSR compliance standards may gradually lower their compliance standards whilst adapting to the already tricky and ineffective CSR compliance situation in China [22], thereby forming a vicious circle of 'competition to the bottom' with domestic companies [23]. Low-quality CSR activities vastly reduce discriminatory barriers and the transparency of international production activities. These activities also result in the indifference of foreign companies towards the welfare of Chinese consumers, the greening of production processes and their fulfilment of environmental responsibility. The resulting 'pollution paradise' effect will also further aggravate the degree of haze pollution [24] and ultimately reduce the overall level of CSR and further aggravate the environmental pollution in China. During the investment process, enterprises need to be jointly driven to fulfil their social responsibilities, strengthen their environmental awareness and promote a balanced development of bilateral economy, ecology and society through moral guidance and environmental regulations.

To cope with the further deterioration of the ecological environment, China actively advocates optimizing the structure and quality of its FDI, enhancing CSR fulfilment and promoting green technology innovation [25]. Nevertheless, the balance in the relationship amongst FDI, CSR and haze pollution remains a practical problem that hinders the low-carbon green development of the country. On the basis of the general equilibrium model of Copeland and Taylor, most domestic and foreign scholars have investigated the internal relationship amongst FDI, CSR and environmental pollution under the constraints of FDI and environmental regulation conditions [15,26]. In terms of research methods, scholars have transitioned from the econometric model with ordinary least square method and simultaneous equations at the core to the endogenous growth model and data envelopment analysis [26,27]. On the basis of the regional differences in practice and development, much achievement has been reported on spatio-temporal evolution analysis and sub-regional testing via micro-scopization [28,29]. In the context of technological innovation leading to low-carbon and green development, individual scholars have introduced environmental technology innovation behavior based on the 'factor–behavior–performance' research idea to understand the role of FDI, environmental regulation and technology innovation in environmental performance paths [30]. To address this problem, this study attempts to integrate FDI, CSR and haze pollution into a unified theoretical framework, explore the nonlinear effects of FDI and CSR on haze pollution, further explore the problem based on regional economic level and resource endowment heterogeneity, and provide a scientific and sound theoretical basis for the haze pollution management and ecological environmental protection decisions in each province.

The theoretical significance of this paper is as follows: Firstly, in the issue of the CSR measurement, this paper innovatively proposes an optimized method that considers the social responsibility carrying capacity of enterprises of different scales and regions and the matching degree of local economic development. Through theoretical deduction and empirical results, it is preliminarily confirmed that CSR has an inhibition function on haze pollution in terms of time and depth, and the research findings provide a reference for relevant theory. Secondly, the action path and boundary conditions of FDI and CSR on haze pollution are proposed based on panel threshold model. This can objectively evaluate the environmental effects of FDI and provide a new perspective for identifying the drivers and governance mechanisms of haze pollution. Thirdly, this paper takes FDI and CSR as threshold variables, preliminary fits the findings through panel regression and analyses the spatial–geographical distribution characteristics of provinces based on different interval thresholds, thereby further verifying the completeness of its findings.

The remainder of the paper is presented as follows: Section 2 provides an overview of methodology and data. Section 3 represents research results and analysis. Section 4 reports the discussion, and Section 5 reports the conclusion.

2. Methodology and Data

2.1. Theoretical Mechanisms and Research Hypotheses

FDI exerts a driving effect on haze pollution mainly through the technology-locking and extrusion effects. The technology-locking effect is manifested in the transfer of high-value industries and the export of clean technologies. The transfer of high-value industries from developed countries brings a large amount of capital to less-developed countries and regions, boosts economic growth and, to some extent, provides financial support for local ecological and environmental pollution management [12,19]. However, the inflow of large amounts of foreign capital can lead to R&D inertia in less-developed countries and regions, thereby curbing the enthusiasm of domestic enterprises in scientific research and innovation and stagnating the number of green capital investments and patent applications, thus leading to the 'technology-locking' phenomenon [31,32]. Meanwhile, developed countries will not directly send their most cutting-edge clean technologies to less-developed countries and regions because they need to maintain their monopoly over clean technology. In addition, the technological gap between less-developed countries and regions and the limitations of many factors, such as digestion and absorption capacity [27,33], eventually triggers a 'technology-locking' effect in these areas. Given the crowding-out effect, the use of FDI from 'quantity-to-quality' momentum conversion is driven by national policy and institutional guidance. To give full play to the economic, technological and environmental effects of FDI, the government formulates various policies and regulations to regulate the market, which reduce its expenditure in other fields [1,32]. Meanwhile, under the thinking of local government officials who are eager for quick success and immediate benefits, and the thinking of only GDP-oriented growth and the current performance appraisal system, local government still takes FDI with solid liquidity and fast returns as their first choice to attract investment [30,31] and excludes those FDI with high environmental protection, clean requirements and a low conversion rate of technological achievements, thereby falling into a vicious circle of pursuing FDI quantity [34], which will eventually aggravate the level of regional haze pollution. Hypothesis 1 is then proposed, that is, that there is a nonlinear double-threshold positive effect between FDI and haze pollution.

CSR takes the spillover and leverage effects as carriers to inhibit haze pollution. Previous studies show that with the improvement of CSR performance, low carbon green development can promote regional ecological environment quality by improving the CSR performance intensity in the region [35]. Although CSR fulfilment increases capital costs in the short term and is not conducive to improving economic performance, CSR fulfilment can promote the customer-oriented strategy of enterprises and improve their business performance in the long run [36]. Therefore, enterprises must be committed to CSR activities through various channels and approaches in a specific country or region and strive to build a 'community of social responsibility with a shared future'. In the form of point-to-line, line-to-surface and surface-to-field connections, and through social responsibility scale effect at the industry level, enterprises balance and address environmental, social and economic benefits to meet the broader social needs for sustainability, including the protection of natural assets, services and functions of ecosystem on which human society ultimately depends [37], thus effectively promoting low-carbon and green development and reducing pollution. The leverage effect is reflected in how, through the formulation and implementation of social strategies, enterprises establish social cooperation and win-win relationship networks with various stakeholders, pool their knowledge to scientifically design and develop low-carbon green products, improve their production process reengineering, promote the use of clean production technologies, reduce their emission of haze pollutants and provide more health products and services for the public [38]. CSR performance also brings economic, social and environmental advantages [33]. Therefore, one can reasonably assume that those enterprises effectively carrying out their social responsibility can maximize their ecological innovation to improve atmospheric conditions and enhance the contributions of their commercial activities to the society and environment [39]. Hypothesis 2 is then proposed, that is, that CSR has a nonlinear double-threshold negative influence on haze pollution.

2.2. Variable Definitions

(1) Haze pollution

Given that haze concentration in China was only monitored in 2012, following the bounds of data selection years in this paper and the views of Yan and Qi and Beatriz et al. [8,15], the satellite observation and chemical migration model from the Center for International Earth Science Information Network (CIESIN) at Columbia University in New York City, USA, which was used to export and convert the data into global PM2.5 average annual concentration monitoring raster data. The monitoring results from this data are roughly the same as those from the domestic environmental protection department, thereby verifying their credibility and applicability.

(2) Corporate social responsibility (CSR)

CSR data were collected from the 'Social Responsibility Report of Listed Companies' published between 2009 and 2018 by Rankins CSR Ratings, RKS [39]. This report measured the degree of CSR fulfilment from four aspects, namely overall, content, technical and industry. Considering the differences in the CSR fulfilment capacities of different regions and industries, the CSR scores of each enterprise were revised using the enterprise scale from the CSMAR database, that database is developed by Xishma Data Technology Co., Ltd., Shenzhen city, China. Afterwards, the average social responsibility scores of each province and region across different years were calculated and used as evaluation indices of regional social responsibility.

(3) Foreign direct investment (FDI)

The FDI data used in this study included the inflow of capital and technology, the transfer of industries and the absorption and utilization of actual capital. In this paper, the actual amount of foreign capital utilization in each province and city was used to measure the FDI for each year. Given the variability of statistical units, the data were converted based on the USD:RMB exchange rate for the current year to obtain the value of FDI in each province and city, and then the logarithm was taken as the measurement variable [5,10].

(4) Control variables

① Environmental regulation (ER). A greater number of environmental regulations and a stricter degree of regulation correspond to a higher motivation for enterprises to invest in environmental pollution control. In this paper, the ratio of the amount of investment in industrial pollution control to the industry-added value in a specific year was used as a proxy for ER [1]. ② Economic growth (EG). In general, a faster economic growth corresponds to a greater demand for FDI and a greater value of GDP per capita. The GDP per capita of each province and city was then used to measure the level of economic growth [4]. ③ Energy structure (ES). China is currently at the stage of high-quality economic development, and its energy consumption is undergoing structural adjustment from coal to clean energy consumption. Therefore, the proportion of natural gas consumption to total energy consumption was used in this paper as a proxy for ES [40]. ④ Industrial structure (IS). Given that the development of primary and secondary industries has a significant impact on the generation of haze pollution, the ratio of the added value of primary and secondary industries to the GDP of each province and city was used in this paper to measure IS [1]. ⑤ Regional innovation ability (IA). Regional innovation focuses on environmental protection, sustainable development of the society and the degree of regional innovation cannot be precisely measured by the number of patent applications or authorizations alone. Therefore, the strength of provincial and municipal innovation ability was measured in this paper using the comprehensive index of IA following the recommendations from the China Regional Innovation Ability Evaluation Report [41].

2.3. The Model

Based on the interaction mechanism amongst the variables, the following panel regression model was constructed to verify the effect of FDI and CSR on haze pollution:

$$PM_{it} = \beta_0 + \beta_1 FDI_{it} + \beta_2 FDI_{it}^2 + \beta_3 CSR_{it} + \beta_4 EG_{it} + \beta_5 ES_{it} + \beta_6 IS_{it} + \beta_7 ER_{it} + \beta_8 IA_{it} + \varepsilon_{it} \quad (1)$$

$$PM_{it} = \beta_0 + \beta_1 CSR_{it} + \beta_2 CSR_{it}^2 + \beta_3 FDI_{it} + \beta_4 EG_{it} + \beta_5 ES_{it} + \beta_6 IS_{it} + \beta_7 ER_{it} + \beta_8 IA_{it} + \varepsilon_{it} \quad (2)$$

The threshold effect model proposed by Hansen was then used to further examine the difference in the fluence degree of FDI and CSR on haze pollution across different threshold domains due to the existence of threshold values. FDI and CSR were taken as threshold variables in turn to establish the following single-threshold effect models for accurately capturing the critical values of explanatory variables and for understanding the nonlinear relationships when the structure changes as follows:

$$PM_{it} = \beta_i + \beta_1 CSR_{it} + \beta_2 EG_{it} + \beta_3 ES_{it} + \beta_4 IS_{it} + \beta_5 ER_{it} + \beta_6 IA_{it} + \beta_7 FDI_{it} \cdot I(FDI \le \gamma) + \beta_8 FDI_{it} \cdot I(FDI > \gamma) + \varepsilon_{it} \quad (3)$$

$$PM_{it} = \beta_i + \beta_1 FDI_{it} + \beta_2 EG_{it} + \beta_3 ES_{it} + \beta_4 IS_{it} + \beta_5 ER_{it} + \beta_6 IA_{it} + \beta_7 CSR_{it} \cdot I(CSR \le \delta) + \beta_8 CSR_{it} \cdot I(CSR > \delta) + \varepsilon_{it} \quad (4)$$

Given that action may have a double-threshold or even a multi-threshold effect due to the existence of multi-stage characteristics, the above single-threshold model was further extended as follows:

$$PM_{it} = \beta_i + \beta_1 CSR_{it} + \beta_2 EG_{it} + \beta_3 ES_{it} + \beta_4 IS_{it} + \beta_5 ER_{it} + \beta_6 IA_{it} + \beta_7 FDI_{it} \cdot I(FDI \le \gamma_1) + \beta_8 FDI_{it} \cdot I(\gamma_1 < FDI \le \gamma_2) + \beta_9 FDI_{it} \cdot I(FDI_{it} > \gamma_2) + \varepsilon_{it} \quad (5)$$

$$PM_{it} = \beta_i + \beta_1 FDI_{it} + \beta_2 EG_{it} + \beta_3 ES_{it} + \beta_4 IS_{it} + \beta_5 ER_{it} + \beta_6 IA_{it} + \beta_7 CSR_{it} \cdot I(CSR \le \delta_1) + \beta_8 CSR_{it} \cdot I(\delta_1 < CSR \le \delta_2) + \beta_9 CSR_{it} \cdot I(CSR_{it} > \delta_2) + \varepsilon_{it} \quad (6)$$

where i represents different provinces; t represents different years; β_0 reflects the individual effect of provincial differences; $I(\bullet)$ is the indicator function; ε_{it} is the random disturbance term; r and δ are the threshold values of FDI and CSR, respectively; β_i is the regression coefficient of each variable; PM, FDI and CSR are the explanatory variables (haze pollution) and threshold variables (FDI and CSR) of the study design; and EG, ES, IS, ER and IA are a set of control variables. After estimating the regression coefficients and the corresponding thresholds for each variable, the significance and authenticity of the thresholds were tested. The regression coefficients and corresponding threshold values were obtained based on the minimum sum of squared residuals of the variables under the given threshold number, and the existence of the threshold effect was tested according to the p-value. The consistency of the obtained threshold values with the actual values was evaluated using the likelihood ratio (LR) statistic.

2.4. Data Sources

The panel data of 30 provinces and cities in China (excluding Hong Kong, Macao, Taiwan and Tibet) from 2009 to 2018 were taken as the sample. EG, FDI and IS data were collected from China Statistical Yearbook and China Urban Statistical Yearbook; ES and ER data were obtained from the China Energy Statistical Yearbook, China Regional Economic Statistical Yearbook and China Environmental Statistical Yearbook; and IA data were derived from the China Regional Innovation Ability Evaluation Report compiled by the China Science and Technology Development Strategy Research Group and the China Innovation and Entrepreneurship Management Research Center of the University of Chinese Academy of Sciences University. Moreover, CSR data were collected from the ratings released by Rankins CSR Ratings, RKS, which were sorted by the authors into provincial and municipal data, and the PM data were collected from the raster data published by the Center for Socioeconomic Data and Applications of Columbia University based on the annual mean global PM2.5 concentrations monitored by satellites.

3. Results

3.1. Collinearity Analysis and Model Selection

Firstly, correlation analysis and variance inflation factor (VIF) were used to verify the multicollinearity amongst the variables. As shown in Table 1, none of the correlation coefficients amongst the variables exceed 0.7, thereby indicating that the correlation coefficients are within a reasonable value range. However, the maximum VIF of 2.41 is much smaller than the reference value of 10, thereby confirming the absence of any severe collinearity amongst the variables. Secondly, F-test and Hausman test were performed to determine the most appropriate estimation method for the model. The panel data estimation methods of OLS regression and random effects were rejected by comparative analysis, thereby confirming that the fixed-effects model was suitable for this study.

Table 1. Correlation and variance inflation factors of variables.

Variable	PM	FDI	CSR	EG	ES	IS	ER	VIF	1/VIF
PM									
FDI	0.456 *							2.32	0.432
CSR	0.183 *	0.151 *						1.32	0.756
EG	0.403 *	0.470 *	0.469 *					2.19	0.457
ES	−0.213 *	−0.246 *	0.083	0.108				1.19	0.838
IS	−0.085	−0.207 *	−0.245 *	−0.476 *	−0.215 *			1.41	0.712
ER	−0.173 *	−0.484 *	0.042	−0.190 *	0.078	0.045		1.73	0.729
IA	0.401 *	0.676 *	0.186 *	0.592 *	−0.056	−0.388 *	−0.402 *	2.41	0.415

Note: * Indicates significance at the 10% level.

3.2. Panel Regression Analysis

The effects of FDI and CSR on haze pollution were initially estimated using Stata 15.1 to fit the panel data with the fixed effects. To further reveal the differential effects produced by the strength of the explanatory variables on the explained variables, the effects of the first and second powers of FDI and CSR on haze pollution were examined in the model after introducing control variables. Table 2 presents the test results. When the explanatory variable is FDI, its first power regression coefficient is negative and significantly correlated at the 1% level ($\beta = -25.560, p < 0.1$), whereas its second power coefficient is positive and passes the 5% significance level test ($\beta = 0.645, p < 0.05$). Therefore, FDI and haze pollution have a U-shaped relationship. With the continuous increase in the total amount of FDI, the haze pollution degree initially decreases and then increases. This trend shows prominent stage characteristics and is constrained by the effect strength of FDI. When CSR is the explanatory variable, the first power coefficient is significantly positive ($\beta = 31.601, p > 0.1$), whereas the second power coefficient is significantly negative ($\beta = -15.202, p < 0.1$), thereby suggesting that CSR and haze pollution have an inverted U-shaped structure. As the degree of CSR fulfilment increases, the degree of haze pollution decreases.

Table 2. Regression results of FDI, CSR and Haze Pollution.

Variable	Haze Pollution							
	Coefficient	Standard Error	t	p	Coefficient	Standard Error	t	p
FDI	−25.560	13.908	−1.84	0.067	3.614	0.774	4.67	0
FDI^2	0.645	0.305	2.11	0.035				
CSR	−13.569	4.745	−2.86	0.005	31.601	29.245	1.08	0.281
CSR^2					−15.202	9.166	−1.66	0.098
EG	−0.628	3.005	−0.21	0.835	−0.412	3.010	−0.14	0.891
IS	5.441	11.165	0.49	0.626	5.194	11.212	0.46	0.644
ES	−38.968	12.402	−3.14	0.002	−42.179	12.490	−3.38	0.001
ER	4.910	3.311	1.48	0.139	5.093	3.318	1.53	0.126
IA	0.247	0.130	1.90	0.058	0.366	0.121	3.04	0.003
R^2(F)		0.328 (17.20)				0.324 (16.89)		

Note: $p < 0.1$ indicates significance at the 10% level, $p < 0.05$ indicates significance at the 5% level, $p < 0.01$ indicates significance at the 1% level.

3.3. Threshold Test

The above panel regression analysis reveals that both FDI and CSR have significant nonlinear effects on haze pollution, which reflects that the different influences of FDI and CSR obviously restrict the degree of haze pollution. Therefore, Bootstrap repeated sampling was performed 300 times to obtain the F statistic and p value as well as the corresponding critical value distribution. Table 3 presents the results. Firstly, the results reveal that both the single and double thresholds of FDI are significant at the 1% level. However, the presence of the triple threshold is not significant, thereby implying that this threshold is invalid. In other words, FDI has a significant double-threshold effect on haze pollution. Similarly, only the single threshold of CSR is significant at the 1% level, and the double and triple thresholds are not significant. In other words, CSR only has a single-threshold effect on haze pollution. Secondly, the threshold value was tested to confirm if it is equivalent to the actual value. Table 3 shows that the two thresholds for FDI are 24.877 and 25.558, and the single threshold for CSR is 1.879. To further verify whether the corresponding estimated threshold value was equal to the actual value, the LR statistic was used to draw the likelihood ratio function graph of each threshold value of FDI (Figure 1) and CSR (Figure 2) under the 95% confidence interval. According to the relationship between the actual LR statistic (lowest point) and the critical value (7.35) at the significance level of 5%, the lowest point of LR statistic is significantly lower than 7.35, thereby confirming the consistency between the threshold value of FDI and CSR and the actual value.

Table 3. Test of FDI and CSR threshold effect results.

Independent Variable	Threshold Variable	Thresholds	F	p	Threshold Value	95% Confidence Interval	Critical Value		
							1%	5%	10%
FDI	FDI	Single	22.02	0.000	24.877	(24.844, 24.900)	28.868	26.807	25.941
		Double	24.74	0.000	25.558	(25.511, 25.561)	14.928	12.926	11.618
		Triple	23.88	1.000	22.302	(21.971, 22.318)	33.099	29.304	27.418
CSR	CSR	Single	17.67	0.080	1.879	(1.826, 1.879)	10.486	9.971	9.026
		Double	16.05	0.3200	1.976	(1.966, 1.988)	9.867	8.646	7.692

Note: $p < 0.1$ indicates significance at the 10% level, $p < 0.05$ indicates significance at the 5% level, $p < 0.01$ indicates significance at the 1% level.

3.4. Analysis of Threshold Effect

Table 4 presents the results of the threshold regression of FDI and CSR on haze pollution. The threshold effect of FDI was initially evaluated. In general, the positive effect of FDI on haze pollution demonstrates the 'pollution paradise' effect because the current use of FDI in China emphasizes quantity over quality. In the context of intensified local competition, local governments blindly expand their use of FDI to promote economic growth. However, those specific industries into which FDI flows are loosely regulated, thereby leading to many FDI flows into industries with high pollution and energy consumption. This behavior also results in the 'market theft' effect of FDI, which further deepens the degree of local haze pollution. When the FDI intensity is lower than 24.877, the impact coefficient is 1.6889, which is significant at the 5% level. Under the early extensive economic development mode, FDI influx plays a special role in promoting environmental pollution. In other words, FDI has a significant positive impact on haze pollution within the first threshold. When the FDI intensity ranges between 24.877 and 25.558, the influence is 2.273 at the significance level of 1%, thereby suggesting that during a process of economic growth that relies on FDI for a long time, the transfer of heavily polluting industries and obsolete technology from developed countries further deepens the environmental deterioration of developing countries. In other words, within the two threshold intervals of FDI, the promoting effect of FDI on haze pollution is significantly enhanced. When the FDI intensity is more significant than 25.558, the impact coefficient decreases to 1.637, which also passes the test at the 5% significance level. Compared with the influence strength of the above two threshold intervals, the positive influence of FDI on haze pollution is the weakest after crossing the second threshold value, thereby suggesting that with the increasing demand for high-quality economic development, both the government and enterprises will adjust their use of FDI, upgrade their industrial structure and use part of their funds for environmental governance. Given that the optimization and adjustment of FDI use are part of a long-term process, the transformation from 'quantity to quality' has not yet achieved the effect of restraining environmental pollution. After FDI crosses the second threshold value, the influence of FDI on haze pollution becomes positive, but its degree is the weakest. Therefore, hypothesis 1 is verified.

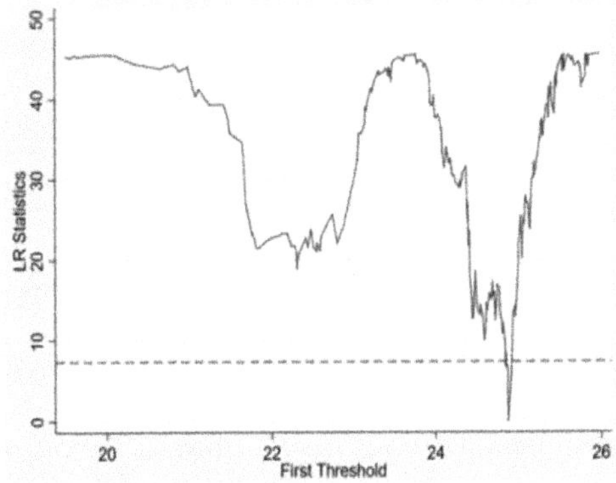

(a) First Threshold

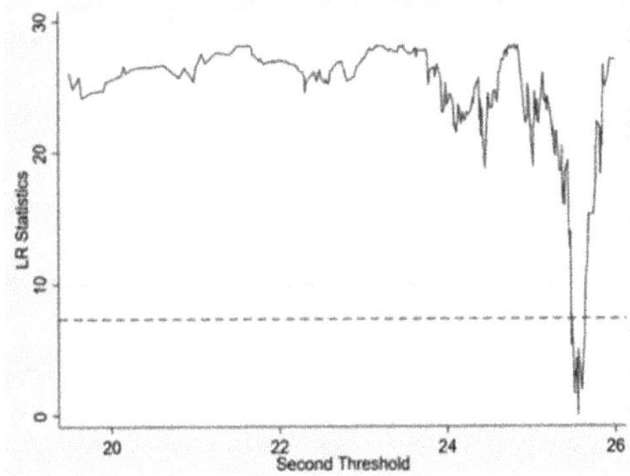

(b) Second Threshold

Figure 1. Plot of threshold estimates and likelihood ratio function for FDI.

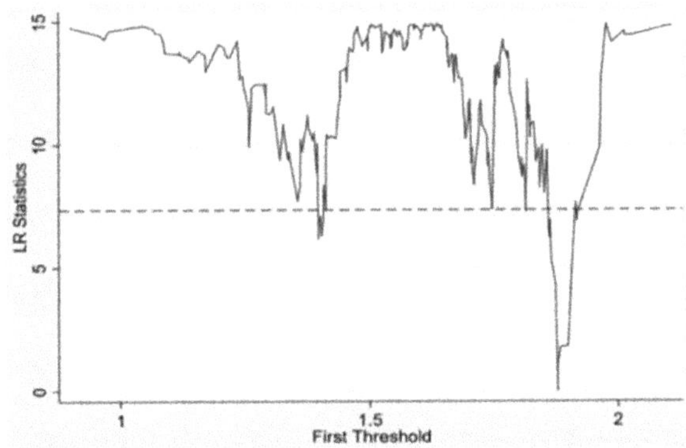

Figure 2. Plot of the threshold estimate of CSR versus the likelihood ratio function.

Table 4. Threshold effect model regression results.

Variable	Haze Pollution							
	Coefficient	Standard Error	t	p	Coefficient	Standard Error	t	p
FDI					3.536	0.768	4.61	0.000
CSR	−11.947	4.236	−2.82	0.005				
EG	−1.325	2.741	−0.48	0.629	−1.254	3.016	−0.42	0.678
IS	15.005	10.3501	1.45	0.148	4.863	11.122	0.44	0.662
ES	−36.896	11.370	−3.25	0.001	−43.796	12.414	−3.53	0.000
ER	4.107	3.054	1.34	0.180	5.153	3.290	1.57	0.118
IA	0.491	0.130	3.79	0.000	0.422	0.122	3.45	0.001
FDI-1	1.689	0.760	2.22	0.027				
FDI-2	2.273	0.730	3.11	0.002				
FDI-3	1.637	0.749	2.18	0.030				
CSR-1					−9.652	5.199	−1.86	0.064
CSR-2					−14.040	4.629	−3.03	0.003
F		24.74				17.67		
R^2		0.442				0.334		

Note: FDI-1, FDI-2, and FDI-3 refer to low-, medium-, and high-intensity intervals of FDI; CSR-1 and CSR-2 refer to the low- and high-intensity intervals of corporate social responsibility, respectively; p value is the result obtained by repeated sampling 300 times with Bootstrap.

The threshold effect of CSR was then analysed. Overall, CSR shows a negative nonlinear effect on haze pollution, and this negative effect is characterized by increasing marginal efficiency. When pursuing economic value, enterprises actively practice social responsibility by paying attention to social harmony and ecological environmental protection, all-around development of environment friendly products and multi-channel innovation of low-carbon green production technology. As a starting point for the sustainable development of enterprises and a support point for the healthy development of the social economy, CSR plays a crucial role in scientific and technological innovation, industrial adjustment and environmental protection, providing a kinetic conversion for haze pollution. When CSR intensity is lower than 1.879, its impact coefficient on haze pollution is −9.652, which is significant at the 10% level. In other words, at the early stage of economic development, the collaborative economic, social and environmental nature of CSR effectively suppress the level of environmental pollution, thereby verifying that CSR is an excellent path to reduce environmental pollution. Meanwhile, when the CSR intensity is higher than 1.879, its impact on haze pollution is significantly more substantial with an impact coefficient of −14.040, which is significant at the 1% level. This finding can be mainly ascribed to the fact that with the increasing seriousness of environmental pollution, the government gradually increases the strength of its environmental regulations, thereby highlighting the necessity and comprehensiveness of CSR performance. Many enterprises incorporate CSR into their development strategies in response to the calls of the government and the public, regard the International Social Responsibility Guide (ISO26000) as a benchmark and fulfil their CSR by reducing their degree of environmental pollution. Therefore, on both sides of the single-threshold value of CSR, the influence of CSR on haze pollution is negative and gradually enhanced. Hypothesis 2 is then verified.

3.5. Further Analysis

There are also changes in the number of provinces across different threshold intervals. According to the different threshold values of FDI and CSR, the 30 provincial samples were divided into 5 intervals to analyse their threshold variability. According to the statistical results in Table 5, FDI and CSR evolve in the direction of adjustment and optimization. From 2009 to 2018, the number of provinces with two variables that are less than the first threshold value showed a downward trend, whilst the number of provinces with two variables that are greater than the first threshold value showed an upward trend. A total of eighteen provinces did not pass the first threshold for FDI in 2018, and these provinces were mainly located in the less-developed regions of central and western China. In the same year, only three provinces passed the second threshold, namely Tianjin, Guangdong and Jiangsu Province. In terms of CSR, due to the influence of China's economic development model and the degree of CSR fulfilment, none of the provinces passed the first threshold from 2009 to 2011. With the transformation of economic development and the popularization of the social responsibility concept, the number of provinces crossing the first threshold gradually increased since 2012, but their number remains relatively small. These results suggest that the development of China's economy and its process of environmental governance are accompanied by the gradual optimization of the FDI structure and the development and implementation of CSR.

Table 5. Statistical results of the number of provinces within different threshold intervals, 2009–2018 (unit: one).

Threshold Interval	2009	2010	2011	2012	2013	2014	2015	2016	2017	2018
FDI ≤ 24.877	15	24	22	22	21	21	20	18	18	18
24.877 < FDI ≤ 25.558	9	3	5	5	6	6	7	10	9	9
FDI > 25.558	6	3	3	3	3	3	3	2	3	3
CSR ≤ 1.879	30	30	30	29	29	26	25	25	23	26
CSR > 1.879	0	0	0	1	1	4	5	5	7	4

An analysis of the geographical characteristics of the number of threshold provinces is also included. The number of provinces where FDI and CSR crossed the threshold shows prominent eastern and western geographical characteristics. From the perspective of FDI, those provinces that exceeded the second threshold in 2009 included Shanghai, Shandong, Guangdong, Jiangsu, Zhejiang and Liaoning. Most of these provinces are located in the eastern coastal region, which is favorable to FDI. Some central provinces such as Sichuan, Anhui, Jiangxi, Henan, Hubei and Hunan are located between the two thresholds due to the constraints in resource endowment and industrial transfer. By contrast, fifteen provinces, including Yunnan, Inner Mongolia, Ningxia, Shanxi, Guangxi, Xinjiang and Guizhou, were less than the first threshold due to geographical location and ecological environment constraints. From the CSR perspective, China still has a long way to go to fulfil its social responsibility. Therefore, none of its provinces passed the first threshold before 2011. According to the measurement standard, it was only after 2012 that the CSR value of Yunnan, Shanghai, Sichuan and Guangdong gradually crossed the first threshold, which, to some extent, indicates that the current social responsibility governance work in China remains challenging and requires further planning and promotion.

4. Discussions

Healthy investment in China is the guarantee of stable economic growth. Different from the developed countries who adopt industrial development model, developing countries have a larger effective utilization gap of capital, more environmental barriers to hurdle and a more imperfect social responsibility environment. As a result, protecting the environment in China and hindrance factors affecting haze pollution are worth studying. This article highlights the economic–social environmental impact, embodied in FDI and CSR.

For FDI, whether it is linear regression or threshold effect regression, the promotion effect of FDI on haze pollution exists. These results not only support most of these previous studies suggesting the linear influence of FDI on haze pollution from the perspective of the pollution paradise and pollution halo effects [4–6], but also expand its nonlinear relationship with stage characteristics. This may be because China attaches importance to the quantity rather than the quality of FDI due to its eagerness to develop the economy, as well as that the relevant environmental protection mechanism and regulatory mechanism are not sound enough, resulting in the consequences of increasing economic aggregate and environmental pollution.

For CSR, we consider the social-responsibility-bearing capacity of enterprises with different scales and from regions and the matching degree of local economic development. Then, our study shows that CSR has a significant inhibitory effect on haze pollution [13,15], whether linear or threshold effect. This is consistent with theoretical deduction, indicating that CSR effect plays a larger part in the threshold affection. It is possible that China's policies towards CSR have gradually increased, and thus, enterprises vigorously implement social responsibility and nurture the concept of low-carbon green development. When foreign capital can adapt to China's institutional environment, a harmonious road between economic development and a beautiful environment may take shape.

When taking FDI and CSR as threshold variables, examining the spatial–geographical distribution characteristics of provinces based on different interval thresholds, the study has found that the number of provinces with FDI and CSR greater than the first threshold has gradually increased, indicating that the Chinese government has gradually attached importance to optimizing and adjusting the structure of FDI use and promoting CSR implementation over time.

5. Conclusions

By analysing the influence mechanism of FDI and CSR on haze pollution, this paper reveals a nonlinear relationship between FDI and haze pollution based on the panel data of 30 provinces and cities across China from 2009 to 2018 and by using the fixed-effects model and threshold regression analysis. This study also comprehensively examines the change in the number of provinces based on the threshold interval and geographical characteristics and draws the following research conclusions. Firstly, there is a significantly positive double-threshold effect between FDI and haze pollution; that is, whether FDI is at the first or second threshold, its influence on haze pollution is significantly positive, and its influence reaches the most substantial level within the two threshold values. Meanwhile, there is a significantly negative single-threshold effect between CSR and haze pollution, that is, the effect of CSR on haze pollution on both sides of the single threshold has the positive effect of increasing marginal efficiency. The management of haze pollution in China is accompanied by optimizing the FDI structure and improving CSR. However, those provinces where each variable crosses different threshold intervals have prominent geographical characteristics. Secondly, from the threshold value and interval distribution perspective, the number of provinces that are below the first threshold value of FDI and CSR decreases yearly. Improving the quality of FDI use and actively carrying out CSR activities have become new approaches to haze pollution control. In terms of the geographical distribution of provinces, the eastern region, with its superior geographical features and developed economy, acts as the main force that crosses the second threshold of FDI and the first threshold of CSR. Meanwhile, the central provinces in the critical period of industrial optimization and investment attraction primarily lie between the two thresholds of FDI.

The above empirical evidence suggests that high-quality FDI and CSR can be used as tools to achieve haze pollution control targets and to construct a green, low-carbon and circular economic system. Policy recommendations are then proposed as followed: Firstly, the quality and structural optimization of FDI should be given priority. Given the need for high-quality economic development, people should adhere to the environmental access threshold of FDI, reduce the entry of enterprises with high energy consumption and pollution, and introduce more clean production and technological innovation enterprises. A group of foreign enterprises that are equipped with technological advantages and are in line with China's economic development should also be introduced to reduce the probability of haze pollution and its negative effects by accumulating and diffusing their capital, technology and knowledge. Secondly, the fulfilment of social responsibilities should be vigorously promoted. Considering their current situation in fulfilling their social responsibilities, foreign and domestic enterprises should be guided to form a development concept that combines high-quality economic development with ecological and environmental protection. These enterprises should also jointly design social responsibility projects with the government and the public to solve social and environmental problems. They should extend their social responsibility to the whole industrial chain to form standard social and environmental value norms, drive chain enterprises to participate in social and environmental governance practices, and coordinate and cooperate with one another to address haze pollution. Thirdly, an environmental governance system shared amongst the government, enterprises and society should be established. The government should not only share its responsibility through environmental regulations and strengthen the restraint mechanism of enterprises' pollution emissions, but also promote a long-term cooperation mechanism with enterprises, social organizations and other actors in environmental governance by taking advantage of the situation. Enterprises should also take an active part in this process by appropriately increasing their R&D investment to innovate green and clean production technologies and by exploring and developing closed-loop value creation systems that reduce emissions and costs, save production materials and recycle energy in circulation. They can significantly contribute to improving the environment by working with third parties, such as universities and research centers, in developing joint business plans, such as eco-patent sharing, to expand the space for collaboration.

Author Contributions: Conceptualization, Z.W. and Y.W.; data curation, S.W. and Y.M.; formal analysis, Y.W., S.W. and Y.M.; investigation, Z.W. and Y.M.; writing—original draft preparation, Z.W. and Y.M.; writing—review and editing, Z.W. and Y.W. All authors have read and agreed to the published version of the manuscript.

Funding: This research was supported by The National Natural Science Fund of China, grant number 72162006; Youth Talent Development Fund of Guizhou Education Department, China, grant number KY Zi [2021]128; the Science and Technology Project of Science and Technology Department of Guizhou Province, grant number ZK [2021]196.

Institutional Review Board Statement: Not applicable.

Informed Consent Statement: Not applicable.

Data Availability Statement: The data presented in this study are available on request from the corresponding author. The data are not publicly available due to student confidentiality and privacy regulations.

Conflicts of Interest: The authors declare that the research was conducted in the absence of any commercial or financial relationships that could be construed as a potential conflict of interest.

References

1. Liu, C.Y.; Xu, Y.Z. How does environmental regulation affect the governance of haze pollution: An empirical study based on the mediating effect method. *J. China Univ. Geosci. Soc. Sci. Ed.* **2017**, *17*, 41–53.
2. Wang, X.L.; He, F.; Zhu, Z.J. Study on the impact of FDI on the environmental quality of Beijing-Tianjin-Hebei. *Res. Financ. Econ. Issues* **2019**, *9*, 45–52.
3. Zhao, D.D.; Zhang, J.S. Foreign direct investment, labor bias and environmental pollution. *Front. Sci. Eng. Manag.* **2022**, *41*, 90–96.
4. Zeng, H.; Shen, J. Spatial correlation analysis between provincial FDI and haze pollution. *Jiangxi Soc. Sci.* **2019**, *39*, 50–60.
5. Kivyiro, P.; Arminen, H. Carbon dioxide emissions, energy consumption, economic growth, and foreign direct investment: Causality analysis for Sub-Saharan Africa. *Energy* **2014**, *74*, 595–606. [CrossRef]
6. Liu, Y.; Hao, Y.; Gao, Y. The environmental consequences of domestic and foreign investment Evidence from China. *Energy Policy* **2017**, *108*, 271–280. [CrossRef]

7. Dong, F.; Yu, B.; Pan, Y.; Hua, Y. What contributes to the regional inequality of haze pollution in China? Evidence from quantile regression and Shapley value decomposition. *Environ. Sci. Pollut. Res.* **2020**, *27*, 17093–17108. [CrossRef]
8. Yan, Y.X.; Qi, S.Z. FDI and haze pollution in China. *Stat. Res.* **2017**, *34*, 69–81.
9. Hassaballa, H. Testing for granger causality between energy use and foreign direct investment inflows in developing countries. *Renew. Sustain. Energy Rev.* **2014**, *31*, 417–426. [CrossRef]
10. He, P.; Liu, Y. The effect of FDI on China's environmental pollution: A research based on instrument of geographical distance. *J. Cent. Univ. Financ. Econ.* **2016**, *6*, 79–86.
11. Nathaniel, S.; Aguegboh, E.; Iheonu, C.; Sharma, G.; Shah, M. Energy consumption, fdi, and urbanization Linkage in Coastal mediterranean countries: Re-assessing the pollution haven hypothesis. *Environ. Sci. Pollut. Res.* **2020**, *27*, 35474–35487. [CrossRef] [PubMed]
12. Hu, J.F.; Wang, Z.; Huang, Q.H. Research on threshold effect of environmental regulation of heterogeneous FDI green technology spillover. *Sci. Technol. Prog. Policy* **2021**, *38*, 20–28.
13. Gong, M.Q.; Liu, H.Y. The influence of two-way FDI coordinated development and industrial structure evolution on environmental pollution in China. *J. Int. Trade* **2020**, *46*, 110–124.
14. Lu, X.H.; Bai, T.T. How can green technology innovation effectively reduce smog pollution? *China Soft Sci.* **2020**, *6*, 174–182+191.
15. Beatriz, J.P.; Daniel, A.M.; José, L.; Godos, D. The influence of corporate social responsibility on air pollution: Analysis of environmental regulation and eco-innovation effects. *Corp. Soc. Responsib. Environ. Manag.* **2018**, *25*, 1–13.
16. Liang, H.; Renneboog, L. On the Foundations of corporate social responsibility. *J. Financ.* **2017**, *72*, 853–910. [CrossRef]
17. Boulouta, I.; Pitelis, C.N. Who needs CSR? The impact of corporate social responsibility on national competitiveness. *J. Bus. Ethics* **2014**, *119*, 349–364. [CrossRef]
18. Costantini, V. Eco-innovation, sustainable supply chains and environmental performance in European industries. *J. Clean. Prod.* **2016**, *155*, 141–154. [CrossRef]
19. Fordham, A.E.; Robinson, G.M. Identifying the social values driving corporate social responsibility. *Sustain. Sci.* **2019**, *3*, 1–16. [CrossRef]
20. Lei, P.; Huang, Q.; He, D. Determinants and welfare of the environmental regulatory stringency before and after regulatory capture. *J. Clean. Prod.* **2017**, *166*, 107–113. [CrossRef]
21. Handayani, R.; Wahyudi, S.; Suharnomo, S. The effects of corporate social responsibility on manufacturing industry performance: The mediating role of social collaboration and green innovation. *Bus. Theory Pract.* **2017**, *18*, 152–159. [CrossRef]
22. Li, D.; Shen, S. Social Environment and Healthy Investment Behavior: Joint Influence of Culture and Institution on China. *Int. J. Environ. Res. Public Health* **2022**, *19*, 607. [CrossRef] [PubMed]
23. Huang, L.Y.; Zheng, S.F.; Wang, Y. Research on the win-win corporation mechanism between multi-national corporation and host country under the background of one belt and one road: Based on the perspective of corporate social responsibility. *Manag. Rev.* **2018**, *30*, 172–182.
24. Zhou, J.Q.; Liang, W.G.; Zhang, Y. Foreign direct investment, environmental regulation and haze pollution: Theoretical analysis and experience from China. *J. Beijing Inst. Technol. Soc. Sci. Ed.* **2019**, *21*, 37–49.
25. Fukuda, K.; Ouchida, Y. Corporate social responsibility (CSR) and the environment: Does CSR increase emissions? *Energ. Econ.* **2020**, *92*, 104933. [CrossRef]
26. Mei, L.A.; Am, B.; Mc, C. Foreign direct investments: The role of corporate social responsibility-Science Direct. *J. Multinatl. Financ. Manag.* **2020**, *59*, 100663.
27. Ye, A.Z.; Zheng, J. Research on the mutual effects among economic growth, DI and environmental pollution in temporal-spatial dimension. *Soft Sci.* **2016**, *30*, 17–21.
28. Zhang, Y.; Jiang, D.C. FDI, environmental regulation and industrial air pollution: An empirical test based on decomposition of industry structure and technology progress. *J. Int. Trade* **2013**, *3*, 102–118.
29. Li, S. An Integration Method for Regional PM2.5 Pollution Control Optimization Based on Meta-Analysis and Systematic Review. *Int. J. Environ. Res. Public Health* **2022**, *19*, 344.
30. Jin, J.Y.; Du, J.G. Research on the mechanism of Chinese FDI enterprises' environmental performance: Double mediating model of different environmental behavior. *Manag. Rev.* **2021**, *33*, 68–79.
31. Zhou, Y.Q.; Xia, N.X.; Liang, W.G. FDI, independent innovation and haze pollution—Evidence from China. *RD Manag.* **2019**, *31*, 78–90.
32. Hao, Y.; Wu, Y.; Wu, H.; Ren, S. How do FDI and technical innovation affect environmental quality? Evidence from China. *Environ. Sci. Pollut. Res.* **2020**, *27*, 7835–7850. [CrossRef] [PubMed]
33. Liu, Y.F.; Gao, L.M. Influence of Chinese provincial FDI on environmental pollution. *Econ. Geogr.* **2019**, *5*, 47–54.
34. Abdo, A.B.; Li, B.; Zhang, X.; Lu, J.; Rasheed, A. Influence of FDI on environmental pollution in selected Arab countries: A spatial econometric analysis perspective. *Environ. Sci. Pollut. Res.* **2020**, *27*, 28222–28246. [CrossRef]
35. Wang, X.L.; Ni, D.Y. Regional economy, corporate social responsibility and carbon emissions. *Mod. Econ. Res.* **2018**, *11*, 87–92.
36. Luo, J.Q.; Kuang, H.B.; Shen, S.W. Impact of corporate social responsibility on financial performance by taking the transportation industry as an example. *Sci. Res. Manag.* **2019**, *40*, 199–208.
37. Bayer, P.; Urpelainen, J. External sources of clean technology: Evidence from the clean development mechanism. *Rev. Int. Organ.* **2013**, *8*, 81–109. [CrossRef]

38. Song, W.; Yu, H. Green innovation strategy and green innovation: The roles of green creativity and green organizational identity. *Corp. Soc. Responsib. Environ. Manag.* **2018**, *25*, 135–150. [CrossRef]
39. Wang, Z.J.; Mai, S. Impact of corporate social strategy on value creation: The moderating role of organic organizational culture. *Sci. Res. Manag.* **2019**, *40*, 266–275.
40. Xu, W.Y.; Xu, Z.X.; Liu, C.J. Energy structure, ecological environment and economic development: Analysis of threshold effect and heterogeneity. *J. Stat. Inf.* **2020**, *35*, 81–89.
41. Xie, Q.Y.; Wang, Y.W. A Study on the Influence Mechanism of Financial Development on the Improvement of Regional Innovation Ability. *Econ. Surv.* **2022**, *39*, 138–149.

International Journal of
Environmental Research and Public Health

MDPI

Article

Air Pollution, Environmental Violation Risk, and the Cost of Debt: Evidence from China

Aiqun Wang, Ming Zhang * and Shuya Zhou

School of Business and Management, Jilin University, Changchun 130012, China; aiqun1964@163.com (A.W.); zhou_shuya0304@163.com (S.Z.)

* Correspondence: vanilla5176@163.com

Abstract: Although a firm's exposure to air pollution-related risk has become an important factor that creditors cannot ignore in the procedure of lending decision making with the aggravation of air pollution, empirical evidence on whether and how air pollution affects the cost of debt has been relatively scarce. Employing a series of Chinese listed firms from the main board of the Shanghai and Shenzhen Stock Exchanges covering 2014 to 2018, our research responds to this research gap by exploring how air pollution-induced environmental violation risk affects the cost of debt by constructing an assessment system of firms' environmental violation risk. The results shed light on an issue that firms exposed to higher concentrations of air pollution may suffer a higher environmental violation risk, resulting in a higher debt cost. In addition, a further analysis shows that environmental regulatory pressure and heavily polluting firms enhance the influence of air pollution on the cost of debt, while state-owned firms and firms' economic contributions weaken the influence of air pollution on the cost of debt. Our research is conducive to highlighting not only the importance of environmental governance for mitigating the cost of debt to the firms exposed to air pollution, but also its importance to creditors exposed to their clients' environmental violation risk and default risk.

Keywords: atmospheric pollution; debt financing cost; environmental penalties; environmental regulatory pressure; heavily polluting firms; state-owned firms; economic contribution

Citation: Wang, A.; Zhang, M.; Zhou, S. Air Pollution, Environmental Violation Risk, and the Cost of Debt: Evidence from China. *Int. J. Environ. Res. Public Health* **2022**, *19*, 3584. https://doi.org/10.3390/ijerph19063584

Academic Editor: Timothy W. Collins

Received: 23 January 2022
Accepted: 14 March 2022
Published: 17 March 2022

1. Introduction

For a long time, some areas in China have sacrificed the environment in exchange for the rapid development of the regional economy [1], causing irreversible damage to the ecological environment. In the last few years, air pollution incidents have occurred frequently in China. In particular, in December 2013, a large-scale air pollution incident occurred in northern China that resulted in large-scale flight delays, the cancellation of outdoor work, and a surge in respiratory diseases [2]. Subsequently, air quality has received increasing public attention as it poses a serious threat to human health, welfare, and psychology [3–5]. How to coordinate the relationship between economic growth and environmental protection has become a crucial issue for China's social development.

With the expansion of the impact of air pollution on the social economy, researchers have gradually become interested in the impact of air pollution on the behavior of capital market participants, which provided us with a crucial and unique idea to explore the relationship between air pollution and the cost of debt. Some frontier literature has noticed that air pollution may increase people's pessimistic emotions, which leads to a series of aberrant behaviors in capital markets, including risk aversion behaviors and attention-driven buying behaviors among investors [6,7], pessimistic earnings forecasts among securities analysts [8], and more audit efforts for auditors [9]. These unique and innovative studies link air pollution to subsequent abnormal behaviors and outcomes in capital markets, but they have not explored the potential influence of air pollution on the decisions of creditors and the cost of corporate debt. Whether and how air pollution affects the cost of debt is a crucial issue, which is related to the economic benefits and sustainable development of firms. Our study explores the effect of air pollution on the cost of debt in depth based on large-sample data to provide empirical evidence for whether and how air pollution affects creditors' decisions.

Public concern regarding the ecological environment has increased along with air pollution, stimulating government regulators to formulate stricter environmental regulations [10]. Firms are not only powerful drivers of economic growth, but also major producers of air pollution. The Chinese government has passed a great deal of powerful environmental regulations aimed at reducing the pollutant gas emissions of firms, including environmental protection interview systems and the Blue Sky Defense Campaign. Undoubtedly, these measures severely punish the polluting behaviors of firms, which leads many heavily polluting firms to face a huge environmental violation risk [11]. Firms' exposure to environmental violation risk may increase the uncertainty risk in its current and future cash flows and, ultimately, increase the likelihood of the default [10,12]. With the increasing pressure of environmental supervision, firms' exposure to environment-related risk has become an important factor that creditors cannot ignore in the procedure of lending decision making [10,13,14], which means that more and more firms are facing the severe challenge of financing constraints caused by air pollution [15,16]. Therefore, whether air pollution affects the cost of debt through its effect on environmental violation risk is worthy of exploration.

Empirical evidence from China can provide an ideal natural setting for this study. First, China's vast geographical area includes different climatic conditions, which can be more conducive for us to identify the potential impact of different levels of air pollution on corporate debt cost. Second, Chinese empirical evidence provides us with data on firms under discrepant conditions of environmental regulation, industry characteristics, ownership, and economic contribution that facilitate the analysis of the moderating roles of these heterogeneous factors in the impact of air pollution on the cost of debt.

Employing the listed firms from the main board of the Shanghai stock exchange (SSE) and Shenzhen stock exchange (SZSE) from 2014 to 2018 as the research sample, we thoroughly explore whether and how air pollution surrounding the cities where the firms are located becomes an important factor affecting the cost of debt. The results show that air pollution in the cities of the firms are significantly positively associated with the cost of debt, and environmental violation risk is the mechanism of air pollution affecting the cost of debt. This reveals that firms exposed to higher concentrations of air pollution may suffer a higher risk of environmental violations, which promotes the creditors' pessimistic assessment of firms' default prospects and results in a higher cost of debt. Our main regression result remains robust after a range of sensitivity tests and endogenous tests. A further analysis shows that the environmental regulatory pressure and heavily polluted firms enhance the influence of air pollution on the cost of debt, while state-owned firms and firms' economic contributions weaken the influence of air pollution on the cost of debt.

The research contributions we consider may include the following parts: First, we fill in the gaps in the existing literature on the relation between air pollution and the cost of debt. This research focuses on the effect of air pollution on the cost of debt, which not only extends the related literature on the economic outcomes of air pollution, but also enriches the research paradigm of "environmental conditions–decision making". Second, our research provides deeper insights into how air pollution affects creditors' decisions and corporate debt cost from the perspective of environmental violation risk. Although some scholars have attempted to analyze the restrictive effect of air pollution on corporate debt financing from the mechanism of credit risk and financial uncertainty [15,16], they ignored the mechanism of environmental violation risk, which is the fundamental mechanism by which air pollution affects the cost of debt. Additionally, few studies have comprehensively measured the environmental violation risk of listed firms. Our research responds to these research gaps by clarifying the internal mechanism of air pollution affecting the cost of debt by constructing an evaluation system of firms' environmental violation risk based on the scoring standard for firms' environmental violations formulated by the Chinese local government. Third, our study provides firms with useful insights for forecasting creditors' responses and decision making in the face of air pollution based on diverse conditions that include regional environmental supervision, industry characteristics, ownership, and firms' economic contribution. Specifically, our research reminds the firms subject to strict environmental supervision, heavily polluting firms, non-state-owned firms, and firms with low economic contributions to be more vigilant against the environmental risk hidden in air pollution, which complements the existing literature on the environmental violation risk of firms to a certain extent [17–19].

2. Literature Review and Hypothesis Development

With the frequent occurrence of global haze events, the severity of air pollution has gradually attracted extensive attention. As a typical indicator of environmental conditions, air pollution observably affects the judgment and decision making of capital market participants, which is gradually being recognized by scholars. Levy and Yagil [6] and Lepori [20] propose that air pollution may deteriorate investors' emotions and intensify their risk aversion, which leads to a negative impact on the subsequent stock returns. Coincidentally, Huang et al. [7] confirmed that air pollution makes investors more prone to the disposition effect, attention-driven buying behavior, and excessive trading, which may contribute to poor trading performance. Wu et al. [1] also verified that the fluctuation of investor sentiment related to air pollution is an important reason for the abnormal stock price. Subsequently, some scholars even further extended the relationship between air pollution and the decision making of capital market participants to analysts' earnings forecast and auditors' pessimistic bias [8,9]. However, few studies discuss whether and how atmospheric pollution influences the judgment and decision making of the lenders compared to stock investors, which is closely related to the debt financing and sustainable development of firms.

In the following analysis, we deeply explored (1) the impact of air pollution on the cost of debt and the mediating effect of environmental violation risk on this impact and (2) the heterogeneous factors affecting the relationship between air pollution and debt cost.

2.1. Influence of Air Pollution on Cost of Debt and Mechanism of Environmental Violation Risk

Recently, the frequent occurrence of serious air pollution, such as high concentrations of haze, has aroused widespread concern and pessimistic expectations among the public regarding a plummeting air quality and environmental livability, since air pollution has gradually become the main culprit of many diseases [21–23]. Meanwhile, these public concerns and pessimistic expectations concerning air pollution are also potential forces to promote government regulators to implement sterner environmental regulatory measures to severely punish environmental violations [24,25]. Firms are not only the main driving force of economic development, but also resource consumers and manufacturers of environmental pollution. In particular, heavily polluting firms may face a greater environmental violation risk due to the penalty of pollution discharge by environmental supervisions in areas with serious air pollution [11], which is a crucial risk signal to creditors and other stakeholders in circumstances of asymmetric environmental information. The financial losses of environmental violation risk for firms are quite serious, which may increase the uncertainty of their future cash flows [10,12,26,27], threaten viability [28], and even increase the default risk or bankruptcy risk of firms [12,18]. Generally, firms with higher exposure to environmental violation risk face higher credit risk [14]. In response, creditors, including banks and other lending institutions, cannot ignore the environmental risk related to air pollution in the process of lending decision making [13,28,29] and prevent the air pollution-related default loss by revising debt contracts, such as raising the bond yields and interest rate as a response [10,12,30]. Tan et al. [16] and Tan et al. [15] also confirmed that firms in areas with serious air pollution face more significant debt financing constraints in the process of negotiation with banks due to a higher credit risk and financial uncertainty related to the environment. Consequently, we expect that firms located in areas with serious air pollution face a higher risk of environmental violations, and, thus, they need to afford higher debt costs in the process of debt financing. The above theoretical analysis led to the following hypothesis:

Hypothesis 1a (H1a). *Higher concentrations of air pollution cause firms to bear a higher cost of debt.*

Hypothesis 1b (H1b). *Environmental violation risk is the mechanism for the effect of air pollution on the cost of debt, that is, air pollution increases the cost of debt through its effect on environmental violation risk.*

2.2. Moderating Role of Environmental Regulatory Pressure

The original pressure for firms to carry out environmental management comes from strict governmental regulations, and its purpose is to maintain legitimacy [31–33]. Strict environmental policies threaten the legitimacy of polluting firms and bring administrative penalties or legal proceedings against polluting firms [34,35], which bring losses to the reputation and market valuation of firms [36]. During the 11th five-year plan period, Chinese local governments continued to strengthen the implementation of environmental policies, which was represented by the inclusion of regional air pollution control in the assessment system of local governments [17,37]. Since 2014, the Ministry of Environmental Protection of China has implemented a new policy of interviewing local government managers. In 2018, the State Council issued the three-year action plan for winning the Blue Sky Defense Campaign, focusing on strengthening the supervision of regional air quality. As a result of a stricter environmental regulation enforcement, the environmental violations of firms due to poor environmental performance have gradually attracted serious attention from investors and have begun to affect investors' assessments of corporate risk [38]. We speculate that the environmental violation risk related to air pollution is higher for firms located in these regions with strict environmental supervision. Creditors are more sensitive to environmental regulatory compliance, environmental violation risk, and the legitimacy status of firms subject to strict environmental supervision. In particular, OECD countries have had a strong risk awareness of climate change for a long time, which has prompted them to adopt strict environmental regulations in response to the threat of climate change [39]. Firms in OECD countries may face a higher environmental violation risk related to air pollution, which results in higher costs of debt.

Hypothesis 2 (H2). *The effect of air pollution on the cost of debt is strengthened by environmental regulatory pressure.*

2.3. Moderating Role of Firms' Industry Characteristics

Prior studies have demonstrated that industry characteristics are a critical factor affecting the investors' attention to firms' environmental issues. Firms in environmentally sensitive industries have a higher tendency to engage in polluting activities and noncompliance with environmental regulations, and investors are more likely to perceive the environmental risk characteristics of these firms [38]. In particular, the announcement of new environmental regulations reduces investors' aspiration to invest in heavily polluting stocks, which eventually leads to a relatively poor stock return performance of heavily polluting firms in a short period [11]. Konar and Cohen [40] demonstrated that, driven by environmental responsibility, some investors have consciously penalized heavily polluting firms by raising the cost of capital. Thus, we expect that when cities are covered in heavy air pollution, creditors would pay more attention to the environmental problems of heavily polluting firms in these areas and worry about the risk of environmental violations of these heavily polluting firms.

Hypothesis 3 (H3). *The effect of air pollution on the cost of debt is strengthened by heavily polluting firms.*

2.4. Moderating Role of Firms' Ownership

Under the special circumstances of China, the ownership of firms is often regarded as an important factor affecting environmental governance issues. In China, state-owned firms often belong to important industries related to the national economy and the people's livelihood and bear the important functions of maintaining national security, economic stability, industrial leadership, and public services. In order to ensure the state-owned economy's control over these key industries, the government must hold a certain proportion of shares in these key firms and appoint specific officials as senior managers of state-owned firms [41]. Thus, ownership is regarded as a strong political link between firms and government [42]. Political connection not only brings more advantages to state-owned firms, such as more investment funds, tax incentives, lower financing constraints, and capital cost [43–46], but also brings some obstacles to the implementation of environmental supervisions by local governments [17,38]. When local governments implement environmental supervision, the political connection can always protect state-owned firms from punishments due to environmental violations to a great extent. Therefore, the regional political protection of heavy polluters has often been accused of being a barrier for the enforcement of environmental regulations [47]. Zhou et al. [18] also demonstrated that creditors more carefully assess the carbon risk of private firms and adopt stricter approval standards in the process of reviewing loans. We expect that when air pollution occurs, political connection can play a role as an umbrella to minimize the penalties for environmental violations doled out to state-owned firms, reducing the sensitivity of creditors to the environmental risk of state-owned firms.

Hypothesis 4 (H4). *The effect of air pollution on the cost of debt is weakened by state-owned firms.*

2.5. Moderating Role of Firms' Economic Contribution

For a long time, it has generally been acknowledged that the importance of economic development is higher than environmental protection in Chinese society, which has led to weak economic punishments for environmental violations among Chinese firms [48]. Under the current administrative system in China, environmental policies are mainly formulated by the central government, while the implementation functions of environmental policies are mostly carried out by local governments, which are the main body of environmental governance and supervision. However, in the process of implementing environmental policies, local governments often fail to supervise the environmental violations of local firms. As Wang and Wheeler [49] noted, the bargaining power of firms is an important factor affecting the effectiveness of governments' environmental supervision. The tax contributions from firms' business activities are conducive to promoting local economic growth and realizing the promotion goals of local officials. If a firm is a major contributor to local economic development, it may have stronger bargaining power when it comes to environmental supervision. Therefore, under the condition of serious information asymmetry in environmental problems, for the consideration of economic development and political performance objectives, local managers are likely to tolerate the environmental violations of heavily polluting firms that provide significant contributions to local economies and taxation at the expense of environmental protection [17]. We expect that an economically important firm has stronger negotiating abilities when it comes to environmental supervision, effectively reducing the administrative penalty they may suffer in air pollution events and reducing the sensitivity of creditors to the environmental violation risk.

Hypothesis 5 (H5). *The effect of air pollution on the cost of debt is weakened by firms' economic contribution.*

3. Research Design

3.1. Sample Selection

We selected a sample of Chinese listed firms from the main board of the SSE and SZSE from 2014 to 2018. The SSE and SZSE are the distinguished stock exchanges in China. Listed firms from the main board of the SSE and SZSE are generally characterized by a high market share, large scale, and strong comprehensive strength, and their environmental performance and environmental violations are more susceptible to the attention of government regulators, creditors, and other stakeholders. Consequently, these firms served as the focus for our research to analyze the possible relationship between air pollution and the cost of debt. The reason for selecting this time period was that China's air quality evaluation system has changed to a great extent since 2014. Therefore, this paper selected the air quality data of 2014 and later to ensure the consistency of the calculation caliber of air pollution degree. Air pollution data of cities in China and financial data of the listed firms were collected from the China Stock Market and Accounting Research (CSMAR) database. We obtained related macroeconomic data, including the regional GDP growth rate and proportion of secondary industry, from the China Statistical Yearbooks. Furthermore, regional climate data, including wind speed and annual rainfall, were manually collected from the meteorological database on the website of greenhouse data sharing platform and the Statistical Yearbooks of China's provinces. Data related to firms' environmental violations were manually collected from the firms' environmental supervision records on the website of the Institute of Public and Environmental Affairs (IPE). Next, we eliminated the sample firms that met the following conditions according to the conventional practice to ensure the reliability of the research results: (i) financial firms (since the accounting standards of the financial industry are quite different from those of other industries, we excluded financial listed firms from the full sample); (ii) firms with missing key financial data; (iii) firms with serious financial abnormalities or delisting risk (ST or *ST). We ultimately retained 3993 firm-year observations after a prudent screening process. Finally, all continuous variables were winsorized at the 1% level to eliminate the effects of extreme values.

3.2. Main Variable Descriptions

3.2.1. Dependent Variable: Cost of Debt (Cost)

In terms of the measurement of the core dependent variable, we used the firm's interest expenditure rate as the proxy of the cost of debt similar to prior studies on the listed firms, which was measured as the total interest expense divided by the average interest-bearing debt [18,50–52].

3.2.2. Independent Variable: Air Pollution (Air)

Before 2013, the Ministry of Environmental Protection of China (MEPC) issued the air pollution index (API), which is an index for calculating and comprehensively evaluating air quality conditions according to the concentration of five main pollutants, such as PM_{10} [1]. Subsequently, after 2013, a severe haze occurred in many cities in China, but the API did not contain the index of $PM_{2.5}$, which is the main pollutant component of haze. Therefore, the MEPC used the air quality index (AQI) to replace the original air pollution index (API). According to the technical regulations of ambient AQI issued by the MEPC, the AQI is evaluated based on six atmospheric pollutants, including SO_2, NO_2, PM_{10}, $PM_{2.5}$, CO, and O_3, which can more comprehensively evaluate the air quality of cities in China. Therefore, similar to prior studies on the Chinese atmospheric pollution [1,8,53], we used the mean value of the daily AQI of the cities where the firms were located in the current year to comprehensively measure the degree of air pollution. Larger values of average daily AQI indicated that firms were exposed to more severe air pollution.

3.2.3. Mediating Variable: Environmental Violation Risk (Violation)

The variable violation refers to the environmental violation risk of firms. Although Dobler et al. [54] and Zhou et al. [18] used ordered variables to measure the environmental violation risk of firms, we still insist that this measurement is too simple to accurately and comprehensively evaluate firms' environmental violation risk, which may lead to biased research conclusions. Some local governments in China have issued the measures for the environmental credit evaluation of firms, which recorded detailed scoring standards for environmental violations of firms, including information on environmental violations and their corresponding scoring. Therefore, combined with the scoring standard of firms' environmental violations attached to the firms' environmental credit evaluation criteria issued by China's local governments, we constructed an assessment system of corporate environmental violation risk as represented in detail in Table 1. The specific process of calculating the environmental violation risk of firms in this research was as follows: First, we summed up the punishment scores corresponding to each environmental violation of a firm in the current year to obtain the severity of environmental violations (Severity). In order to more accurately identify the impact of local air pollution concentration on the environmental violation risk of firms, we excluded the records of environmental violations of affiliated firms whose regions were inconsistent with where the listed firms' offices were located. Secondly, Ln (Severity + 1) was calculated to measure the firms' environmental violation risk (Violation).

Table 1. Assessment system of firms' environmental violation risk.

Items	Category of Punishment for Environmental Violations		Score Value
1	Warning		1
2	Order to make corrections or make corrections within a time limit		1
3	Penalty	The penalty is less than 10,000 yuan	1
		The penalty is more than 10,000 yuan and less than 50,000 yuan	2
		The penalty is more than 50,000 yuan and less than 100,000 yuan	3
		The penalty is more than 100,000 yuan and less than 200,000 yuan	4
		The penalty is more than 200,000 yuan	6
4	Order to stop construction	Construction projects of registration form	3
		Construction projects of report form	6
		Construction projects of report	12
5	Order to restrict production		6
6	Order to stop production for rectification		12
7	Seal up and detain		6
8	Confiscation of illegal income and illegal property		6
9	Temporary seizure of permits or other documents		6
10	Revocation of licenses or other certificates		12
11	Environmental violation cases of administrative detention		12
12	Cases suspected of environmental crimes		12

3.2.4. Control Variables

Referring to Zhou et al. [18] and Shailer and Wang [51], we considered a series of control variables that were used in the literature and were related to the cost of debt. First, we controlled for size (Size), which was calculated as the natural logarithm of the firm's total assets. Larger firms may have more mortgage assets, which means they have a stronger risk resistance and lower debt cost. The second control variable was ownership of firms (State). State-owned firms usually have debt guaranteed and financial support provided by the government, and their default risk and debt cost are correspondingly lower. The third control variable was leverage (Lev), which was calculated as the ratio of total debt to total assets. A higher asset liability ratio means a higher possibility of corporate debt default, and creditors need more risk premiums as compensation. The fourth control variable was the interest coverage ratio (Ic), which was calculated as the ratio of earnings before interest and tax (EBIT) to interest expense. The interest coverage ratio represents the firm's ability to pay interest, and the interest coverage ratio is negatively associated with the cost of debt. The fifth control variable was the fixed assets ratio (Fix), which was calculated as the ratio of total fixed assets to total assets. Firms with more fixed assets usually have a lower asset liquidity, lower capital turnover rates, and weaker operating capacities, which result in a higher cost of debt financing. The sixth control variable is the rate of return on total assets (Roa), which represents the profitability of firms. Firms with a higher Roa have a stronger profitability and solvency. Thus, Roa is negatively associated with corporate debt cost. The seventh control variable was growth opportunity (Growth), which was measured as the revenue growth rate. Firms with higher growth opportunities are generally expected to have a higher default risk. The eighth control variable was the operating cash flow of firms (Cfo), which was calculated as the ratio of operating cash flow to total assets. Firms with an adequate cash flow from operations usually have a lower debt cost. We also controlled some regional macroeconomic variables expected to impact the cost of debt: the provincial GDP growth rate (Gdp) and proportion of provincial secondary industry (Second). Specifically, the pursuit of GDP growth is an important reason for the government to reduce environmental supervisions for firms. At the same time, the proportion of secondary industry is a reflection of industrialization, which is typically accompanied by the environmental deregulation of local governments [55]. From this, GDP growth and the proportion of secondary industry may be associated with the environmental violation risk of firms and could ultimately affect the cost of debt. Furthermore, we also considered the possible impact of annual, industrial, and provincial fixed effects on the empirical results. Detailed variable definitions are listed in Table 2.

Table 2. Definitions of variables.

Variables	Definition
Dependent variable	
Cost	Cost of debt, measured as the ratio of total interest expense to average interest-bearing debt
Independent variable	
Air	Air pollution, measured as the mean value of the daily AQI of the cities where the firms were located
Mediating variable	
Violation	Environmental violation risk of firms, measured as Ln (Severity + 1)
Control variables	
Size	Natural logarithm of total assets at the end of year
State	0 if the affiliation of the actual controller of a firm is the state, and 1 otherwise
Lev	Ratio of total liabilities to total assets at the end of the year
Ic	Interest coverage ratio, measured as the ratio of EBIT to interest expense
Fix	Ratio of total fixed assets to total assets at the end of year
Roa	Return on assets, measured as the ratio of EBIT to average total assets
Growth	Growth rate of a firm's total revenue from year t − 1 to year t
Cfo	Ratio of net cash flow from operating activities to total assets at the end of year
Gdp	Growth rate of a province's total GDP from year t − 1 to year t
Second	Proportion of secondary industry in the province where the firms were located

Notes: AQI was the air quality index; Severity was the severity of environmental violations; EBIT was the earnings before interest and tax; GDP was the gross domestic product.

3.3. Research Model

In order to verify the main theoretical hypothesis that air pollution surrounding cities where the firms were located increased their cost of debt, we used the OLS method to construct the multiple linear regression model seen in Equation (1). The principal coefficient of interest was α_1, which was expected to be positive. In the robust test, endogeneity-corrected regression methods (two-stage least squares regressions and the propensity score matching method) were used.

$$\begin{aligned} Cost_{i,t} = {} & \alpha_0 + \alpha_1 Air_{i,t} + \alpha_2 Size_{i,t} + \alpha_3 State_{i,t} + \alpha_4 Lev_{i,t} + \alpha_5 Ic_{i,t} + \alpha_6 Fix_{i,t} + \alpha_7 Roa_{i,t} \\ & + \alpha_8 Growth_{i,t} + \alpha_9 Cfo_{i,t} + \alpha_{10} Gdp_{i,t} + \alpha_{11} Second_{i,t} + \sum Year + \sum Industry + \sum Province + \varepsilon \end{aligned} \quad (1)$$

According to the previous theoretical analysis, we conjectured that air pollution increases the negative expectations of the public for atmospheric quality, which then encourages the government regulatory authorities to formulate stricter environmental regulatory policies and punish the environmental violations of firms. Firms exposed to heavy air pollution are confronted with more severe environmental violation risks, which increases the debt cost of firms. Referring to the mediating effect test of Baron and Kenny [56], we constructed Equations (2) and (3) by the OLS method to test whether air pollution increased the cost of debt through its effect on environmental violation risk.

$$\begin{aligned} Violation_{i,t} &= \alpha_0 + \alpha_1 Air_{i,t} + \alpha_2 Size_{i,t} + \alpha_3 State_{i,t} + \alpha_4 Lev_{i,t} + \alpha_5 Ic_{i,t} + \alpha_6 Fix_{i,t} + \alpha_7 Roa_{i,t} \\ &+ \alpha_8 Growth_{i,t} + \alpha_9 Cfo_{i,t} + \alpha_{10} Gdp_{i,t} + \alpha_{11} Second_{i,t} + \sum Year + \sum Industry + \sum Province + \varepsilon \end{aligned} \quad (2)$$

$$\begin{aligned} Cost_{i,t} &= \alpha_0 + \alpha_1 Air_{i,t} + \alpha_2 Violation_{i,t} + \alpha_3 Size_{i,t} + \alpha_4 State_{i,t} + \alpha_5 Lev_{i,t} + \alpha_6 Ic_{i,t} \\ &+ \alpha_7 Fix_{i,t} + \alpha_8 Roa_{i,t} + \alpha_9 Growth_{i,t} + \alpha_{10} Cfo_{i,t} + \alpha_{11} Gdp_{i,t} + \alpha_{12} Second_{i,t} + \sum Year \\ &+ \sum Industry + \sum Province + \varepsilon \end{aligned} \quad (3)$$

In order to further test the moderating role of environmental regulatory pressure on the relationship between air pollution and the cost of debt, we introduced the intersection term of environmental regulatory pressure and air pollution as shown in Equation (4). Since 2008, the IPE and NRDC (Natural Resources Defense Council) have continuously published the Pollution Information Transparency Index in China (PITI), which is an index system for evaluating the information disclosure of pollution source supervision in key environmental protection cities in China. Specifically, the PITI takes more than 100 critical cities in China as the evaluation object every year and comprehensively scores and evaluates the environmental supervision pressure of these cities according to eight items, such as the publicity of the daily exceeding standard of pollution sources and illegal record information, the publicity of centralized remediation information of pollution sources, and the publicity of cleaner production audit information. As such, we used the PITI to measure the environmental regulatory pressure suffered by the sample firms. The coefficient α_3 in Equation (4) was expected to be significantly positive if Hypothesis 2 was confirmed.

$$\begin{aligned} Cost_{i,t} &= \alpha_0 + \alpha_1 Air_{i,t} + \alpha_2 PITI_{i,t} + \alpha_3 AQI \times PITI_{i,t} + \alpha_4 Size_{i,t} + \alpha_5 State_{i,t} + \alpha_6 Lev_{i,t} + \alpha_7 Ic_{i,t} \\ &+ \alpha_8 Fix_{i,t} + \alpha_9 Roa_{i,t} + \alpha_{10} Growth_{i,t} + \alpha_{11} Cfo_{i,t} + \alpha_{12} Gdp_{i,t} + \alpha_{13} Second_{i,t} + \sum Year \\ &+ \sum Industry + \sum Province + \varepsilon \end{aligned} \quad (4)$$

In order to further test the moderating role of firms' industry characteristics on the relationship between air pollution and the cost of debt, we introduced the intersection term of firms' industry characteristics and air pollution as shown in Equation (5). Specifically, we extracted the list of heavily polluting firms according to the classified management directory of environmental protection verification industry of listed firms issued by the MEPC. Next, we created a classified variable, Polluted, which represented the firms' industry characteristics and assigned it to 1 if the firm belonged to a heavily polluting industry, while the others were 0. The coefficient α_3 in Equation (5) was expected to be significantly positive if Hypothesis 3 was confirmed.

$$\begin{aligned} Cost_{i,t} &= \alpha_0 + \alpha_1 Air_{i,t} + \alpha_2 Polluted_{i,t} + \alpha_3 Air \times Polluted_{i,t} + \alpha_4 Size_{i,t} + \alpha_5 State_{i,t} + \alpha_6 Lev_{i,t} \\ &+ \alpha_7 Ic_{i,t} + \alpha_8 Fix_{i,t} + \alpha_9 Roa_{i,t} + \alpha_{10} Growth_{i,t} + \alpha_{11} Cfo_{i,t} + \alpha_{12} Gdp_{i,t} + \alpha_{13} Second_{i,t} + \sum Year \\ &+ \sum Industry + \sum Province + \varepsilon \end{aligned} \quad (5)$$

In order to further test the moderating role of firms' ownership on the relationship between air pollution and the cost of debt, we introduced the intersection term of firms' ownership and air pollution as shown in Equation (6). As mentioned above, we assigned the variable State to 0 if the affiliation of the actual controller of a firm was the state, and 1 otherwise. The coefficient α_3 in Equation (6) was expected to be significantly positive if Hypothesis 4 was confirmed.

$$\begin{aligned}Cost_{i,t} &= \alpha_0 + \alpha_1 Air_{i,t} + \alpha_2 State_{i,t} + \alpha_3 Air \times State_{i,t} + \alpha_4 Size_{i,t} + \alpha_5 Lev_{i,t} + \alpha_6 Ic_{i,t} \\ &+ \alpha_7 Fix_{i,t} + \alpha_8 Roa_{i,t} + \alpha_9 Growth_{i,t} + \alpha_{10} Cfo_{i,t} + \alpha_{11} Gdp_{i,t} + \alpha_{12} Second_{i,t} + \sum Year \\ &+ \sum Industry + \sum Province + \varepsilon\end{aligned} \tag{6}$$

In order to further test the moderating role of firms' economic contribution on the relationship between air pollution and the cost of debt, we introduced the intersection term of firms' economic contribution and air pollution as shown in Equation (7). Following Liu et al. [17], we measured the economic contribution of a firm by its tax expenditure as a percentage of the total tax revenue in that province. The variable Tax refers to a firm's economic contribution. The coefficient α_3 in Equation (7) was expected to be significantly negative if Hypothesis 5 was confirmed.

$$\begin{aligned}Cost_{i,t} &= \alpha_0 + \alpha_1 Air_{i,t} + \alpha_2 Tax_{i,t} + \alpha_3 Air \times Tax_{i,t} + \alpha_4 Size_{i,t} + \alpha_5 State_{i,t} + \alpha_6 Lev_{i,t} \\ &+ \alpha_7 Ic_{i,t} + \alpha_8 Fix_{i,t} + \alpha_9 Roa_{i,t} + \alpha_{10} Growth_{i,t} + \alpha_{11} Cfo_{i,t} + \alpha_{12} Gdp_{i,t} + \alpha_{13} Second_{i,t} \\ &+ \sum Year + \sum Industry + \sum Province + \varepsilon\end{aligned} \tag{7}$$

4. Empirical Results

4.1. Descriptive Statistics

Table 3 presents a sample distribution by the provinces where the firms were located and mean values of the AQI during our sample period. In our research sample, the province with the largest number of sample firms was Beijing, followed by Zhejiang and Guangdong. The province with the worst air quality was Hebei, with a mean AQI of 124.1973 during the sample period, and the province with the best air quality was Hainan, with a mean AQI of 46.4630 during the sample period. Combined with the technical regulation on ambient AQI issued by the MEPC in 2012, an AQI exceeding 100 was considered polluted, indicating that some central and eastern provinces in China had an unhealthy air quality.

Table 3. Air pollution conditions of the provinces where the sample firms were located.

Province	Mean AQI	N	Province	Mean AQI	N
Hebei	124.1973	99	Hunan	85.2114	111
Henan	115.0412	146	Jilin	84.2591	77
Beijing	111.4179	354	Inner Mongolia	83.9621	59
Shaanxi	109.8092	69	Heilongjiang	82.3223	70
Tianjin	107.3100	60	Chongqing	82.2441	80
Xinjiang	105.8833	84	Shanghai	82.0048	307
Shandong	104.0479	267	Zhejiang	78.8685	351
Shanxi	101.2360	108	Jiangxi	71.6547	72
Gansu	97.1313	49	Guangxi	66.4132	60
Hubei	96.4351	150	Guizhou	63.4468	44
Ningxia	96.0047	29	Tibet	63.0077	16
Jiangsu	90.9829	304	Guangdong	62.9685	343
Sichuan	90.9101	176	Fujian	57.5179	101
Qinghai	86.8484	32	Yunnan	56.5209	61
Anhui	86.8321	146	Hainan	46.4630	34
Liaoning	86.2856	134	Total	88.7361	3993

Notes: Mean AQI was the mean values of the daily air quality index during the sample period.

Table 4 shows the descriptive statistical results of the main research variables. The cost of debt financing varied from 0.0004 to 0.0773, which showed that there were prominent differences in debt financing cost between different firms from the sample. Moreover, the mean and median values of debt financing cost from our sample were 0.0320 and 0.0314, respectively, which were much lower than 0.065 and 0.061 [18], indicating that the debt cost of high-carbon firms was higher to a certain extent. The mean value of our sample cities' daily AQI was 88.7361, indicating that the air quality of the cities where most of the sample firms were located was good according to the MEPC. However, the maximum value of AQI was 142, indicating that some firms were still exposed to a polluted atmosphere. In addition, the mean value of the leverage (Lev) was 0.5173, confirming that most of the firms from the sample preferred debt financing.

Table 4. Variable descriptive statistics.

Variables	Obs.	Mean	Std. Dev.	Min	Median	Max
Cost	3993	0.0320	0.0157	0.0004	0.0314	0.0773
Air	3993	88.7361	20.9627	46.4630	86.7135	142.000
Size	3993	22.9095	1.3719	20.1914	22.7688	26.8076
State	3993	0.3869	0.4871	0	0	1
Lev	3993	0.5173	0.1824	0.1207	0.5179	0.9450
Ic	3993	16.2714	49.4600	−16.7000	4.0300	388.0000
Fix	3993	0.2967	0.1916	0.0103	0.2673	0.7888
Roa	3993	0.0520	0.0562	−0.1487	0.0468	0.2313
Growth	3993	0.1625	0.4536	−0.4920	0.0849	3.0500
Cfo	3993	0.0474	0.0626	−0.1325	0.0459	0.2331
Gdp	3993	0.0769	0.0432	−0.2240	0.0823	0.1459
Second	3993	0.4104	0.0874	0.1863	0.4399	0.5313

Notes: Cost was the cost of debt; Air was the air pollution; Size was the size of firms; State was the ownership of firms; Lev was the leverage of firms; Ic was the interest coverage ratio of firms; Fix was the fixed assets ratio of firms; Roa was the return on total assets; Growth was the revenue growth rate of firms; Cfo was the operating cash flow of firms; Gdp was the provincial GDP growth rate; Second was the proportion of provincial secondary industry.

The results in Table 5 show that the highest correlation coefficient was 0.4049 between Roa and Cfo. In addition, given that the coefficients were all less than 0.8, it could be considered that there was no serious multicollinearity among regression model variables.

Table 5. Pearson correlations.

Variables	Cost	Air	Size	State	Lev	Ic	Fix	Roa	Growth	Cfo	Gdp	Second
Cost	1.0000											
Air	0.0084	1.0000										
Size	0.0734 ***	0.1175 ***	1.0000									
State	−0.0101	0.1862 ***	0.2975 ***	1.0000								
Lev	0.1345 ***	0.1202 ***	0.3795 ***	0.2388 ***	1.0000							
Ic	0.3158 ***	−0.0232	0.0753 ***	0.0651 ***	0.2649 ***	1.0000						
Fix	0.2994 ***	0.0391 **	0.1378 ***	0.2073 ***	0.0701 ***	0.1231 ***	1.0000					
Roa	−0.0401 **	0.0476 ***	0.0584 ***	0.1450 ***	0.3310 ***	0.3035 ***	−0.0321 **	1.0000				
Growth	0.0214	−0.0240	0.0036	0.0954 ***	0.0462 ***	0.0508 ***	0.0870 ***	0.2420 ***	1.0000			
Cfo	0.0793 ***	−0.0224	0.1299 ***	−0.0197	0.1379 ***	0.1313 ***	0.3237 ***	0.4049 ***	−0.0109	1.0000		
Gdp	0.0827 ***	0.0571 ***	−0.0135	0.0577 ***	0.0624 ***	0.0045	0.0800 ***	0.0610 ***	0.0308 *	0.0162	1.0000	
Second	0.0784 ***	0.0871 ***	0.2299 ***	0.0801 ***	−0.0208	−0.0312 **	0.1013 ***	0.0188	−0.0089	0.0128	−0.0265 *	1.0000

Notes: The explanation for abbreviations of all variables used inside Table 5 were mentioned in the footer of Table 4. *, **, and *** reflect p-values of the correlation coefficients between variables were less than 0.1, 0.05, and 0.01, respectively.

4.2. Regression Results of the Influence of Air Pollution on the Cost of Debt

Columns (1) and (2) of Table 6 present the regression results of the influence of air pollution on the cost of debt. Column (1) shows the regression results without control variables of firms' characteristics and regional economic characteristics. As predicted in the prior hypothesis, the coefficient of air pollution (Air) in column (1) was positive and statistically significant at the 5% level, supporting our hypothesis that firms exposed to more serious air pollution in their cities have to bear higher debt costs. Column (2) added control variables representing firms' characteristics and regional economic characteristics. The coefficient of Air in column (2) still remained statistically significant at the 1% level, demonstrating that air pollution was significantly positively associated with the cost of debt (Cost) after firms' characteristics and regional economic characteristics were considered. The positive effect of air pollution surrounding the cities where the firms were located on the cost of debt demonstrated that firms exposed to more serious air pollution were more likely to face a higher risk of environmental violations, which correspondingly increased the default risk of firms. Under such circumstances, creditors can only raise the debt cost of firms to reduce air pollution-related default losses. Thus, the empirical results supported Hypothesis 1a. It is worth noting that the cost of debt includes multiple influencing factors that cannot be captured completely. Based on the existing literature, the control variables in our research models largely included the factors that may affect the cost of debt. The adjusted-R^2 value in model (2) equaled 0.3318, which was similar to the 0.334 of Tan et al. [16] and the 0.373 of Zhou et al. [18], and was slightly higher than the 0.283 of Jung et al. [10] and the 0.105 of Chen et al. [9]. It indicated that the fitting degree of the model was acceptable. In addition, we also employed an analysis of variance to evaluate the overall goodness-of-fit of models (1) and (2). The results of the analysis of variance in Table 7 revealed that the F-statistics of models (1) and (2) were all significant at the 1% level, which indicated that the regression models had statistical significance as a whole.

Table 6. Influence of air pollution on cost of debt and mechanism of environmental violation risk.

Variables	(1) Cost	(2) Cost	(3) Violation	(4) Cost
Air	0.0000 ** (2.42)	0.0001 *** (3.02)	0.0044 *** (3.38)	0.0001 *** (2.73)
Violation				0.0006 ** (2.06)
Size		0.0003 (1.60)	0.0834 *** (6.01)	0.0002 (0.69)
State		0.0036 *** (7.06)	−0.0867 ** (−2.43)	0.0044 *** (7.91)
Lev		0.0085 *** (5.90)	0.0425 (0.44)	0.0084 *** (5.50)
Ic		−0.0001 *** (−17.28)	−0.0001 (−0.19)	−0.0001 *** (−14.71)
Fix		0.0119 *** (7.63)	0.4438 *** (4.10)	0.0104 *** (6.11)
Roa		0.0078 (1.64)	−0.1507 (−0.46)	0.0093 * (1.82)
Growth		0.0025 *** (5.32)	−0.0087 (−0.27)	0.0027 *** (5.41)
Cfo		0.0033 (0.83)	0.0770 (0.28)	0.0058 (1.35)
Gdp		−0.0015 (−0.23)	0.2689 (0.61)	−0.0015 (−0.21)
Second		0.0021 (0.13)	−0.1988 (−0.18)	0.0076 (0.44)
_Cons	0.0205 *** (4.00)	0.0030 (0.29)	−1.0131 (−1.43)	0.0011 (0.10)
Year	Yes	Yes	Yes	Yes
Industry	Yes	Yes	Yes	Yes
Province	Yes	Yes	Yes	Yes
F-statistics	15.57 ***	22.08 ***	9.22 ***	18.76 ***
Adjusted-R^2	0.2346	0.3318	0.1832	0.3288
N	3993	3993	3372	3372
Sobel-test			1.7600 *	

Notes: Columns (1) and (2) represent the regression results on the influence of air pollution on the cost of debt. Columns (3) and (4) represent the regression results on the mechanism of environmental violation risk. Violation was the environmental violation risk of firms. The explanation for abbreviations of other variables used inside Table 6 were mentioned in the footer of Table 4. The numbers in brackets are *t*-values. *, **, and *** reflect *p*-values of the coefficients were less than 0.1, 0.05, and 0.01, respectively.

Table 7. Analysis of variance (ANOVA).

Model (1)					
Source	**SS**	**df**	**MS**	**F-Statistics**	**Sig.**
Model	0.2452	84	0.0029	15.57	0.0000
Residual	0.7329	3908	0.0002		
Total	0.9781	3992	0.0002		
Model (2)					
Source	**SS**	**df**	**MS**	**F-Statistics**	**Sig.**
Model	0.3399	94	0.0036	22.08	0.0000
Residual	0.6383	3898	0.0002		
Total	0.9781	3992	0.0002		

Consistent with Shailer and Wang [51], the coefficient of firms' ownership (State) was statistically significant, confirming that the ownership of firms is a critical factor affecting the cost of debt in the Chinese context. In other words, the cost of debt financing of firms under government control was significantly lower than that of firms under private control. The coefficients for Lev, Ic, Fix, and Growth were all significant at the 1% level, confirming that firms with higher leverage, a lower interest coverage, higher fixed assets ratio, and higher growth opportunities were more likely to be able to afford higher costs of debt.

4.3. Mechanism of Environmental Violation Risk

Columns (3) and (4) of Table 6 reflect the regression results on the mechanism of environmental violation risk. The coefficient of Air in column (3) of Table 6 was positive and statistically significant at the 1% level, confirming that air pollution surrounding the cities where the firms were located significantly increased the firm's environmental violation risk (Violation). The coefficient of Violation in column (4) of Table 6 was also significantly positive, which meant that air pollution increased the firm's cost of debt through its effect on the environmental violation risk. Therefore, the mediating effect test confirmed Hypothesis 1b (environmental violation risk was the mechanism of air pollution affecting the cost of debt). In addition, the coefficient of Air in column (4) remained significant, indicating that the environmental violation risk was the partial mediating variable between air pollution and the cost of debt. To guarantee the reliability of the conclusion, our research also carried out a Sobel test. The Z statistic was 1.7600 and was significant at the level of 10%, which proved the inference of the mediating effect of environmental violation risk again.

4.4. Robustness Test

4.4.1. Sensitivity Test for Measurement of Air Pollution and the Cost of Debt

In the previous regression analysis, we used the AQI to measure the air pollution concentrations of the cities where the firms were located; here, we needed to use other indicators to replace the AQI so as to ensure the robustness of the regression results. Firstly, considering that $PM_{2.5}$ was the most important component of atmospheric pollutants, we used the method developed by Chen et al. [9] to measure the air pollution by average $PM_{2.5}$ concentration (PM_2.5) in cities where the firms were located. Next, MEPC divided the AQI once into six levels according to the severity of pollution: excellent ($0 < AQI \leq 50$), good ($50 < AQI \leq 100$), slightly polluted ($100 < AQI \leq 150$), moderately polluted ($150 < AQI \leq 200$), heavily polluted ($200 < AQI \leq 300$), and severely polluted ($AQI > 300$). Therefore, referring to Dong et al. [8], we changed the AQI into a classified variable (AQI_grade) according to the classification standard of MEPC for the air quality grade. We employed PM_2.5 and AQI_grade to remeasure air pollution. The results of the sensitivity test of air pollution are presented in columns (1) and (2) of Table 8. The coefficients of PM_2.5 and AQI_grade were 0.0001 and 0.0016, which were significantly positively associated with the Cost, and they were coincident with the results shown in

Table 6. This showed that our inference remained robust after replacing the measurement of air pollution.

Table 8. Sensitivity test for measurement of air pollution and the cost of debt.

Variables	(1) Cost	(2) Cost	(3) Expense	(4) Expense	(5) Expense
Air			0.0000 *** (2.83)		
PM_2.5	0.0001 *** (3.57)			0.0001 *** (3.25)	
AQI_grade		0.0016 ** (2.39)			0.0015 *** (2.72)
Size	0.0003 * (1.79)	0.0003 (1.61)	0.0003 ** (2.15)	0.0004 ** (2.19)	0.0003 ** (2.16)
State	0.0039 *** (7.44)	0.0036 *** (6.94)	0.0034 *** (8.12)	0.0037 *** (8.53)	0.0034 *** (8.03)
Lev	0.0074 *** (5.00)	0.0085 *** (5.90)	0.0101 *** (8.59)	0.0096 *** (7.84)	0.0101 *** (8.58)
Ic	−0.0001 *** (−17.05)	−0.0001 *** (−17.34)	−0.0001 *** (−18.09)	−0.0001 *** (−17.84)	−0.0001 *** (−18.16)
Fix	0.0115 *** (7.28)	0.0119 *** (7.59)	0.0136 *** (10.62)	0.0130 *** (10.01)	0.0136 *** (10.59)
Roa	0.0068 (1.40)	0.0079 * (1.67)	0.0051 (1.30)	0.0042 (1.05)	0.0052 (1.33)
Growth	0.0026 *** (5.28)	0.0026 *** (5.36)	0.0017 *** (4.38)	0.0017 *** (4.26)	0.0017 *** (4.42)
Cfo	0.0032 (0.78)	0.0033 (0.83)	0.0008 (0.26)	0.0012 (0.35)	0.0009 (0.26)
Gdp	−0.0020 (−0.31)	−0.0009 (−0.14)	−0.0018 (−0.33)	−0.0019 (−0.35)	−0.0015 (−0.28)
Second	0.0066 (0.40)	−0.0001 (−0.01)	−0.0003 (−0.03)	0.0013 (0.10)	−0.0024 (−0.18)
_Cons	−0.0008 (−0.07)	0.0061 (0.58)	−0.0021 (−0.24)	−0.0034 (−0.39)	−0.0002 (−0.02)
Year	Yes	Yes	Yes	Yes	Yes
Industry	Yes	Yes	Yes	Yes	Yes
Province	Yes	Yes	Yes	Yes	Yes
F-statistics	21.54 ***	22.03 ***	25.80 ***	24.88 ***	25.79 ***
Adjusted-R^2	0.3357	0.3312	0.3687	0.3701	0.3686
N	3822	3993	3993	3822	3993

Notes: Columns (1) and (2) represent the regression results on the sensitivity test of air pollution. Columns (3), (4), and (5) represent the regression results on the sensitivity test of the cost of debt. Cost was the ratio of total interest expense to average interest-bearing debt; Expense was the ratio of total interest expense to average total debt; Air was the average air quality index; PM_2.5 was the average $PM_{2.5}$ concentration; AQI_grade was the grade of average air quality index. The explanation for abbreviations of other variables used inside Table 8 were mentioned in the footer of Table 4. The numbers in brackets are *t*-values. *, **, and *** reflect *p*-values of the coefficients were less than 0.1, 0.05, and 0.01, respectively.

Following previous studies on the Chinese firms [57,58], we applied the total interest expense divided by the average total debt (Expense) as an alternative measurement for the cost of debt and re-estimated Equation (1). The results of the sensitivity test of the cost of debt are presented in columns (3), (4), and (5) of Table 8. The coefficients of Air, PM_2.5, and AQI_grade were all significantly positively associated with the Expense, and they were consistent with those shown in Table 6.

4.4.2. Two-Stage Least Squares (2SLS) Analysis

From a more rigorous perspective, this study may have had some endogenous problems that needed to be solved. The first problem was that there may have been a two-way causal relationship between air pollution and the cost of debt. Previous studies have shown that debt financing can have an impact on corporate performance or corporate value [59–61], and a higher cost of debt capital inevitably increases the operating pressure and earnings pressure of firms. Firms with earnings pressure are more motivated to emit polluting gases [17], which, inevitably, aggravates air pollution. In addition, there may be the problem of missing variables in the research, since there were other potential factors affecting the cost of debt, including the regional legal environment [62–64], monetary policy [65], internal control quality [66], and other factors. Consequently, we employed a 2SLS regression with instrumental variables to mitigate potential endogenous problems.

Past studies have found that air pollution is closely related to meteorological factors, including rainfall, wind speed, wind direction, air temperature, etc. [67,68]. At the same time, these meteorological factors were unlikely to have an influence on the cost of debt. Therefore, rainfall (Rain) and wind speed (Wind) in the cities where the firms were located were selected as instrumental variables of air pollution in this paper. As reflected in column (1) of Table 9, Rain and Wind were both significantly related to Air. Under the 2SLS method, the coefficient of Air in column (2) was still significantly positive, which reflected that the regression results of 2SLS also supported the main assumptions of this paper. Furthermore, in order to examine the effectiveness of instrumental variables, this paper implemented a weak instrumental variable test and over-identification test. The LM statistic was significant at the 1% level, demonstrating that the problem of the underestimation of instrumental variables could be eliminated. The *p*-value of the Sargan statistic was 0.2462, which indicated that there was no overestimation problem either. In addition, the Shea's partial R^2 was 0.0290, and its F statistic and the *p*-value of its F statistic were 36.6104 (more than 10) and 0.0000, respectively. We could, thus, insist that there was no weak instrumental variable. Overall, this result re-confirmed that air pollution enhanced the cost of corporate debt.

Table 9. Endogenous test of 2SLS analysis.

Variables	**(1) Step 1 Air**	**(2) Step 2 Cost**
Air		0.0003 *** (2.75)
Rain	−0.0018 ** (−2.26)	
Wind	−5.2605 *** (−8.38)	
Size	−0.0174 (−0.11)	0.0003 (1.53)
State	−1.9900 *** (−4.79)	0.0042 *** (7.21)
Lev	0.2585 (0.21)	0.0084 *** (5.10)
Ic	0.0001 (0.03)	−0.0001 *** (−14.65)
Fix	−0.6003 (−0.48)	0.0122 *** (6.79)
Roa	0.9495 (0.22)	0.0077 (1.33)
Growth	0.0934 (0.24)	0.0025 *** (3.58)
Cfo	0.1855 (0.06)	0.0033 (0.72)
Gdp	31.2401 *** (5.36)	−0.0095 (−1.28)
Second	−9.8322 (−0.58)	0.0026 (0.15)
_Cons	113.4483 *** (11.68)	−0.0215 (−1.36)
Year	Yes	Yes
Industry	Yes	Yes
Province	Yes	Yes
F-statistics	200.27 ***	
Wald		10187.56 ***
R^2	0.7562	0.3176
N	3993	3993

Notes: Table 9 reflects the results on the endogenous test of 2SLS analysis. Rain was the rainfall of the cities; Wind was the wind speed of the cities. The explanation for abbreviations of other variables used inside Table 9 were mentioned in the footer of Table 4. In particular, Rain and Wind were tool variables used in 2SLS analysis. ** and *** reflect *p*-values of the coefficients were less than 0.05 and 0.01, respectively.

4.4.3. Propensity Score Matching Estimation (PSM)

Finally, for the purpose of alleviating the endogenous problems caused by selection bias, we also used the propensity score matching (PSM) method to construct new samples to re-test our major assumptions. Specifically, we first selected the experimental group and the control group. We classified the samples with an Air value higher than the median Air of the full sample as the experimental group and other samples were classified as the control group. Next, we used the indicators of the firm size, ownership, leverage, interest coverage ratio, fixed assets ratio, return on total assets, growth opportunity, cash flow, provincial GDP growth rate, and proportion of provincial secondary industry to carry out a one-to-one matching of the nearest neighbors within the caliper radius (0.01), screening out the corresponding counterfactual samples, and carried out a regression. After the balanced test, the deviation of the most control variables was greatly reduced after matching, and the t-test results did not reject the original hypothesis that there was no systematic difference between the two groups. The coefficients for Air presented in Table 10 were all positive and significant at the 5% level, which proved once again that the results of the previous regression analysis were robust.

Table 10. Endogenous test of PSM estimation.

Variables	(1) Cost	(2) Cost
Air	0.0001 ** (2.30)	0.0001 ** (2.53)
Size		0.0001 (0.50)
State		0.0043 *** (6.12)
Lev		0.0100 *** (4.98)
Ic		−0.0001 *** (−11.17)
Fix		0.0102 *** (4.62)
Roa		0.0038 (0.57)
Growth		0.0025 *** (3.93)
Cfo		0.0110 ** (2.01)
Gdp		0.0022 (0.24)
Second		0.0069 (0.30)
_Cons	0.0254 *** (3.30)	0.0078 (0.52)
Year	Yes	Yes
Industry	Yes	Yes
Province	Yes	Yes
F-statistics	9.72 ***	12.89 ***
Adjusted-R^2	0.2568	0.3456
N	2095	2095

Notes: Table 10 reflects the results on the endogenous test of PSM estimation. The explanation for abbreviations of all variables used inside Table 10 were mentioned in the footer of Table 4. ** and *** reflect p-values of the coefficients were less than 0.05 and 0.01, respectively.

5. Further Analysis

5.1. The Moderating Role of Environmental Regulatory Pressure

Next, we explored the moderating role of environmental regulatory pressure on the relationship between air pollution and the cost of debt. As reflected in column (1) of

Table 11, the coefficient on Air × PITI was positive and statistically significant at the 1% level. From this result, we concluded that environmental regulatory pressure on firms enhanced the influence of air pollution on the cost of debt. Moreover, we designated the samples into the "high PITI" partition if environmental regulatory pressure on firms was higher than the median PITI in the full sample and re-regressed Equation (1) into "high PITI" and "low PITI". The results showed that the coefficient Air presented a significant positive correlation at the level of 1% in the "high PITI" partition in column (2), while the coefficient Air presented an insignificant correlation in the "low PITI" partition in column (3). The regression results of sub samples also showed that the impact of air pollution on the cost of debt was more significant when firms faced stronger environmental regulatory pressure.

Table 11. Moderating effect of firm's environmental regulatory pressure and industry characteristic.

Variables	(1) Cost	(2) Cost	(3) Cost	(4) Cost	(5) Cost	(6) Cost
	Full Sample	High PITI	Low PITI	Full Sample	High Polluted	Low Polluted
Air	0.0001 *** (3.43)	0.0001 *** (3.26)	0.0000 (0.89)	0.0001 *** (2.76)	0.0001 *** (3.31)	0.0000 (1.32)
Air * PITI	0.0003 *** (2.93)					
Air * Polluted				0.0000 ** (2.25)		
PITI	−0.0029 (−0.96)					
Polluted				0.0184 *** (4.04)		
State	0.0037 *** (7.14)	0.0021 *** (2.91)	0.0051 *** (6.62)	0.0036 *** (7.05)	0.0036 *** (4.92)	0.0036 *** (4.95)
Size	0.0003 (1.51)	0.0003 (1.28)	0.0002 (0.66)	0.0003 (1.51)	0.0004 (1.27)	0.0001 (0.52)
Lev	0.0084 *** (5.85)	0.0100 *** (4.81)	0.0074 *** (3.62)	0.0086 *** (5.95)	0.0109 *** (5.59)	0.0068 *** (3.18)
Ic	−0.0001 *** (−17.31)	−0.0001 *** (−12.05)	−0.0001 *** (−11.91)	−0.0001 *** (−17.19)	−0.0001 *** (−12.28)	−0.0001 *** (−11.68)
Fix	0.0119 *** (7.59)	0.0093 *** (4.14)	0.0150 *** (6.68)	0.0117 *** (7.44)	0.0132 *** (6.19)	0.0088 *** (3.66)
Roa	0.0078 (1.64)	0.0069 (1.03)	0.0120 * (1.77)	0.0077 (1.62)	0.0096 (1.53)	0.0016 (0.22)
Growth	0.0025 *** (5.34)	0.0039 *** (5.92)	0.0014 ** (2.04)	0.0026 *** (5.38)	0.0041 *** (5.89)	0.0011 (1.63)
Cfo	0.0030 (0.76)	−0.0045 (−0.81)	0.0094 * (1.67)	0.0035 (0.88)	0.0108 * (1.95)	−0.0016 (−0.28)
Gdp	−0.0004 (−0.07)	−0.0072 (−0.58)	−0.0037 (−0.41)	−0.0013 (−0.20)	−0.0035 (−0.43)	−0.0007 (−0.07)
Second	0.0077 (0.46)	0.0962 (1.94)	−0.0030 (−0.15)	0.0012 (0.08)	0.0182 (0.86)	−0.0151 (−0.58)
_Cons	0.0057 (0.54)	−0.0273 (−1.02)	0.0105 (0.79)	0.0179 * (1.84)	0.0049 (0.37)	0.0240 (1.57)
Year	Yes	Yes	Yes	Yes	Yes	Yes
Industry	Yes	Yes	Yes	Yes	Yes	Yes
Province	Yes	Yes	Yes	Yes	Yes	Yes
F-statistics	21.78 ***	13.74 ***	11.97 ***	21.93 ***	14.01 ***	14.57 ***
Adjusted-R^2	0.3332	0.3330	0.3329	0.3324	0.2983	0.3317
N	3993	1991	2002	3993	2051	1942

Notes: Columns (1) to (3) of this table represent the results on the moderating role of environmental regulatory pressure. Columns (4) to (6) of this table represent the results on the moderating role of industry characteristic. PITI was the Pollution Information Transparency Index; Polluted was the firms' industry characteristics. The explanation for abbreviations of other variables used inside Table 11 were mentioned in the footer of Table 4. The numbers in brackets are *t*-values. *, **, and *** reflect *p*-values of the coefficients were less than 0.1, 0.05, and 0.01, respectively.

5.2. The Moderating Role of Firms' Industry Characteristics

Next, we tested the moderating role of firms' industry characteristics. As presented in column (4) of Table 11, the coefficient of Air × Polluted was positive and statistically significant at the 5% level, which confirmed Hypothesis 3, which states that heavily polluting firms enhance the influence of air pollution on the cost of debt. Furthermore, we divided the full sample into a high polluted group and a low polluted group and re-regressed Equation (1) in both the high polluted group and the low polluted group. The coefficient on Air in column (5) presented a significant positive correlation at the level of 1% for the high polluted subsample. Conversely, the coefficient on Air in column (6) was statistically insignificant at conventional levels for the low polluted subsample. The regression results of sub samples also suggested that the influence of air pollution on the cost of debt was more pronounced for heavily polluting firms.

5.3. The Moderating Role of Firms' Ownership

Our next test was the moderating role of firms' ownership. As shown in column (1) of Table 12, the coefficient on Air × State was also positive and statistically significant at the 10% level, confirming Hypothesis 4, which states that being a state-owned firm lessens the influence of air pollution on the cost of debt. Furthermore, we divided the baseline sample into state-owned firms versus non-state-owned firms and re-regressed Equation (1) in these subsamples. The coefficient of Air in column (2) was nonsignificant in the state-owned subsample. Comparatively, the coefficient of Air in column (3) was statistically significant at the 1% level in the non-state-owned subsample. These results also confirmed that the influence of air pollution on the cost of debt was more pronounced for non-state-owned firms.

Table 12. Moderating effect of firm's ownership and economic contribution.

Variables	(1) Cost	(2) Cost	(3) Cost	(4) Cost	(5) Cost	(6) Cost
	Full Sample	State-Owned	Non-State-Owned	Full Sample	High Tax Contribution	Low Tax Contribution
Air	0.0001 ***	0.0000	0.0001 ***	0.0001 ***	−0.0000	0.0001 ***
	(3.06)	(1.61)	(2.79)	(2.91)	(−0.38)	(2.60)
Air * State	0.0000 *					
	(1.75)					
Air * Tax				−0.0031 **		
				(−2.48)		
Tax				−0.0887 ***		
				(−2.35)		
State	0.0037 ***			0.0037 ***	0.0030 ***	0.0046 ***
	(7.16)			(7.20)	(4.14)	(6.17)
Size	0.0003	0.000 *	0.0001	0.0006 ***	−0.0004	0.0015 ***
	(1.58)	(1.91)	(0.14)	(2.90)	(−1.30)	(3.64)
Lev	0.0085 ***	0.0079 ***	0.0118 ***	0.0080 ***	0.0133 ***	0.0084 ***
	(5.93)	(4.59)	(4.52)	(5.55)	(6.44)	(4.11)
Ic	−0.0001 ***	−0.0001 ***	−0.0001 ***	−0.0001 ***	−0.0001 ***	−0.0001 ***
	(−17.30)	(−12.08)	(−10.18)	(−17.21)	(−11.86)	(−11.92)
Fix	0.0119 ***	0.0086 ***	0.0208 ***	0.0121 ***	0.0139 ***	0.0126 ***
	(7.62)	(4.67)	(6.75)	(7.78)	(6.49)	(5.31)
Roa	0.0077	0.0146 **	0.0001	0.0080 *	0.0173 **	0.0068
	(1.63)	(2.44)	(0.01)	(1.69)	(2.47)	(1.03)
Growth	0.002 ***	0.0026 ***	0.0023 ***	0.0025 ***	0.0050 ***	0.0004
	(5.25)	(4.47)	(3.00)	(5.18)	(7.99)	(0.50)
Cfo	0.0034	0.0133 ***	−0.0084	0.0044	0.0070	0.0003
	(0.86)	(2.77)	(−1.26)	(1.10)	(1.25)	(0.05)
Gdp	−0.0013	0.0035	−0.0073	−0.0012	−0.0021	−0.0004
	(−0.20)	(0.50)	(−0.54)	(−0.18)	(−0.29)	(−0.03)
Second	0.0021	0.0032	−0.0068	0.0003	0.0238	−0.0231
	(0.13)	(0.17)	(−0.24)	(0.02)	(1.26)	(−0.83)
_Cons	0.0100	0.0192	−0.0108	0.0021	−0.0138	0.0029
	(0.97)	(1.51)	(−0.57)	(0.19)	(−1.00)	(0.16)
Year	Yes	Yes	Yes	Yes	Yes	Yes
Industry	Yes	Yes	Yes	Yes	Yes	Yes
Province	Yes	Yes	Yes	Yes	Yes	Yes
F-statistics	21.89 ***	18.63 ***	9.29 ***	21.86 ***	17.56 ***	10.38 ***
Adjusted-R^2	0.3321	0.3881	0.3235	0.3341	0.4276	0.3018
N	3993	2448	1545	3993	1996	1997

Notes: Columns (1) to (3) represent the regression results on the moderating role of ownership. Columns (4) to (6) represent the regression results on the moderating role of economic contribution. Tax was the economic contribution of firms. The explanation for abbreviations of other variables used inside Table 12 were mentioned in the footer of Table 4. The numbers in brackets are *t*-values. *, **, and *** reflect *p*-values of the coefficients were less than 0.1, 0.05, and 0.01, respectively.

5.4. The Moderating Role of Firms' Economic Contribution

Next, we tested the moderating role of firms' economic contribution. As shown in column (4) of Table 12, the coefficient on Air × Tax was negative and statistically significant at the 5% level, confirming Hypothesis 5, which states that a firm's economic contribution lessens the influence of air pollution on the cost of debt. Then, we divide the full sample into high tax contribution versus low tax contribution according to the median value of their tax contribution in the full sample and re-regressed Equation (1) in these subsamples. As shown in columns (5) and (6) of Table 12, the coefficient of Air was nonsignificant in the high tax contribution subsample, while the coefficient of Air was statistically significant at the 1% level in the low tax contribution subsample. These results revealed that the influence of air pollution on the cost of debt was more prominent for firms with low tax contributions.

6. Conclusions, Implications, Limitations, and Future Prospects

6.1. Conclusions

Employing a large sample of Chinese listed firms from the main board of the SSE and SZSE under different air quality conditions covering 2014 to 2018, we explored whether and how air pollution affects the cost of debt through environmental violation risk. In addition, we deeply explored the moderating role of environmental regulatory pressure, firms' industry characteristics, firms' ownership, and firms' economic contribution on the relationship between air pollution and the cost of debt. Our study drew the following conclusions.

First, the air pollution of the cities where the firms were located had a positive impact on the cost of debt, and the environmental violation risk of firms was the mechanism of air pollution affecting the cost of debt. Our empirical results confirmed a positive relationship between air pollution and the cost of debt after controlling for various firm-level and provincial-level characteristics that affected the cost of debt and implementing a series of sensitivity and endogenous tests. In addition, we attempted to use the mediating effect test to confirm that environmental violation risk was the mechanism by which air pollution affected the cost of debt, which was distinct from the insights into the mechanism of credit risk and financial uncertainty [16]. Specifically, air pollution-induced environmental violation risk caused pessimistic assessments by creditors on the default risk of firms, which prompted them to raise the interest rates in terms of the loan contract to reduce the air pollution-related default loss. Overall, higher levels of air pollution led to a higher cost of debt tolerated by firms through their effects on environmental violation risk.

Second, the empirical results confirmed that environmental regulatory pressure and heavily polluting firms could strengthen the relationship between air pollution and the cost of debt. Specifically, the influence of air pollution on the cost of debt was more prominent for firms subjected to strict environmental supervision and heavily polluting firms. Strict environmental supervision and heavily polluting industry characteristics aggravated the environmental violation risk of firms caused by air pollution, eventually leading to a higher probability of debt default. Creditors chose to increase the cost of debt for firms subjected to strict environmental supervision and heavily polluting firms to mitigate the default loss associated with air pollution. This conclusion puts forward some conditions regarding increases in the adverse impact of air pollution on debt cost from the perspective of environmental supervision and corporate environmental performance that have not been tested in prior research.

Third, the empirical results confirmed that state-owned firms and firms' economic contribution could weaken the relationship between air pollution and the cost of debt. In other words, the influence of air pollution on the cost of debt was more prominent for non-state-owned firms and firms with lower economic contributions. Since state-owned firms and regional major economic contributors have stronger bargaining power in environmental supervision, they can reduce the environmental violation risk, administrative penalty, and legal proceedings caused by air pollution to a certain extent; thus, mitigating the creditors' concern about the debt default of these firms. This study was the first to examine the mechanism of mitigating the adverse impact of air pollution on debt cost from the perspective of ownership and firms' economic contributions, which differs from insights on the mitigating mechanism of monitoring and the economic environment [16].

6.2. Implications

The results reported in our research may have critical implications for different groups. First, our research was conducive to highlight not only the importance of environmental governance for mitigating the cost of debt to firms that are exposed to air pollution, but also its importance to creditors exposed to their clients' environmental violation risk and default risk. Specifically, firms should attach importance to environmental violation risk related to air pollution and strengthen the ability of environmental governance to reduce the cost of debt; creditors should incorporate the environmental risk signal transmitted by air pollution into their decision making to mitigate their default losses. Second, the moderating effect analysis in our study suggested that firms should judge the possible constraints of severe air pollution on debt financing based on their circumstances and heterogeneity in a timely manner. In particular, this implies that with the enhancement of environmental regulations, firms subjected to strict environmental supervision, heavily polluting firms, non-state-owned firms, and firms with low economic contributions need to be more cautious about environmental risks related to air pollution. These firms should take the initiative to strengthen their environmental governance and reduce the risk of environmental violations and pessimistic expectations of creditors for default risk through low-carbon innovation and green development, which, ultimately, reduce the cost of debt and financing constraints. Additionally, increasingly severe air pollution should suggest to government regulators that state-owned firms and regional major economic contributors may bring resistance to environmental supervision to some extent. Government regulators should strengthen supervision for environmental violations of state-owned firms and economic contributors so as to promote the long-term development of the regional economy.

6.3. Limitations and Future Prospects

Similar to other academic research, our empirical research also had certain limitations, which are worthy of further research. First, Zhou et al. [18] and Xu et al. [69] mentioned that media coverage is an important external mechanism affecting the response of investors or creditors to the risk of firms' environmental violations. Media coverage intensity and media information sources regarding air pollution are likely to affect the sensitivity of creditors to the environmental violation risk of firms, which was not discussed in this research. Future research could further explore the moderating effect of media coverage on the relationship between air pollution and the cost of debt. Second, our research only evaluated the environmental violation risk of firms according to the severity of environmental violations. Creditors may attach greater importance to changes in corporate cash flow caused by environmental violations [10]. Future research could further measure the environmental violation risk based on the amounts of environmental penalties. Third, Wu et al. [1] tried to explore the effect of the six specific pollutants that constitute AQI on stock returns and trading activities. Each air pollutant may also have a different impact on the cost of debt, which was not discussed in our research. Considering that there may exist correlations between these pollutants, future research could further employ statistical methods, such as a factor analysis, principal component analysis, and cluster analysis, to merge these pollutant factors. Fourth, although our research used the OLS model that has been widely applied in many empirical analyses, parametric tests still have inherent limitations. Parameter tests need to estimate the population parameters using the population distribution and sample information, which is only applicable when the function that describes the relationship between the dependent and independent variables is known [70]. Some scholars have found that the application of parametric tests may cause model misspecification [71], which cannot correctly reveal the substantive relationship between the variables. In contrast, nonparametric tests can be applied to assess the effects of independent variables on the dependent variables in a nonlinear fashion without imposing a specific functional form, which allows the data themselves to reveal the functional relationship between the variables [70]. Future studies should further consider using nonparametric approaches to test the impact of air pollution on debt cost.

Author Contributions: Conceptualization, A.W. and M.Z.; methodology, M.Z. and S.Z.; software, M.Z.; validation, M.Z.; formal analysis, A.W. and M.Z.; investigation, M.Z.; resources, M.Z.; data curation, M.Z.; writing—original draft preparation, M.Z.; writing—review and editing, M.Z. and S.Z.; supervision, M.Z. All authors have read and agreed to the published version of the manuscript.

Funding: This research received no external funding.

Institutional Review Board Statement: Not applicable.

Informed Consent Statement: Not applicable.

Data Availability Statement: The data shown in this research are available on request from the corresponding author.

Conflicts of Interest: The authors declare no conflict of interest.

References

1. Wu, Q.; Hao, Y.; Lu, J. Air pollution, stock returns, and trading activities in China. *Pac.-Basin Financ. J.* **2018**, *51*, 342–365. [CrossRef]
2. Ji, D.; Li, L.; Wang, Y.; Zhang, J.; Cheng, M.; Sun, Y.; Liu, Z.; Wang, L.; Tang, G.; Hu, B.; et al. The heaviest particulate air-pollution episodes occurred in northern China in January, 2013: Insights gained from observation. *Atmos. Environ.* **2014**, *92*, 546–556. [CrossRef]
3. Gu, H.; Cao, Y.; Elahi, E.; Jha, S.K. Human health damages related to air pollution in China. *Environ. Sci. Pollut. Res.* **2019**, *26*, 13115–13125. [CrossRef] [PubMed]
4. Qiu, Q.; Wang, Y.; Qiao, S.; Liu, R.; Bian, Z.; Yao, T.; Thai Son, N. Does air pollution affect consumer online purchasing behavior? The effect of environmental psychology and evidence from China. *J. Clean. Prod.* **2020**, *260*, 120795. [CrossRef]
5. Song, Y.; Huang, B.; He, Q.; Chen, B.; Wei, J.; Mahmood, R. Dynamic assessment of PM2.5 exposure and health risk using remote sensing and geo-spatial big data. *Environ. Pollut.* **2019**, *253*, 288–296. [CrossRef] [PubMed]
6. Levy, T.; Yagil, J. Air pollution and stock returns in the US. *J. Econ. Psychol.* **2011**, *32*, 374–383. [CrossRef]
7. Huang, J.; Xu, N.; Yu, H. Pollution and performance: Do investors make worse trades on hazy days? *Manag. Sci.* **2020**, *66*, 4455–4476. [CrossRef]
8. Dong, R.; Fisman, R.; Wang, Y.; Xu, N. Air pollution, affect, and forecasting bias: Evidence from Chinese financial analysts. *J. Financ. Econ.* **2021**, *139*, 971–984. [CrossRef]
9. Chen, H.; Tan, X.; Cao, Q. Air pollution, auditors' pessimistic bias and audit quality: Evidence from China. *Sustain. Account. Manag. Policy J.* **2021**, *12*, 74–104. [CrossRef]
10. Jung, J.; Herbohn, K.; Clarkson, P. Carbon risk, carbon risk awareness and the cost of debt financing. *J. Bus. Ethics* **2018**, *150*, 1151–1171. [CrossRef]
11. Guo, M.; Kuai, Y.; Liu, X. Stock market response to environmental policies: Evidence from heavily polluting firms in China. *Econ. Model.* **2020**, *86*, 306–316. [CrossRef]
12. Schneider, T.E. Is environmental performance a determinant of bond pricing? Evidence from the US pulp and paper and chemical industries. *Contemp. Account. Res.* **2011**, *28*, 1537–1561. [CrossRef]
13. Coulson, A.B.; Monks, V. Corporate environmental performance considerations within bank lending decisions. *Eco-Manag. Audit.* **1999**, *6*, 1–10. [CrossRef]
14. Weber, O. Environmental credit risk management in banks and financial service institutions. *Bus. Strategy Environ.* **2012**, *21*, 248–263. [CrossRef]
15. Tan, J.; Zhang, X.; Zhang, P.; Chan, K.C. Does air pollution matter in a supplier's trade credit strategy? Evidence from an emerging market. *Borsa Istanb. Rev.* **2021**, *21*, S70–S79. [CrossRef]
16. Tan, J.; Chan, K.C.; Chen, Y. The impact of air pollution on the cost of debt financing: Evidence from the bond market. *Bus. Strategy Environ.* **2022**, *31*, 464–482. [CrossRef]
17. Liu, Z.; Shen, H.; Welker, M.; Zhang, N.; Zhao, Y. Gone with the wind: An externality of earnings pressure. *J. Account. Res.* **2021**, *72*, 101403. [CrossRef]
18. Zhou, Z.; Zhang, T.; Wen, K.; Zeng, H.; Chen, X. Carbon risk, cost of debt financing and the moderation effect of media attention: Evidence from Chinese companies operating in high-carbon industries. *Bus. Strategy Environ.* **2018**, *27*, 1131–1144. [CrossRef]
19. Zou, H.L.; Zeng, R.C.; Zeng, S.X.; Shi, J.J. How do environmental violation events harm corporate reputation? *Bus. Strategy Environ.* **2015**, *24*, 836–854. [CrossRef]
20. Lepori, G.M. Air pollution and stock returns: Evidence from a natural experiment. *J. Empir. Financ.* **2016**, *35*, 25–42. [CrossRef]
21. Molina, M.J.; Molina, L.T. Megacities and atmospheric pollution. *J. Air Waste Manag. Assoc.* **2004**, *54*, 644–680. [CrossRef] [PubMed]
22. Brauer, M.; Freedman, G.; Frostad, J.; Van Donkelaar, A.; Martin, R.V.; Dentener, F.; Dingenen, R.V.; Estep, K.; Amini, H.; Apte, J.S. Ambient air pollution exposure estimation for the global burden of disease 2013. *Environ. Sci. Technol.* **2016**, *50*, 79–88. [CrossRef] [PubMed]

23. Cohen, A.J.; Brauer, M.; Burnett, R.; Anderson, H.R.; Frostad, J.; Estep, K.; Balakrishnan, K.; Brunekreef, B.; Dandona, L.; Dandona, R. Estimates and 25-year trends of the global burden of disease attributable to ambient air pollution: An analysis of data from the Global Burden of Diseases Study 2015. *Lancet* **2017**, *389*, 1907–1918. [CrossRef]
24. Pu, S.; Shao, Z.; Fang, M.; Yang, L.; Liu, R.; Bi, J.; Ma, Z. Spatial distribution of the public's risk perception for air pollution: A nationwide study in China. *Sci. Total Environ.* **2019**, *655*, 454–462. [CrossRef]
25. Lan, G.-L.; Yuan, Z.-K.; Maddock, J.E.; Cook, A.; Chu, Y.-Y.; Pan, B.-B.; Tu, H.; Fan, S.; Liao, X.; Lu, Y. Public perception of air pollution and health effects in Nanchang, China. *Air Qual. Atmos. Health* **2016**, *9*, 951–959. [CrossRef]
26. Sharfman, M.P.; Fernando, C.S. Environmental risk management and the cost of capital. *Strateg. Manag. J.* **2008**, *29*, 569–592. [CrossRef]
27. Chen, L.H.; Silva Gao, L. The Pricing of Climate Risk. Available online: https://ssrn.com/abstract=1940727 (accessed on 7 October 2011).
28. Thompson, P. Bank lending and the environment: Policies and opportunities. *Int. J. Bank Mark.* **1998**, *16*, 243–252. [CrossRef]
29. Thompson, P.; Cowton, C.J. Bringing the environment into bank lending: Implications for environmental reporting. *Br. Account. Rev.* **2004**, *36*, 197–218. [CrossRef]
30. Du, X.; Weng, J.; Zeng, Q.; Chang, Y.; Pei, H. Do lenders applaud corporate environmental performance? Evidence from Chinese private-owned firms. *J. Bus. Ethics* **2017**, *143*, 179–207. [CrossRef]
31. Diabat, A.; Govindan, K. An analysis of the drivers affecting the implementation of green supply chain management. *Resour. Conserv. Recycl.* **2011**, *55*, 659–667. [CrossRef]
32. Zhang, B.; Bi, J.; Yuan, Z.; Ge, J.; Liu, B.; Bu, M. Why do firms engage in environmental management? An empirical study in China. *J. Clean. Prod.* **2008**, *16*, 1036–1045. [CrossRef]
33. Hofmann, K.H.; Theyel, G.; Wood, C.H. Identifying firm capabilities as drivers of environmental management and sustainability practices–evidence from small and medium-sized manufacturers. *Bus. Strategy Environ.* **2012**, *21*, 530–545. [CrossRef]
34. Stewart, R.B. Environmental regulation and international competitiveness. *Yale Law J.* **1992**, *102*, 2039–2106. [CrossRef]
35. McGuire, W. The effect of ISO 14001 on environmental regulatory compliance in China. *Ecol. Econ.* **2014**, *105*, 254–264. [CrossRef]
36. Wang, Z.; Dear, K. Region and firm level determinants of environmental regulation violations: An empirical study in Chongqing, China. *J. Clean. Prod.* **2017**, *141*, 1011–1022. [CrossRef]
37. Jin, Y.; Andersson, H.; Zhang, S. Air pollution control policies in China: A retrospective and prospects. *Int. J. Environ. Res. Public Health* **2016**, *13*, 1219. [CrossRef]
38. Zou, H.; Zeng, S.; Zhang, X.; Lin, H.; Shi, J.J. The intra-industry effect of corporate environmental violation: An exploratory study. *J. Clean. Prod.* **2015**, *107*, 428–437. [CrossRef]
39. Sadik-Zada, E.R.; Ferrari, M. Environmental policy stringency, technical progress and pollution haven hypothesis. *Sustainability* **2020**, *12*, 3880. [CrossRef]
40. Konar, S.; Cohen, M.A. Does the market value environmental performance? *Rev. Econ. Stat.* **2001**, *83*, 281–289. [CrossRef]
41. Tian, L.; Estrin, S. Retained state shareholding in Chinese PLCs: Does government ownership always reduce corporate value? *J. Comp. Econ.* **2008**, *36*, 74–89. [CrossRef]
42. Faccio, M.; Masulis, R.W.; McConnell, J.J. Political connections and corporate bailouts. *J. Financ.* **2006**, *61*, 2597–2635. [CrossRef]
43. Duchin, R.; Sosyura, D. The politics of government investment. *J. Financ. Econ.* **2012**, *106*, 24–48. [CrossRef]
44. Adhikari, A.; Derashid, C.; Zhang, H. Public policy, political connections, and effective tax rates: Longitudinal evidence from Malaysia. *J. Account. Public Policy* **2006**, *25*, 574–595. [CrossRef]
45. Chan, K.S.; Dang, V.Q.; Yan, I.K. Chinese firms' political connection, ownership, and financing constraints. *Econ. Lett.* **2012**, *115*, 164–167. [CrossRef]
46. Boubakri, N.; Guedhami, O.; Mishra, D.; Saffar, W. Political connections and the cost of equity capital. *J. Corp. Financ.* **2012**, *18*, 541–559. [CrossRef]
47. Eaton, S.; Kostka, G. Central protectionism in China: The "central SOE problem" in environmental governance. *China Q.* **2017**, *231*, 685–704. [CrossRef]
48. Xu, X.; Zeng, S.; Tam, C.M. Stock market's reaction to disclosure of environmental violations: Evidence from China. *J. Bus. Ethics* **2012**, *107*, 227–237. [CrossRef]
49. Wang, H.; Wheeler, D. Equilibrium pollution and economic development in China. *Environ. Dev. Econ.* **2003**, *8*, 451–466. [CrossRef]
50. Pittman, J.A.; Fortin, S. Auditor choice and the cost of debt capital for newly public firms. *J. Account. Res.* **2004**, *37*, 113–136. [CrossRef]
51. Shailer, G.; Wang, K. Government ownership and the cost of debt for Chinese listed corporations. *Emerg. Mark. Rev.* **2015**, *22*, 1–17. [CrossRef]
52. Bliss, M.A.; Gul, F.A. Political connection and cost of debt: Some Malaysian evidence. *J. Bank. Financ.* **2012**, *36*, 1520–1527. [CrossRef]
53. Wang, Y.; Lu, T.; Qiao, Y. The effect of air pollution on corporate social responsibility performance in high energy-consumption industry: Evidence from Chinese listed companies. *J. Clean. Prod.* **2021**, *280*. [CrossRef]
54. Dobler, M.; Lajili, K.; Zéghal, D. Environmental performance, environmental risk and risk management. *Bus. Strategy Environ.* **2014**, *23*, 1–17. [CrossRef]

55. Pan, Y.; Chen, Q.; Zhang, P. Does policy uncertainty affect corporate environmental information disclosure: Evidence from China. *Sustain. Account. Manag. Policy J.* **2020**, *11*, 903–931. [CrossRef]
56. Baron, R.M.; Kenny, D.A. The moderator-mediator variable distinction in social psychological research: Conceptual, strategic, and statistical considerations. *J. Personal. Soc. Psychol.* **1986**, *51*, 1173–1182. [CrossRef]
57. Li, S.; Fu, H.; Wen, J.; Chang, C.-P. Separation of ownership and control for Chinese listed firms: Effect on the cost of debt and the moderating role of bank competition. *J. Asian Econ.* **2020**, *67*, 101179. [CrossRef]
58. Lim, C.Y.; Wang, J.; Zeng, C. China's "mercantilist" government subsidies, the cost of debt and firm performance. *J. Bank. Financ.* **2018**, *86*, 37–52. [CrossRef]
59. Jensen, M.C. Agency costs of free cash flow, corporate finance, and takeovers. *Am. Econ. Rev.* **1986**, *76*, 323–329.
60. Hart, O.; Moore, J. A theory of debt based on the inalienability of human capital. *Q. J. Econ.* **1994**, *109*, 841–879. [CrossRef]
61. Myers, S.C. Determinants of corporate borrowing. *J. Financ. Econ.* **1977**, *5*, 147–175. [CrossRef]
62. Diamond, D.W. Presidential address, committing to commit: Short-term debt when enforcement is costly. *J. Financ.* **2004**, *59*, 1447–1479. [CrossRef]
63. Laeven, L.; Majnoni, G. Does judicial efficiency lower the cost of credit? *J. Bank. Financ.* **2005**, *29*, 1791–1812. [CrossRef]
64. Jappelli, T.; Pagano, M.; Bianco, M. Courts and banks: Effects of judicial enforcement on credit markets. *J. Money Credit Bank.* **2005**, *37*, 223–244. [CrossRef]
65. Mojon, B.; Smets, F.; Vermeulen, P. Investment and monetary policy in the euro area. *J. Bank. Financ.* **2002**, *26*, 2111–2129. [CrossRef]
66. Dhaliwal, D.; Hogan, C.; Trezevant, R.; Wilkins, M. Internal control disclosures, monitoring, and the cost of debt. *Account. Rev.* **2011**, *86*, 1131–1156. [CrossRef]
67. Fortelli, A.; Scafetta, N.; Mazzarella, A. Influence of synoptic and local atmospheric patterns on PM10 air pollution levels: A model application to Naples (Italy). *Atmos. Environ.* **2016**, *143*, 218–228. [CrossRef]
68. He, H.-D. Multifractal analysis of interactive patterns between meteorological factors and pollutants in urban and rural areas. *Atmos. Environ.* **2017**, *149*, 47–54. [CrossRef]
69. Xu, X.; Zeng, S.; Zou, H.; Shi, J.J. The impact of corporate environmental violation on shareholders' wealth: A perspective taken from media coverage. *Bus. Strategy Environ.* **2016**, *25*, 73–91. [CrossRef]
70. Sadik-Zada, E.R.; Loewenstein, W. Drivers of CO_2-Emissions in Fossil Fuel abundant settings:(Pooled) mean group and nonparametric panel analyses. *Energies* **2020**, *13*, 3956. [CrossRef]
71. Hsiao, C. Panel data analysis—Advantages and challenges. *Test* **2007**, *16*, 1–22. [CrossRef]

MDPI
St. Alban-Anlage 66
4052 Basel
Switzerland
Tel. +41 61 683 77 34
Fax +41 61 302 89 18
www.mdpi.com

International Journal of Environmental Research and Public Health Editorial Office
E-mail: ijerph@mdpi.com
www.mdpi.com/journal/ijerph

www.ingramcontent.com/pod-product-compliance
Lightning Source LLC
Chambersburg PA
CBHW061824210926
51850CB00006B/796
9783036579436